Structure and Activity
of Natural Peptides

Structure and Activity of Natural Peptides

Selected Topics

Proceedings of the Fall Meeting
Gesellschaft für Biologische Chemie
Tübingen, Germany, September 1979

Editors
Wolfgang Voelter · Günther Weitzel

Walter de Gruyter · Berlin · New York 1981

Editors:

Wolfgang Voelter, Prof., Dr. rer. nat.
Universität Tübingen
Abteilung für Physikalische Biochemie
Hoppe-Seyler-Straße 1
D-7400 Tübingen

Günther Weitzel, Prof., Dr. rer. nat., Dr. med.
Universität Tübingen
Physiologisch-chemisches Institut
Hoppe-Seyler-Straße 1
D-7400 Tübingen

CIP-Kurztitelaufnahme der Deutschen Bibliothek

Structure and activity of natural peptides: Selected topics; Tübingen, Germany, September 1979/ed. Wolfgang Voelter; Günther Weitzel. -
Berlin; New York: de Gruyter, 1981.
(Proceedings of the Fall Meeting / Gesellschaft für Biologische Chemie; 1979)
ISBN 3-11-008264-0
NE: Voelter, Wolfgang [Hrsg.]; Gesellschaft für Biologische Chemie; Chemie: Proceedings of the . . .

Library of Congress Cataloging in Publication Data

Structure and activity of natural peptides.
Bibliography: p., Includes index.
1. Peptides-Congresses.
2. Structure-activity relationship (Pharmacology)-Congresses.
I. Voelter, W. II. Weitzel, Günther, 1915-
III. Gesellschaft für Biologische Chemie.
QP552.P4S87 612'.015756 81-3288
ISBN 3-11-008264-0 AACR2

 Printing: Color Druck, Berlin. – Binding: Dieter Mikolai, Berlin. Printed in Germany.

To Margarethe

PREFACE

The book "Structure and Activity of Natural Peptides" is the result of selected contributions presented at the Fall Meeting of the Gesellschaft für Biologische Chemie in Tübingen in 1979. Excellent survey lectures as well as many talks on unpublished research results by internationally recognized authorities were the initiators of an idea conceived during the Congress: that the information of highest interest, carefully selected, should become available in a Proceedings monograph to the scientific world.

Coverage of this volume ranges from traditional peptide chemistry to physically and biologically-oriented areas. Survey articles of different aspects on gastrointestinal hormones (E. Wünsch), amatoxins, phallotoxins (T. Wieland), insulin (R. Geiger), thymic peptides (K. Folkers), alamethicin, trichotoxin (G. Jung) or ACTH (K. H. Voigt) are of introductory character for those not directly involved in the special field. The introductory contributions on field desorption mass spectrometry and crystal structure analyses of polypeptides may cause peptide chemists to apply these techniques to their special field. In part II of the book, a series of new peptides isolated from different natural sources with interesting biological activities is described, and these peptides may open up new areas for synthetic chemists, biochemists, physiologists or even physicians. In chapter III of the book new methods for purification and characterization of peptides are introduced, which are of interest to every peptide and protein laboratory. Part IV of the book is dedicated to the synthetic peptide chemist. Examples for new synthetic routes and new protective groups are discussed; furthermore a new system for planning of peptide synthesis with the aid of electronic data-processing is introduced. The monograph ends with the discussion of biological effects of peptides in different areas such as the hypothalamus, pituitary, gonads or thymus.

We would like to take this opportunity of expressing our thanks to Mrs. Hella Schmid, who, after editing the text, typed most of the monograph's pages. Our appreciation is also extended to Mrs. Isa Jantzen for her assistance in preparing the indices. Finally, we would like to thank our Publisher, Walter de Gruyter, especially Dr. Weber and Mrs. Glowka, for their cooperation.

Tübingen, February, 1981

Wolfgang Voelter, Günther Weitzel

CONTENTS

I. SURVEYS OF SELECTED TOPICS

ISOLATION, STRUCTURE ELUCIDATION AND SYNTHESIS OF GASTROINTESTINAL HORMONES - CRITICAL ASPECTS

Erich Wünsch
Max-Planck-Institut für Biochemie, Abteilung für Peptidchemie
D-8033 Martinsried, Germany

Introduction

The isolation and the structure elucidation of naturally occurring polypeptides, and finally their synthesis as ultimate proof of the proposed primary structure, constitute one single research project, complete in itself. The fast development, however, with consequently high specialization in these three research areas generated a diverging process which prevents already since a longer period of time to tackle such a complete project with one single research team.

In fact, the isolation relies upon a group of highly specialized researchers well acquainted with the most advanced purification techniques and perfectly trained in precise biological and immunological assay systems;

on the other hand, the sequence analysis has to be performed by a qualified research team with the most sophisticated techniques in the effort to propose the primary structure for the isolated natural product usually available only in minimal amounts;

finally, the synthesis can only be carried out by a team of experts in the preparative organic chemistry, capable of building high molecular weight polypeptides in the desired sequence order and of isolating in the final stage from the synthetic material the endproduct at the highest possible degree of purity as judged by the analytical methods available today.

Structure and Activity of Natural Peptides

In this context the collaboration of the "isolation team" is absolutely necessary for the comparative analysis of the biological and immunological properties of the synthetic material relative to those of the natural preparation. The collaboration of the "sequencing team" is desirable, sometimes essential, even if the peptide chemists usually are well trained in the modern analytical techniques needed for the purity tests and generally are able to use them in the obligatory critical sense.

As a resume these three specialized research groups should fuse into a research community in the common effort for a maximum success.

Results and Discussion

1) Critical aspects of isolation and sequence analysis of peptide hormones. The isolation of natural peptides, especially of gastrointestinal hormones, has clearly revealed that in most of the cases it is very difficult, if not impossible, to obtain these active factors as homogeneous materials. Even by means of very advanced procedures the resulting peptide preparations are contaminated by other active substances, which can sometimes not be removed completely. As an example, the pancreas stimulating hormone secretin is, according to our estimation (comparison of natural and synthetic secretin by amino acid analysis), still contaminated, even in its qualitatively pure form, by 3 to 4% of motilin, although the sequences of secretin and motilin are very different (1).

It is very well known that also other natural polypeptides can not be isolated as homogeneous materials: in fact, glucagon is contaminated by varying amounts of insulin, pancreozymin by GIP, etc.

For the sequence analysis of natural peptide factors, however, these contaminants, present to a few percent or even up to 20%, do not interfere. The analytical methods used, both, in the sequencing work and in the evaluation of the experimental data do not account for these impurities. So W.W.Bromer et al. (2) succeeded in the sequence determination of the cristalline natural glucagon and to propose the correct primary structure for this insulin antagonist, even if it is usually contaminated by randomly desamidated glucagon (3) and proglucagon fragment (4).

On the other side, the modern methods of sequence analysis are still not free of errors. In the case of human ACTH it has been the extensive studies of peptide chemists to finally allow the correction of the previously proposed primary structure and to resolve the problems concerning the sequence of this important hormone. The sequence errors in the region 25-30 were clearly demonstrated by a comparative investigation of the synthetic and natural product by means of different analytical tests (5). For motilin, too, the comparative analysis of the synthetic product with the natural preparation led to a minor correction of the sequence.

J.Brown, in fact, proposed the sequence of motilin as occupied in positions 14 and 15 by glutamic acid residues (6). Using the synthetic material, prepared in our laboratory, i.e. [Nle^{13}, Glu^{14}]-motilin (7), J.Brown carried out a comparative analysis of the fragments derived from enzymatic degradation. The results strongly suggested that the position 14 is occupied by a glutamine residue (possibly, even in position 15 of the motilin sequence a glutamine is present) (8). It has been, however, a very fortuitous case that the exchange of glutamine by a glutamic acid residue had no effect on the biological (7) and immunological (8) response of this hormone. Certainly, this may not be true in other cases. In fact desamidination of the glutamine residue in position 3 of the glucagon sequence produces a dramatic decrease of the biological activity to zero %.

As last example, synthetic human big-gastrin I, prepared in

collaboration with Kenner's group (10) and subsequently by an alternative route in our laboratory (11) according to the primary structure proposed by Gregory, Kenner and Harris (12) raised

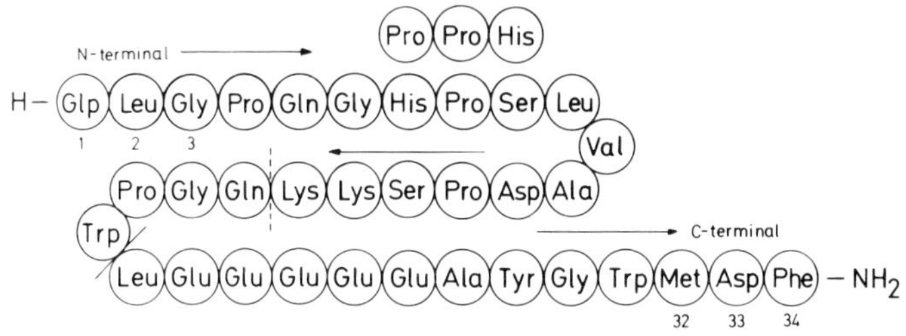

Fig. 1. The primary structure of human big-gastrin I as proposed by R.A.Gregory and H.J.Harris (12) and the corrected structure suggested by R. Ramage (16).

questions about possible errors in the sequence analysis: the synthetic human big-gastrin I was shown to exhibit a biological activity equal, if not superior, to that of 15-leucine-human gastrin I (13) which is considered as a fully active human little-gastrin I analogue; in comparison to the natural big-gastrin I preparation the synthetic product showed the expected activity (14). The immunoreactivity, however, was observed to be different for the two preparations (15).

Extensive immunological studies on several synthetic peptides related to the N-terminal sequence of porcine big-gastrin I allowed to detect the errors in the primary structure proposed for this hormone and consequently to synthetize successfully according to the new, corrected structure porcine big-gastrin I with immunological properties equal to those of the natural preparation (16). Similar preliminary studies on human big-gastrin I strongly suggest that corrections in the sequence-region 7-9 should resolve the existing discrepancies. Most likely the tripeptide sequence His-Ser-Pro (positions 7-9) has to be replaced by Pro-Pro-His (16).

A gastrointestinal factor isolated and sequenced by J.Brown et al. (17,18) as a linear polypeptide chain of 43 amino acid residues has been termed "gastric inhibitory polypeptide" (GIP) because of its characteristic activity profile.

Two tentative syntheses of this tritetracontapeptide (19,20) - in one case reported even as the total synthesis of GIP (19) - have to be considered as failed: the biological tests of both synthetic preparations were carried out by J.Brown in Vancouver and revealed no detectable biological potencies (J.Brown, personal communication). Additionally it has been observed that the synthetic GIP materials examined were not at all homogeneous in the electrophoretic tests (J.Brown and A.Moody, personal communications).

Thus the following questions are still open:

a) is the proposed primary structure of GIP erroneous?
b) have the two syntheses been performed as inefficiently as not to be used for the final proof of the proposed primary structure, whereby the lack of biological activity would become meaningless?

According to findings of J.Brown, a carboxypeptidase A degradation of natural porcine GIP produced an octatriacontapeptide corresponding to the sequence 1-38 of GIP and characterized by the typical activity profile of the parent hormone. In the light of this interesting findings, we have attempted to contribute to the GIP problem by devoting our attention to the synthesis of this octatriacontapeptide (21).

For this purpose the seven fragments shown in fig. 2 were synthetized by conventional procedures in solution and subsequently condensed in sequence order (fig. 3).

I) H-Ser(tBu)-Asp(OtBu)-Trp-Lys(BOC)-OtBu	H- [35-38] -OtBu
II) NPS-Lys(BOC)-Gly-Lys(BOC)-Lys(BOC)-OH	NPS- [31-34] -OH
III) Z-Ala-Gln-Gln-OH	Z- [28-30] -OH
IV) Z-Phe-Val-Asn-Trp-Leu-Leu-OH	Z- [22-27] -OH
V) NPS-Met-Asp(OtBu)-Lys(BOC)-Ile-Arg(HBr)-Gln-Gln-Asp(OtBu)-OH	NPS- [14-21] -OH
VI) NPS-Asp(OtBu)-Tyr(tBu)-Ser(tBu)-Ile-Ala-OH	NPS- [9-13] -OH
VII) BOC-Tyr(tBu)-Ala-Glu(OtBu)-Gly-Thr(tBu)-Phe-Ile-Ser(tBu)-OH	BOC- [1-8] -OH

Fig. 2. Suitably protected fragments for the synthesis of porcine GIP-[1-38].

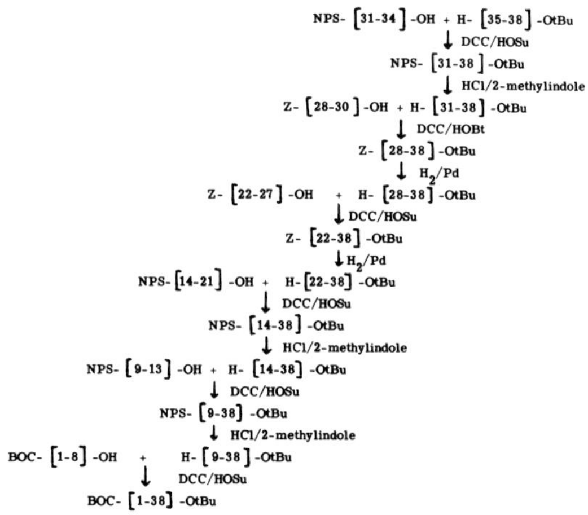

Fig. 3. Scheme of fragment condensations for the synthesis of fully protected porcine GIP-[1-38].

The synthetic route was elaborated according to the Schwyzer-Wünsch strategy, and certainly has to be ulteriorly optimized in several details in order to be proposed as a well defined total synthesis. The synthetic GIP-[1-38] material finally obtained as apparently homogeneous material (fig. 4), exhibited 100% insulinotropic activity, if compared with the natural porcine material; a surprisingly poor inhibitory potency of only 4%, however, was detected by J.Brown.

BOC - [1-38] -OtBu

↓ 1) CF_3COOH/anisole/1,2-ethanedithiol
2) ion-exchange chromatography on enzacryl-COOH, free flow electrophoresis

H-Tyr-Ala-Glu-Gly-Thr-Phe-Ile-Ser-Asp-Tyr-Ser-Ile-Ala-Met-Asp-Lys-Ile-Arg-Gln-Gln-Asp-Phe-Val-Asn-Trp-Leu-Leu-Ala-Gln-Gln-Lys-Gly-Lys-Lys-Ser-Asp-Trp-Lys-OH

Fig. 4. Scheme of the final steps for the synthesis of porcine GIP-[1-38].

These contrasting findings raised the question of whether:

i) the synthesis of the GIP-octatriacontapeptide failed: this question is justified, since we encountered great difficulties in the determination of the correct values of asparagine 24 and aspartic acid 36 on one side, and for tryptophan 25 and 37 on the other side. The low recovery of these amino acids upon acid hydrolysis in presence of thioglycolic acid is surprising and has to be investigated on model peptides;

ii) the primary structure proposed by J.Brown is not correct: according to investigations of A.Moody (personal communication) the sequence analysis of J.Brown may possibly have revealed one glutamine residue in excess in position 29-30 of the peptide chain. In fact the amino acid analysis of the acid hydrolysate of natural GIP preparations yields lower glutamic acid contents than expected from the proposed sequence[x)]. Additionally, cancelling one glutamine residue from positions 29-30, whereby GIP would result as a linear polypeptide of 42 amino acid residues, an alignment of the sequences 29-37 of GIP, 29-37 of proglucagon fragment and 92-100 of gut GLI-I reveals the expected strong sequence homology, especially if an exchange of residue 33 with 34 should also appear to be necessary in the proglucagon fragment (fig. 5);

---

x)These findings have recently been confirmed in our laboratory.

```
                          29
Glucagon       - Met - Asn - Thr -OH
                          29                                                       37
Proglucagon    - Met - Asn - Thr -         Lys - Arg - Asn - Asn - Lys - Asn - Ile - Ala -
Fragment                  92                                                       100
GUT GLI-1      - Met - Asn - Thr -         Lys - Arg - Asn - Lys - Asn - Asn - Ile - Ala -
                          29                                                       38
GIP            - Leu - Ala - Gln -(Gln) -  Lys - Gly - Lys - Lys - Ser - Asp - Trp - Lys -
```

Fig. 5.

iii) the natural preparation of GIP is not pure: the gastric inhibitory factor could be represented by a highly active contaminant; in this case GIP would not be GIP, but IRF (insulin-releasing factor).

2) Critical aspects of synthesis and purity tests of peptide hormones.

As mentioned above, the synthesis may also not be free of errors. A critical attitude has not to be taken up only towards solid phase synthesis, which in the higher molecular weight ranges (above 15 amino acid residues chains) undoubtedly produces mixtures of very closely related peptides; these mixtures can hardly be purified to the desired extent, even using the most modern and advanced techniques. The HPLC, which is praised at the moment as the most efficient method, has its limits, too, as all previous analytical tools: the accuracy and efficiency strongly depends upon the experimental conditions.

A classical example, in this connection, is represented by the low molecular weight peptide factor Kentsin.
H.A.Kent has recently reported the isolation of a contraceptive tetrapeptide (named Kentsin) from hamster zygotes (22) and proposed the sequence H-Thr-Pro-Arg-Lys-OH for this peptide factor (23). Of five synthetic preparations tested by H.A. Kent (24) only two were found to be active, i.e. the synthetic Kentsin of R.Roeske and K.U.Prasad (prepared by the solid phase procedure), and that produced by Calbiochem according to conventional methods. The latter Kentsin preparation has been analyzed and compared with a Kentsin synthetized in our laboratory as reference substance(25).

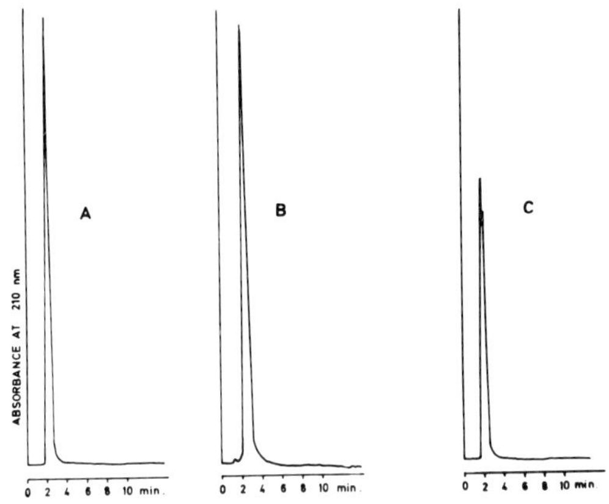

Fig. 6. HPLC [μ-Bondapak C18; acetonitrile/0.01 M ammonium acetate buffer, pH 4.0; 56:34 v/v; isocratic elution] of A) reference Kentsin; B)purified Calbiochem Kentsin; C) mixture of reference and purified Calbiochem Kentsin.

Both preparations yielded the amino acid ratios expected by theory upon acid hydrolysis. With aminopeptidase-M only the reference Kentsin was fully digested, whereas the AP-M digest of the Calbiochem product was uncomplete. Additionally the two synthetic preparations behaved as completely different materials on thin-layer chromatography and on thin-layer electrophoresis (pyridinium acetate buffer, pH 4.9; 10V/cm), but showed identical retention times in HPLC using methanol/water (7:3 v/v) as eluent.

All chromatographic tests of the Calbiochem preparation indicated the presence of small amounts of impurities; thus it was further purified by us via ion-exchange chromatography on SP-Sephadex C-25 using 0.05M ammonium acetate (pH 5) as eluent. The isolated main product was compared with our Kentsin preparation on HPLC using acetonitrile/0.01M ammonium acetate, pH 4.0 (56:34 v/v) as eluent (fig. 6); even in this solvent system a clear differentiation of the two Kentsins could not be achieved. The UV absorption and the ^{1}H NMR spectra of the commercial Calbiochem Kentsin indicate, however, the presence of an aromatic, respectively benzyl group. Finally the mass spectra of the suitably derivatized compounds allowed us to identify the commercial Calbiochem product as a N^{α}- or O-benzyl-threonyl-prolyl-arginyl-lysine (25). In the light of these findings the degree of purity of the biologically active Roeske's solid phase Kentsin has severly to be questioned, too, as well as the chemical structure of Kentsin, since our H-Thr-Pro-Arg-Lys-OH preparation was inactive in the biological test.

The definition of purity for a synthetic compound has always been a very difficult task. Every result derived from an analytical test can be considered positive and of real value, if it reveals and identifies more than one component. On the contrary, the findings one peak, one single band, one single spot, etc. in single assay systems are not indicative. Only the sum of a series of positive results allows according to the rules of the greatest probability to draw conclusions about the homogeneity of the analyzed material - and this only up to the moment, a new analytical procedure will demonstrate the opposite.

The biological activity as a criterium of purity is also meaningless in the cases where peptide analogues, failure sequencesor truncated sequences are more active than the natural peptide factor. In this connection it is noteworthy, that a failure sequence of eledoisin displayed 500% biological activity if compared with that of the natural product. It is also well-known that in the case of somatostatin replacement of the L-tryptophan residue by D-tryptophan is accompanied by a strong increase in biological potency (26). Since racemization is one of the side-reactions difficult to control in peptide synthesis, the related byproducts in a synthetic somatostatin could simulate excellent biological properties even for impure preparations. Therefore this peptide factor belongs to that group of substances, whose synthesis require particular attention: the biological activity cannot be used as a probe for the purity.

In our tentative efforts to elaborate indicative analytical criteria for purity of somatostatin we found that subsequent to enzymatic hydrolysis by aminopeptidase-M the amino acid analysis yielded more representative values for pure samples than after acid hydrolysis under different standard conditions (25). Furthermore thin-layer chromatography on normal precoated silica gel plates -60 (E.Merck, Darmstadt) (tlc) was not as efficient as on high-performance silica gel -60 plates (E.Merck, Darmstadt) (hptlc) for the detection of impurities. With free flow electrophoresis additional trace impurities not revealed by tlc and hptlc could be clearly visualized (25). The trypsin and dipeptidylpeptide hydrolase digest were ideally suited for comparative analysis of different somatostatin samples. A comparison of the UV spectra of synthetic somatostatin and of an artifical mixture of the amino acids (1 Trp + 3 Phe) in the aromatic region was found to be equally suitable; gel filtrations on the other side allow to detect polymeric forms of somatostatin as possible by-products. The high-performance liquid chromatography has extensively been applied for somatostatin purity controls in the last years. It constitutes a very sensitive analytical tool if the operating conditions are optimized in a series of experiments. In fact, in a synthetic somatostatin preparation, which behaved as practically homogeneous material using acetonitrile/0.01M ammonium acetate, pH 4.0 (35:65v/v), three contaminants were revealed by gradually decreasing the acetonitrile content to a final concentration of 32.9% as shown in fig. 7. The experience we gained with HPLC confirms the efficacy of this method, if the optimal conditions are elaborated for every batch by varying eluent, column loading and flow rate. So it has been possible to follow the purification of synthetic somatostatin by HPLC (fig.8).

Already in the case of our total synthesis of glucagon (27) and than especially of 15-leucine-human-little gastrin I (28), we observed a great loss of material at the final purification step because of a side-product supposed to be related to a modified tryptophan residue.

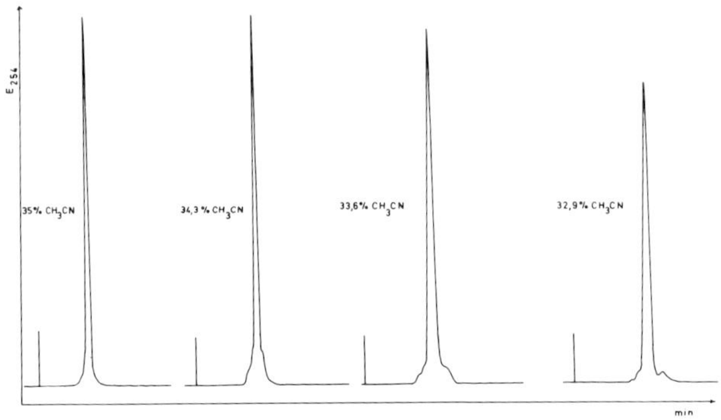

Fig. 7. HPLC [µ-Bondapak C18; acetonitrile /0.01M ammonium acetate buffer, pH 4.0; 1,7 ml/min; isocratic elution] of a synthetic somatostatin preparation at different acetonitrile concentrations.

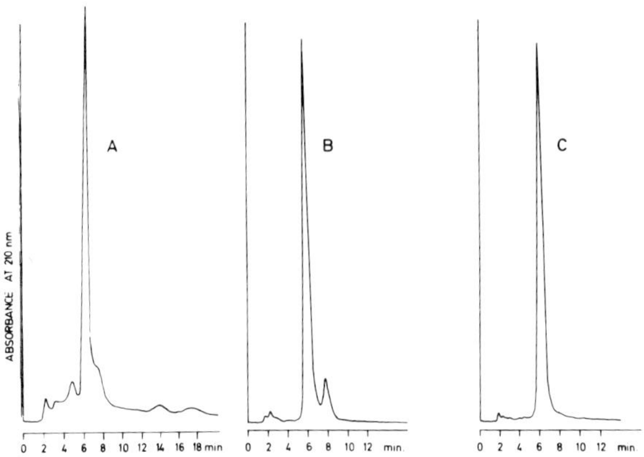

Fig. 8. HPLC [µ-Bondapak C18; acetonitrile/0.01M ammonium acetate buffer, pH 4.0; 32.5:67.5 v/v; 1.4ml/min; isocratic elution] of a crude synthetic somatostatin preparation (A); after LH 20-gel filtration (B) and after additional Biogel P6-gel filtration (C).

An exact identification of these side-products and thus of the nature of the side-reactions, which had occurred, was extremely difficult because of the presence of two tryptophan residues in the peptide chain of the human little gastrin I analog.

More success we had in the case of the synthesis of "11-leucine-human minigastrin I" x) (29). A "minigastrin I" side-fraction of more than 10% of the total crude product containing a modified tryptophan was isolated and thus became available for analytical studies (fig. 9).

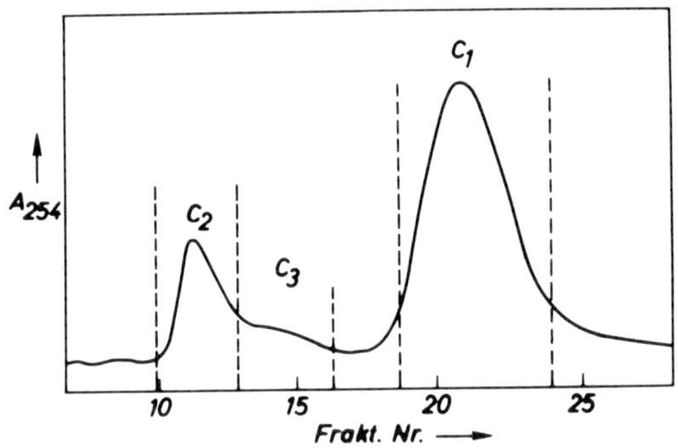

Fig. 9. Purification of 11-leucine-human-"minigastrin I" by partition chromatography on Sephadex G-25 (eluent: 2-butanol/0.1M ammonium acetate buffer, pH 6.0); fraction C_2 contains the tryptophan modified side-fraction.

This side-product could be identified as "10-N^{in}-tert-butyl-tryptophan-11-leucine-minigastrin I". The definitive characterization of this side-product was achieved via a series of experiments on other tryptophan-containing peptides and several model compounds. In this context, simulating the deprotection conditions with liquid hydrogen fluoride - the reagent prevalently used in the Merrifield synthesis - the total amount of tryptophan-modified side-product reached the enormous values of about 70% as found in model experiments by our Hungarian colleagues (32). This side-reaction cannot be markedly minimized or totally suppressed by means of particular scavengers.

x) 11-Leucine-human-minigastrin I was synthetized according to the primary structure proposed by R.A.Gregory and H.J.Tracy (30). Recently this structure has been corrected (31).

Recently, Sakakibara (33) reported data which disagree with the findings we observed in Munich and our colleagues in Budapest. With his experiments Sakakibara confirms, that the tert -butylation reaction occurs prevalently at the indole NH function of the tryptophan residue (followed in successive steps by tert -butylation at other positions of the ring); however, he affirms that scavengers of the anisole type efficiently suppress these side-reactions. As model for his studies Sakakibara investigated the desacylation of tert -butyloxycarbonyl-tryptophan in presence and in absence of scavangers using trifluoroacetic acid and liquid hydrogen fluoride as acids.

Already several years ago during the synthetic work connected with the total synthesis of glucagon we were able to demonstrate, that the electrophilic substitution (2-nitrophenylsulfenylation) at the tryptophan indole ring - which represents also the reaction mechanism of the tert -butylation (32,34) - strongly depends in its extent upon the chemical topography of the tryptophan residues, i.e. i) if the tryptophan is unsubstituted, ii) if this amino acid is substituted at the amino - or carboxyl-function and iii) if it is incorporated into a peptide chain as shown in fig. 10 (35).

In fact, the reaction rate at the tryptophan and at the glucagon-pentapeptide Phe-Val-Gln-Trp-Leu-OH differs by a factor of more than 100. Thus one can expect that the use of scavengers may suppress completely the tert -butylation reaction in the case of free tryptophan and only to a certain extent or even in undetectable amount in the case of tryptophan peptides. Therefore conclusions drawn from simple model experiments may undoubtedly lead to wrong general statements.

We investigated the effect of different kinds of scavangers in the acidolytic deprotection step and observed that this side-reaction cannot be suppressed totally. Infact tert -butylated side products were also found in the synthesis of human big-gastrin I

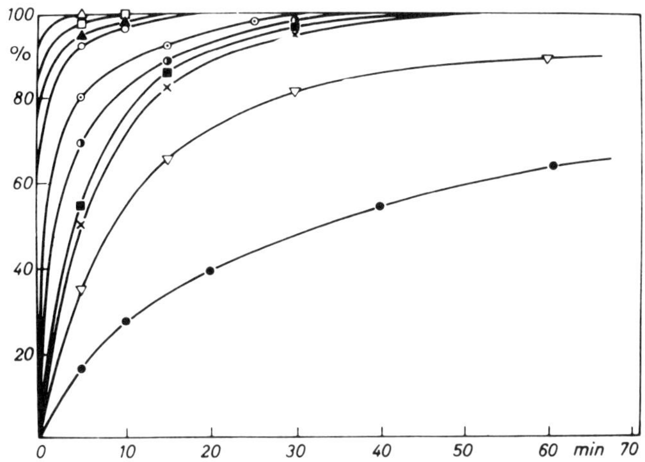

Fig. 10. 2-Nitrophenylsulfenylation, expressed as percentage of conversion, of N-methyl-pyrrole (⊙), indole (▲), 1-methyl-indole (△), 2-methyl-indole (□), 3-methyl-indole (○), ß-(3-indolyl)-propionic acid (◑), tryptophan (●), Z-Trp-OH (▽), Z-Phe-Trp-OMe (×) and H-Phe-Val-Gln-Trp-Leu-OH (■).

using ethanedithiole, anisole, thioanisole, 2-methyl-indole and their combinations as scavangers (36). Only in one case we did not observe this tryptophan-related side-reaction, i.e. in the final trifluoroacetic acid treatment of fully protected somatostatin.

With 90% trifluoroacetic acid as proposed by Kenner's group for peptides related to gastrins (31) no sensible improvement was detected. In fact, the deprotection of a CCK-PZ-nonapeptide analogue with 90% trifluoroacetic acid and 2-methylindole as scavanger led again to tert -butyl-tryptophan-derivatives (fig. 11). This CCK-PZ-nonapeptide analogue has been synthetized by a new route using suitably protected tyrosine-O-sulfate as starting material (37,38).

The stability of the tyrosine-O-sulfate ester in the normal operating conditions of peptide synthesis as derived from this model study should allow the synthesis of even more complicated pep-

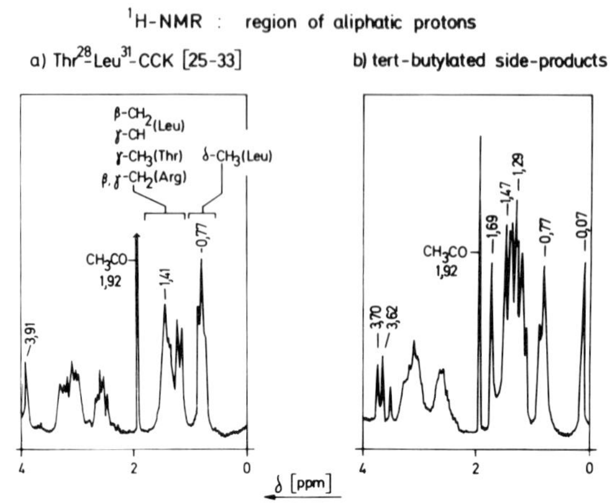

Fig. 11. ^{1}H NMR spectrum of the Trp-modified side-products isolated during the purification of Thr^{28}, Leu^{31}-CCK-PZ-[25-33].

tides, such as the entire CCK-PZ-molecule or the sulfated gastrins.

Conclusions

The examples discussed above strongly suggest that the methodology applied nowadays in the isolation, sequence analysis and synthesis of peptide factors has still to be carefully developed and optimized in order to assure the success of the research projects so extremely important for ulterior knowledges about the biological and immunological events of the living matter. In the area of molecular biology the biological experiments should be performed using only substances at the highest degree of purity, whereby the definition of purity remains relative, since restricted within the limits of error of the analytical methods available today. It may be irresponsable not to attend to this elementary rule.

Thus, we have decided to perform on the synthetic materials an extensive analytical investigation according to a test program based on physio-chemical, biological and immunological assays (35).

References

1. Wünsch, E.: Z.Gastroenterologie 14, 116 (1976).
2. Bromer, W.W., Sinn, G.L., Behrens, O.K.: J.Am.Chem.Soc. 79, 2807 (1957).
3. Bromer, W.W., Boucher, M.E., Patterson, J.M., Pekar, A.H., Frank, B.H.: J.Biol.Chem. 247, 2581 (1972).
4. Tager, H.S., Steiner, D.F.: Proc.Nat.Acad.Sci USA 70, 2321 (1973).
5. Riniker, B., Sieber, P., Rittel, W.: Nature New Biology 235, 114(1972).
6. Brown, J.C., Cook, M.A., Dryburgh, J.R.: Can.J.Biochem. 51, 533 (1973).
7. Wünsch, E., Brown, J.C., Deimer, K.-H., Drees, F., Jaeger, E., Musiol, J., Scharf, R., Stocker, H., Thamm, P., Wendlberger, G.: Z.Naturforsch. 28c, 235 (1973).
8. Schubert, H., Brown, J.C.: Can.J.Biochem. 92, 7 (1974).
9. Brown, J.C., Dryburg, J.-R.: Gut Hormones (ed. S.R.Bloom), Churchill Livingstone, Edinburgh, p. 327 (1978).
10. Choudhury, A.M., Kenner, G.W., Moore, S., Ramage, R., Richards, P.M., Thorpe, W.D., Moroder, L., Wendlberger, G., Wünsch, E.: Peptides 1976 (ed. A.Loffet). Editions de l'Universite de Bruxelles, Bruxelles, p. 257 (1976).
11. Wünsch, E., Wendlberger, G., Hallett, A., Jaeger, E., Knof, S., Moroder, L., Scharf, R., Schmidt, I., Thamm, P., Wilschowitz, L.: Z.Naturforsch. 32c, 495 (1977).
12. Gregory, R.A., Tracy, H.J.: Gastrointestinal Hormones (ed. J.C. Thompson), University of Texas Press, Austin, USA, p. 13 (1975).
13. Konturek, S.J., Tasler, J., Cieszkowski, M., Dobrzanska, M., Wünsch, E.: Am.J.Physiol. 253, E 170 (1977).
14. Kenner, G.W.: Biochem.Soc.Meeting, Edinburg (1977).
15. Gregory, R.A.: Personal communication.
16. Ramage, R.: Biochem.Soc.Meeting, Gregynog (1979).
17. Brown, J.C., Mutt, V., Pederson, R.A.: J.Physiol. (London) 209, 57 (1970).

18. Brown, J.C., Dryburgh, J.R.: Can. J. Biochem., 49, 867 (1971).

19. Yajima, H., Ogawa, H., Kubota, M., Tobe, T., Fujimura, M., Henmi, K., Torizuka, K., Adachi, H., Imura, H., Taminato, T.: J.Am. Chem. Soc. 97, 5593 (1975).

20. Yanaihara, N., Mochizuki, T., Yanaihara, C., Sakura, N., Hashimato, T., Sakagami, M., Keneko, T., Kanetoh, A., Brown, J.C.: Gut Hormones (ed. S.R.Bloom) Churchill Livingstone, Edinburgh, p. 28 (1978).

21. Moroder, L., Hallett, A., Thamm, P., Wilschowitz, L., Goehring, W., Brown, J.C., Wünsch, E.: Peptides 1978 (eds. J.Z. Siemion and Kupryszewski), Wrocław University Press, Wrocław, Poland, p. 585 (1979).

22. Kent, H.A.Jr.: Am. J. Anat., 144, 509 (1975).

23. Kent, H.A.Jr.: Biol. Reprod. 12, 504 (1975).

24. Kent, H.A.Jr.: Biol. Reprod. 19, 785 (1978).

25. Wünsch, E.: Hormone Receptors in Digestion and Nutrition (eds. G.Rosselin, P. Fromageot and S. Bonfils), Elsevier/North-Holland Biomedical Press, Amsterdam, p. 45 (1979).

26. Rivier, J., Brown, M., Rivier, C., Ling, N., Vale, W.: Peptides 1976 (ed. A.Loffet), Editions de l'Université de Bruxelles, Bruxelles, p. 427 (1976).

27. Wünsch, E.: Z.Naturforsch., 22b, 1269 (1967).

28. Wünsch, E., Jaeger, E., Deffner, M., Scharf, R.: Z.Physiol. Chem. 353, 1716 (1972).

29. Jaeger, E., Thamm, P., Schmidt, I., Knof, S., Moroder, L., Wünsch, E.: Z.Physiol.Chem. 359, 155 (1978).

30. Gregory, R.A., Tracy, H.J.: Gut, 15, 683 (1974).

31. Gregory, R.A., Tracy, H.J., Harris, I., Runswick, M.J., Moore, S., Kenner, G.W., Ramage, R.: Z.Physiol.Chem. 360, 73 (1979).

32. Löw, M., Kisfaludy, L., Sohar,P.: Z.Physiol.Chem. 359, 1643 (1978).

33. Chino, N., Masui, Y., Sakakibara, S.: Peptide Chemistry 1977 (ed. T.Shiba), Protein Research Foundation, Osaka, Japan, p. 27 (1977).

34. Lundt, B.F., Johansen, N.L., Vølund, A., Makussen, J.: Int. J. Peptide Protein Res. 12, 258 (1978).

35. Wünsch, E., Fontana, A., Drees, F.: Z. Naturforsch. 22 b, 607 (1967).

36. Wünsch, E.: Gut Hormones (ed. S.R. Bloom) Churchill Livingstone, Edinburgh, p. 28 (1978).

37. Moroder, L., Wilschowitz, L., Jaeger, E., Knof, S., Thamm, P., Wünsch E.: Hormone Receptors in Digestion and Nutrition (eds. G.Rosselin, P.Fromageot, S.Bonfils), Elsevier North-Holland Biomedical Press, Amsterdam, p. 129 (1979).

38. Moroder, L., Wilschowitz, L., Jaeger, E., Knof, S., Thamm, P., Wünsch, E.: Z. Physiol. Chem. 360, 787 (1979).

AMATOXINS AND PHALLOTOXINS - STRUCTURE AND TOXICITY

Theodor Wieland

Max-Planck-Institut für medizinische Forschung,
Abteilung Naturstoff-Chemie, Jahnstraße 29
D-6900 Heidelberg, Germany

Introduction

The poisonous mushrooms Amanita phalloides, A. verna and A. virosa contain two families of toxic compounds, the phallotoxins and the amatoxins (1), A. virosa an additional one, the virotoxins (2). Since the phallotoxins and the virotoxins are 10 times less toxic than the amatoxins and, in lethal doses (> 2 mg per kg body weight) will lead to death already within several hours, at least in the white mouse, we assume that they will not contribute to the deadly intoxication of men after mushroom ingestion. However, the phallotoxins have proven to be interesting compounds from the chemical point of view as well as from their biochemical implications. At first the chemistry and toxicology of the amatoxins will be treated.

A. The Amatoxins

Toxicological aspects

The amatoxins cause death within several days by an inhibition of the protein synthesis mainly in liver and kidneys leading to a necrosis of the organs. As early as in 1954 together with K. Dose (3) I could demonstrate by paper electrophoresis in the blood of frogs that predominantly one protein component was missing after 48 hours intoxication. An electron-microscopical

Structure and Activity of Natural Peptides

observation of alterations in the nuclei of kidney and liver cells of amanitin-intoxicated mice by L. Fiume and R. Laschi in 1965 (4) gave the signal for the crucial findings of several European and American groups (5) of the high and specific sensitivity of the DNA-dependent RNA-polymerase II (or B) for the amatoxins. Since this enzyme is responsible for the synthesis of m-RNA's a progressive deficit of proteins in liver and kidneys will occur thus leading to necrotic degeneration of the organ. The site of attack of the enzyme is its subunit SB3 as has been found by Brodner using a radioactive affinity labelling derivative of the toxin (6). The inhibitory constant of the α-amanitin is rather low, $K_I = 0.5 \times 10^{-8}$M. To give an idea of that tiny concentration, it means that the enzyme will be inhibited by 50% in a solution of 0.5 mg of the toxin in 100 l of water! This is true for all RNA polymerases II (or B) of vertebrate origin; the analogous enzyme of yeast is ca. 1000 times less susceptible, other RNA polymerases are also sensitive, but much less so (figure 1). Due to its high specificity α-amanitin can be used as an indicator in all those processes where an involvement of RNA polymerase II (or B) is in question.

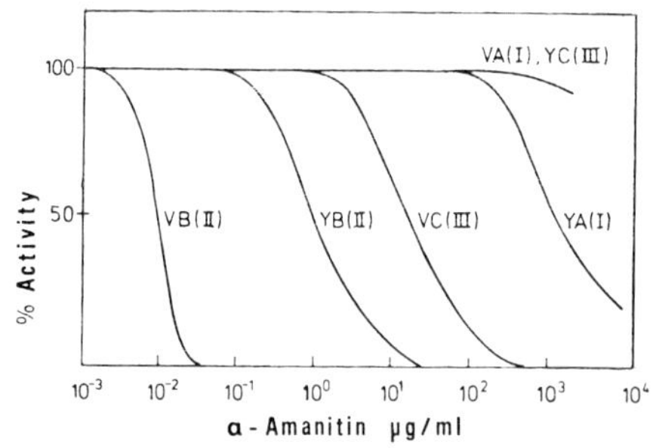

Figure 1. Inhibition curves of the different RNA polymerases A (or I), B (or II), and C (or III) of vertebrates (V) and yeast (Y) by α-amanitin.

Chemistry of amatoxins

The family of the amatoxins consists of 8 defined members all of which have in common a cyclic peptide skeleton consisting

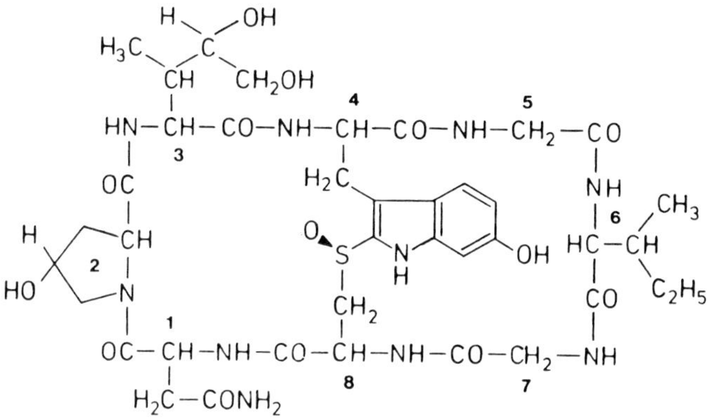

α-Amanitin. In β-amanitin: COOH instead of $CONH_2$ in 1.

of 8 (L) amino acids which is divided by a sulfur-containing bridge. The amatoxins whose main representatives, α- and β-amanitin, are only shown, are composed of the same parent amino acids, the differences provided only by substitutions in the side chains. β-Amanitin is the carboxylic acid of which α-amanitin is the amide derivative. The main differences of the amatoxins rest in their content of more or less oxygen atoms, proamanullin having the lowest and α-amanitin the highest number of hydroxyls. The oxygen atom at the sulfur bridge has been found to be arranged in the configuration of an (R)-sulfoxide. The hydroxyl group at the indole nucleus renders the molecules capable to give a deep violet colour reaction with cinnamic aldehyde in the presence of strong hydrochloric acid and a similar one with lignin components of newsprint (7).

Colour reaction

The newsprint colour reaction allows to detect not less than 0.02 mg of an amatoxin per ml solution. Since in the sap of a mushroom the concentration is ca. 10 times higher it is feasible to recognize toxic Amanita mushrooms by this test: a drop of the sap is dried on an empty place of a newsprint sheet and thereafter spotted with a few drops of 8-10 N hydrochloric acid. A blue or blue greenish spot will appear within 5 to 15 minutes if amatoxins are present (8).

Three-dimensional formulae

A space filling structural formula of α-amanitin (figure 2b) has been constructed after an x-ray analysis in the laboratory of W.N. Lipscomb (9). At the same time a ^{1}H NMR investigation was performed of α-amanitin in dimethylsulfoxide solution (10). The space model designed using the data obtained (figure 2a) is very well agreeing with that from the crystal structure analysis.

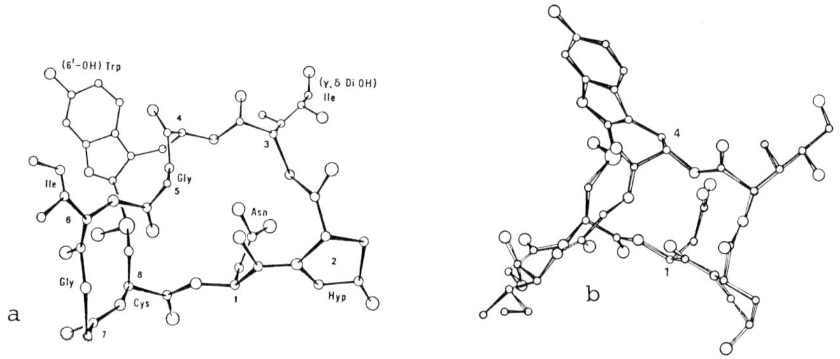

Figure 2a. Conformation of α-amanitin as determined by ^{1}H NMR analysis. 2b. Structure of β-amanitin as determined by x-ray analysis.

Structure and toxicity of the amatoxins

The structural features responsible for high toxicity have been explored not only by comparison of the naturally occuring toxins, but also by checking the toxicities and affinities for RNA polymerase II of derivatives obtained by chemical modifications of amatoxins.
Most important is the bicyclic structure of the octapeptides: after removal of the sulfoxide bridge by treating with Raney-nickel there result entirely non-toxic dethio compounds.

The hydroxylated isoleucine side-chain. The role of the branched side-chain in positon 3 has been investigated thoroughly. As nature shows, at least one hydroxyl group is necessary there in order to cause firm adaption to the polymerase enzyme. Amanullin which lacks any OH in the side-chain is less inhibitory and non-toxic. On oxidation by KIO_4^- the side-chain no. 3 of 6'-O-methyl-α-amanitin can be split to yield the one C-atom less-containing aldo-compound. This compound differing slightly in its CD spectrum from the amatoxins shows no toxicity and has a much higher K_I (table 1). The toxic effect is restored partially after hydrogenation of the aldehyde group with $NaBH_4$ yielding dehydroxymethyl-O-methyl-α-amanitin. This points to an essential role of at least one OH group but additionally

Table 1. Toxicity and inhibitory constants for RNA polymerase B (II) of amatoxins modified in the side-chain of the 3-position.

Side chain of residue no.3	Compound	Toxicity	K_I [M]
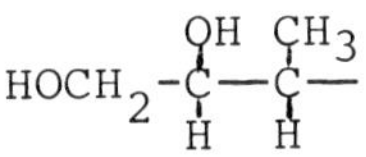	α-Amanitin	++	0.5×10^{-8}
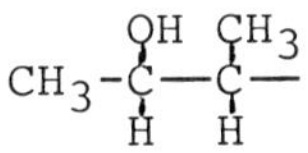	γ-Amanitin	++	0.5×10^{-8}
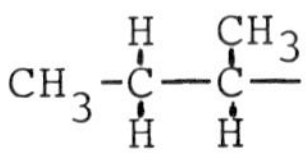	Amanullin	-	$\sim 10^{-8}$
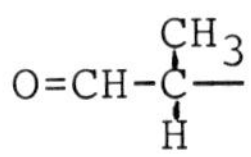	Aldo-amatoxin	-	$> 10^{-6}$
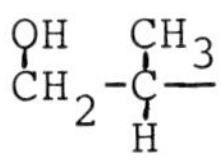	Dehydroxymethyl-α-amanitin	+	$\sim 10^{-8}$
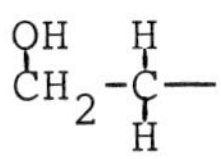	Dinor-S-deoxo-compound	-	$> 10^{-6}$

shows that also a number of four carbon atoms gives rise to a stronger binding than a three carbon containing one. Additional information came from a product of total synthesis, dinor-S-desoxoamaninamide (11). This compound with a side-chain of only two C-atoms showed a 100 times less inhibitory effect than the toxic amatoxins and so demonstrates the necessity of a β-branched side-chain for strong binding to the polymerase enzyme.

Indispensable is also the OH group of the hydroxyproline building block of the amatoxins. Proamanullin, a naturally occuring peptide containing proline instead of hydroxyproline has a 100 times smaller affinity to the enzyme than the strong inhibitors have.

<u>The sulfoxide part.</u> The oxygen atom of the (R)-sulfoxide moiety of all amatoxins is dispensable. The methylether of α-amanitin was deoxygenated at its sulfoxide moiety and so transformed to the thioether S-deoxo-O-methyl-α-amanitin, which

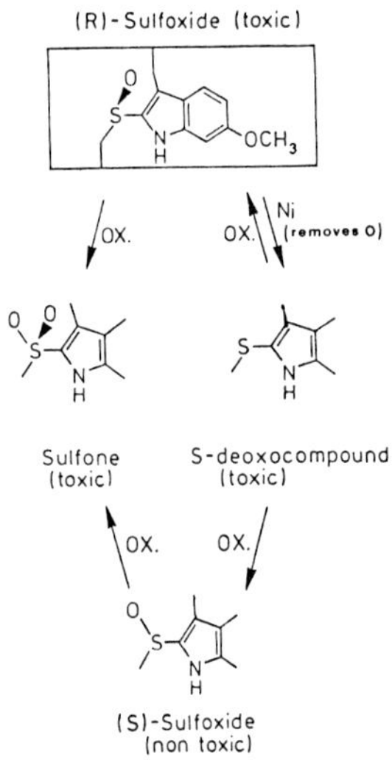

Scheme I.

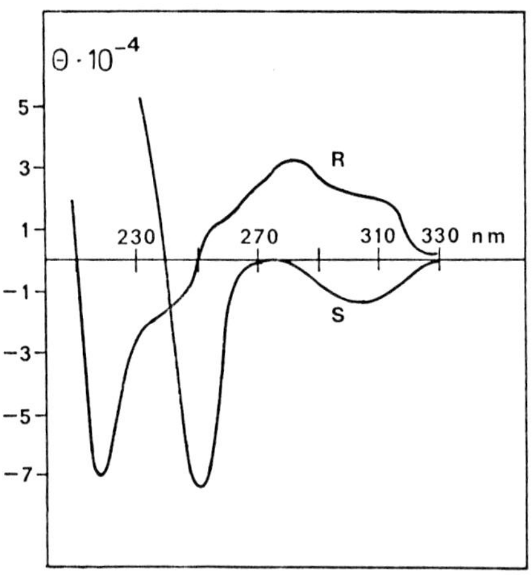

Figure 3. CD-Spectra in water of the R- and S-sulfoxides of 6'-O-methyl-α-amanitin.

was as toxic as the sulfoxide (Scheme I). On oxidation of the thioether with peroxyacetic acid besides the original (R)-sulfoxide also the diastereomeric (S)-compound was obtained which proved non-toxic (12). Accordingly was the inhibitory effect on RNA polymerase II (or B) with the (S)-sulfoxide ten times less than with the (R)-sulfoxide. The sulfone which was formed on further oxidation of both of the sulfoxides showed full toxicity and strong inhibition of the polymerase.

The weaker binding of the (S)-sulfoxide to the enzyme is most probably due to an altered conformation. This is reflected by the different shapes of the CD curves of the (S)- and the (R)-compound as shown in figure 3.

B. The Phallotoxins

The phallotoxins are bicyclic heptapeptides also containing a sulfur bridge. As depicted in the formulae the phallotoxins offer a similar picture as the amatoxins in that also here the extent of hydroxylation characterizes the individual members. From prophalloin, a non-toxic putative precursor with L-proline instead of allo-hydroxy-L-proline up to phallisin

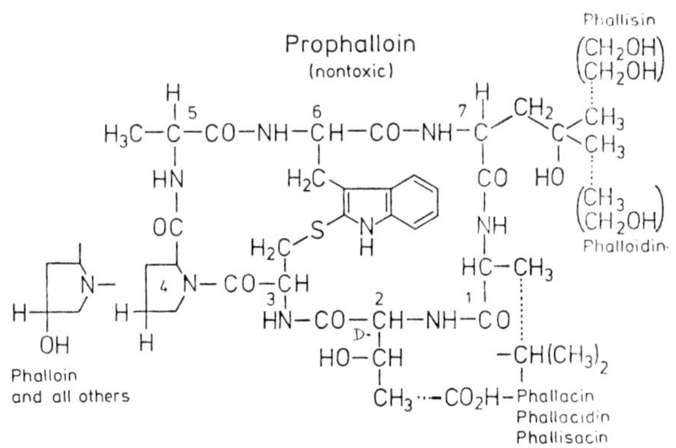

The naturally occurring phallotoxins

with three hydroxyl groups in the L-leucine side chain, we have all degrees of hydroxylation. Other natural varieties are an erythro-β-hydroxy-D-α-aminobutyric acid residue in the acidic phallotoxins instead of D-threonine in the neutral ones. The formula of phalloidin, the main representative of the phallotoxin family, is represented alone in figure 4 (a) together with a drawing of the conformation of phalloin (b) resulting form ^{1}H NMR data and minimal energy calculations (13).

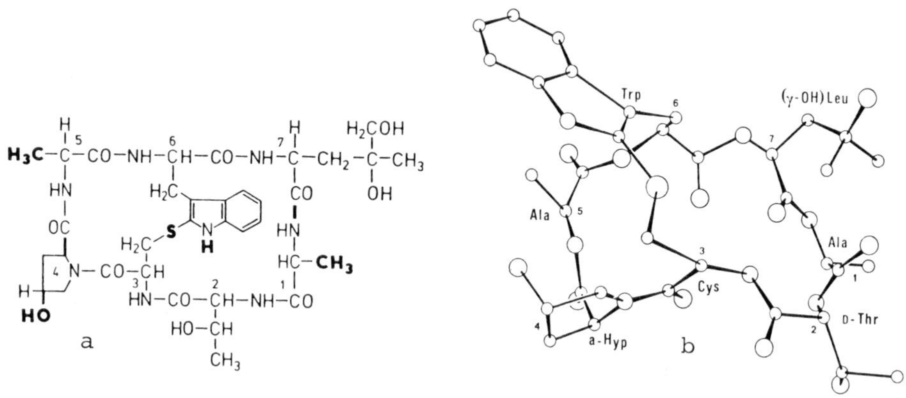

Figure 4 a. Formula of phalloidin. Bold-typed are the parts essental for toxicity; b. Structural formula of phalloin as determined by ^{1}H NMR analysis.

Structure and toxicity of the phallotoxins

Monocyclic derivatives of phalloidin are non-toxic. This was stated for dethiophalloidin obtained from phalloidin by treatment with Raney nickel as well as for seco-phalloidin, the monocyclic product of treating phalloidin with cold acids. Acids will cleave preferably the peptide bond between residues 7 and 1 due to the neighbouring γ-OH group, which easily forms a lactone with the carboxyl originally involved in the peptide bond (see below). The non-toxicity found for both monocyclic derivatives may be understood on the basis of changes in the

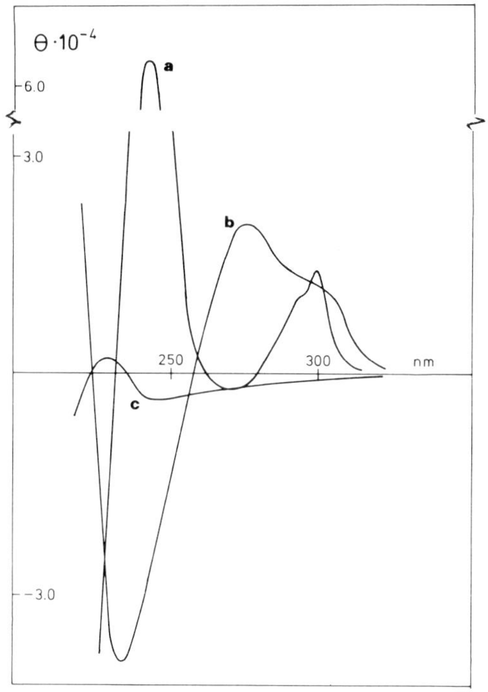

Figure 5.CD-Spectra in water of phalloidin (a), seco-phalloidin (b) and dethiophalloidin (c).

conformation as documented by the CD-curves (figure 5). They differ strongly from that of phalloidin indicating that in the seco-compound the chromophoric moiety has changed its spatial arrangement.

Attempts to recognize the contribution to toxicity of the various side-chains we started with the oxidative degradation of the 1,2-glycol structure in phalloidin by periodate (5). The product, ketophalloidin, was as toxic as the mother compound and it served as the starting material for many further derivatives. For example, hydrogenation of ketophalloidin by $NaBH_4$ yielded demethylphalloin, which is as toxic as phalloin. Its reaction with 1,2-ethanedithiol gave the toxic dithiolane, which after treatment with Raney nickel led to norphalloin. The latter derivative was even a little more toxic than phalloidin. These derivatives indicate that most of the modifications in the leucine-7 side chain do not affect toxicity (Scheme II).

Recently a dithiolane was prepared whose 5-membered ring bears an aminomethyl group (14). On reaction with fluoresceine isothiocyanate a fluorescent phallotoxin was obtained. Some applications in cell biology are described at the end of this article. It has already been mentioned that a hydroxy group in the proline residue is an essential condition of toxicity.

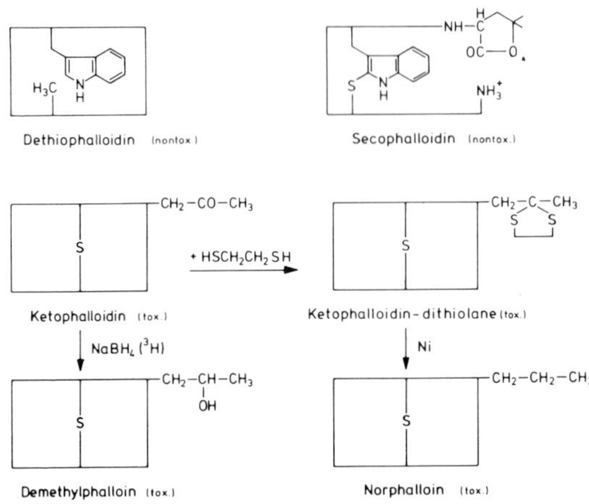

Scheme II. Nontoxic and toxic derivatives of phalloidin.

This group must be arranged in the cis-position to carboxyl (allo) since a synthetic product with trans-hydroxyproline was non-toxic. From the non-toxicity of Gly^5-norphalloin it follows that also the methyl group of alanine-5 is indispensable. Of the other analogs prepared by synthesis (5) only those will be mentioned additionally, which differ from the natural pro-

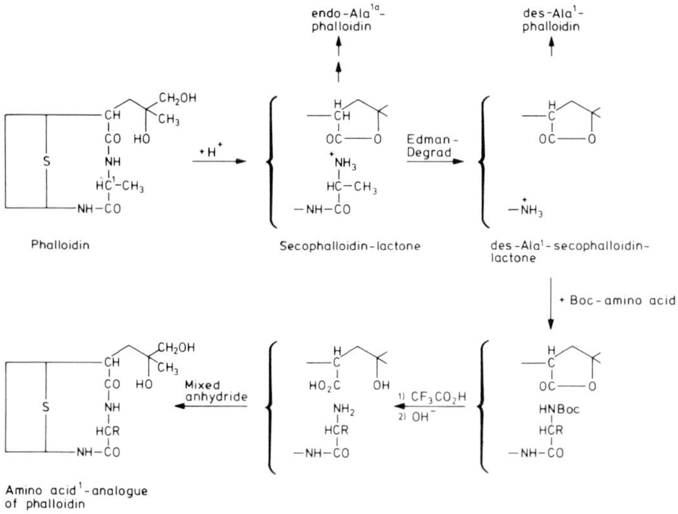

Scheme III. Replacement of alanine-1 in phalloidin by other amino acids (E. Munekata (15).

ducts by the nature of amino acid residue no. 1. They have been obtained by Edman degradation of seco-phalloidin, followed by coupling of various amino acids and recyclization (15) (see Scheme III).

This enabled us to prepare analogues of phalloidin substituted in position 1. Gly^1-phalloidin is 3-4 times less toxic than phalloidin and Phe^1-phalloidin even 10 times while Val^1-phalloidin is as toxic as phalloidin. These results suggest an optimum length of this side chain of 1-3 carbon atoms. D-Alanine in position 1 gives an entirely non-toxic analog.

Symptoms of intoxication

Most of the studies have been done in the rat by intraperitoneal application. Here, phalloidin when given in a dose of > 1 mg per kg body weight causes a severe swelling of the liver; at the end the weight of the liver is 2-3 times that of a normal rat. Such swelling is due to an excessive accumulation of blood in this organ, the blood content being five times higher than that in controls. The resulting depletion of fluid in the periphery probably leads to death of the animals in a hemorrhagic shock.

Such toxic lesions were observed exclusively in the liver. This is a consequence of the specific accumulation of the toxin in this organ (16). In less than ten minutes, the rat liver takes up more then 70 % of the toxin present in the circulatory system. The mechanism of such rapid and specific uptake is still unknown (17). For comparison, the rate of uptake for amatoxins is about 20 times slower. In addition to this, phallotoxins cannot be excreted into the bile like the amatoxins, because the bile flow stops a few minutes after administration of phalloidin.

In search of the cellular events preceding the accumulation of blood in the liver, a study by electron microscopy established that the hepatocytes become interspersed with vacuoles. These vacuoles develop from invaginations of the membranes near the sinusoidal space as indicated by erythrocytes and other blood components present in the interior of the vacuoles. An event very similar to this occurs, when isolated hepatocytes are incubated with phalloidin. In this case exvaginations develop, instead of invaginations giving rise to numerous bulbous protrusions or blebs on the cell surface (18). In both cases the plasma membrane of the hepatocytes may be regarded as to have lost its inherent elasticity. As a consequence, the membrane can be deformed by even low pressure gradients such as the moderate pressure of the portal veins in vivo or a slight intracellular pressure apparently present in the isolated hepatocytes.

Interaction of phallotoxins with actin

Looking for a possible impairment of the hepatocyte membrane isolated plasma membranes from livers of rats poisoned with phalloidin were examined by electron microscopy (19). The membrane preparation contained bundles of microfilaments, which were associated with the membrane fragments. Such filaments were only occasionally visible in the membrane preparations of control animals. The filaments behaved unlike actin in that they withstood treatment with chaotropic ions, such as 0.6 M KI. Nevertheless, they were actin, as documented by their reaction with heavy meromyosin fragments forming the typical arrowhead like structures (20). These experiments gave a first hint that phalloidin had interfered with the actin of rat liver cells.

Phallotoxins bind to actin

Rabbit muscle actin in presence of phalloidin likewise became stable against chaotropic ions. We learned that such stabilization represented the specific effect of phalloidin on actin. The fact that liver cell actin and muscle actin behaved very similar to the toxin prompted us to further investigate this interaction with rabbit muscle actin as a model protein. Most of the results obtained with muslce actin have been reviewed in several papers (2) and will be summarized here only briefly.

Phalloidin binds firmly to actin, the K_D of the toxin being as low as 3.6×10^{-8}M (22). It binds to polymeric or oligomeric actin only, forming a 1:1 complex with the protomers in the filaments. No binding has been observed so far with actin monomers. Binding can be followed by difference spectroscopy in the u.v. The difference spectrum has three maxima at 288, 295 and 305 nm; the latter strongly suggests the participation of the thioether moiety, while the two others may also originate from tyrosine of tryptophan residues in the protein by interaction with the toxin.

In the electron microscope, there is no difference between normal actin filaments and those having complexed phalloidin. However, the adduct phalloidin-F-actin (Ph-actin) differs from F-actin in that it is much more stable against various physical or chemical treatments, which normally depolymerize or destroy filamentous actin.

Among these treatments is 0.6 M KI, in whose presence no depolymerization was observed by viscosimetry (23). Phalloidin even admits a slow polymerisation reaction of G-actin in 0.6 M KI which is not observed in absence of the toxin. Other destructing agents are proteolytic enzymes. Subtilisin as well as trypsin cannot digest F-actin if phalloidin has been added together with K^+-ions to the solution (24).

The question whether affinity to actin runs parallel with the toxicity of the phallotoxins can be answered unambiguously.

Table 2. Toxicity (LD_{50}, mg/kg in white mice) and relative affinity to F-actin of several amino acid-1 analogues of phalloidin.

Amino acid in position 1	Toxicity	Affinity to F-actin (phalloidin = 1.00)
Ala [a]	2.0	1.00
D-Ala	non-toxic [b]	< 0.02
Gly	7.5	-
Val	2.5	0.72
Leu	2.5	0.28
Phe	20.0	< 0.02
des-Ala	non-toxic [b]	< 0.02
Ala-Ala	non-toxic [b]	< 0.02

a = phalloidin; b = tested in doses up to 30 mg/kg.

Table 2 shows e.g. that of the phallotoxin analogs possessing exchanged amino acids in position 1 only those are considerably toxic which bind to F-actin with an affinity comparable to that of phalloidin.

Some applications of phallotoxins in cell biology

Since the phallotoxins react with actin they can be used as tools in cell research. Today the general opinion is that cell actin causes all kinds of motility and participates in processes where contractions occur e.g. secretion of cell products. So the influence of phalloidin and several derivatives on the secretion of triglycerides from hepatocytes was studied by a Swiss group (25). In table 3 it can be seen that this process is markedly inhibited by phalloidin and the toxic R-sulfoxide but much less or virtually not by non-toxic derivatives.

The specific affinity of phallotoxins to F-actin has been used for the visualization of actin in various cells. A fluorescent derivative of ketophalloidin (FL-PH) mentioned above

Table 3. Relationships between stabilization of F-actin against heat denaturation (26), affinity for F-actin, inhibition of triglyceride secretion of rat hepatocytes (25) and LD_{50} in the white mouse of various phallopeptides.

Substance	Heat denatured actin after 3 min at 70°C, %	Affinity for actin (PHD = 1.0)	Triglyceride secretion µmol/g wt wt.	LD_{50} (white mouse) mg/kg body wt.
none	100	-	3.24±0.11	-
Secophalloidin	96	0	2.97±0.14	> 50
S-Phalloidin-sulfoxide	67	< 0.01	2.80±0.17	> 20
R-Phalloidin-sulfoxide	24	0.1	1.54±0.22	2.5
Phalloidin	12	1.0	1.34±0.02	2.0

had an affinity high enough to stain F-actin in various cell preparations: Sections from rat liver were fixed with formaldehyde and incubated with a solution of FL-PH and rinsed (26). The fluorescence microscopic picture (figure 6a) shows that F-actin is mainly localized along the inside of the cytoplasmic membrane and around the bile canaliculi of the paren-

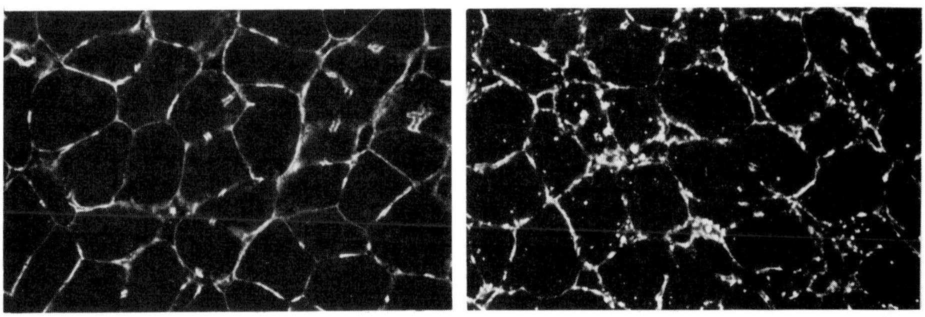

a b

Figure 6. Cryo-sections of rat liver stained with fluorescent phallotoxin. a: Normal rat liver; F-actin along the cell borders and around bile canaliculi. b: Liver perfused for 2 hours with a medium containing 0.5 mg of phalloidin. Layer along cell borders disrupted, part of F-actin concentrated in clusters (27). Magnification 600-fold.

chymal cells (where it presumably has a contracting, secretory function).

In figure 6b an analogous preparation of a liver is shown which has been perfused with phalloidin containing medium a few hours before. The sections after fixation were extracted with dimethylsulfoxide-water in order to remove the bound toxin and then stained with FL-PH as before. Clearly F-actin now is accumulated in discrete clusters, the even layer along normal cytoplasmic membranes being destroyed. This seems to be the reason why the liver cell membrane loses its "elasticity" on intoxication of the animal.

The fluorescent phallotoxin has also been used for the visualization of F-actin in cultures of various vertebrate cells (28). In figure 7 a rat-kangaroo fibroblast cell in the interphase is shown which after fixation with formaldehyde has been stained with FL-PH.

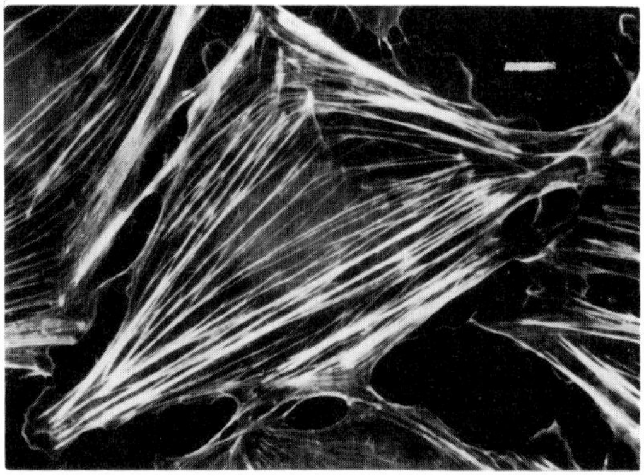

Figure 7. Rat kangaroo fibroblast cell stained with fluorescent phallotoxin after fixation with 4 percent formaldehyde (28). Length of the bar upper right = 10 μm.

Brightly fluorescent are the long parallel fibers of F-actin traversing the whole cell and situated submembranously on the adhesive side of the cell.

F-Actin has been visualized already previously in other laboratories using immuno fluorescent techniques. The FL-PH, however, may have some advantages in penetrating due to its relative small size.

References

1. Wieland, Th.: Chemie in unserer Zeit 13, 56-63 (1979).
2. Faulstich, H., Georgopoulos, D., Bloching, M., Wieland, Th.: Z.Naturforsch. 29c, 86-88 (1974).
3. Wieland, Th., Dose, K.: Biochem.Z. 325, 439-447 (1954).
4. Fiume, L., Laschi, R.: Sperimentale 115, 288-297 (1965).
5. For a review see: Wieland, Th., Faulstich, H.: Amatoxins, Phallotoxins, Phallolysin and Antamanide: The Biologically Active Components of Poisonous Amanita Mushrooms. Crit. Rev.Biochem. 5, 185-260 (1978).
6. Brodner, O.G., Wieland, Th.: Biochemistry 15, 3480-3484 (1976).
7. Wieland, Th., Wirth, L., Fischer, E.: Liebigs Ann.Chem. 654, 152-160 (1949).
8. Zeitungspapier-Test für Giftpilze: Umschau 78, 611 (1978).
9. Kostansek, E.C., Lipscomb, W.N., Yocum, R.R., Tiessen, W.E.: J.Amer.Chem.Soc. 99, 1273-1274 (1977); Biochemistry 17, 3790-3795 (1978).
10. Tonelli, A.E., Patel, D.J., Wieland, Th., Faulstich, H.: Biopolymers 17, 1973-1986 (1978).
11. Buku, A., Altmann, R., Wieland, Th.: Liebigs Ann.Chem. 417-431 (1976).
12. Buku, A., Altmann, R., Wieland, Th.: Liebigs Ann.Chem. 1580-1586 (1974).
13. Patel, D.J., Tonelli, A.E., Pfaender, P., Faulstich, H., Wieland, Th.: J.Mol.Biol. 79, 185-196 (1973).
14. Wieland, Th., Deboben, A., Faulstich, H.: Liebigs Ann. Chem., in press (1980).
15. Munekata, E., Faulstich, H., Wieland, Th.: J.Amer.Chem. Soc. 99, 6151-6153 (1977).
16. Faulstich, H., Wieland, Th., Schimassek, H., Walli, A.J., Ehler, N.: Mechanisms of phalloidin intoxication in: Membrane alterations as basis of liver injury. In: Falk-Symposium 22 (H.Popper, L.Bianci, W.Reuter, eds.) MTP Press, Lancaster 1977

17. For a recent hypothesis see: Frimmer, M.: Biologie in unserer Zeit 9, 147-152 (1979).

18. Weiss, E., Sterz, I., Frimmer, M., Kroker, R.: Beitr. Pathol. 150, 345-356 (1973).

19. Govindan, V.M., Faulstich, H., Wieland, Th., Agostini, B., Hasselbach, W.: Naturwissenschaften 59, 521-522 (1972).

20. Lengsfeld, A.M., Löw, I., Wieland, Th., Dancker, P., Hasselbach, W.: Proc.Natl.Acad.Sci. 71, 2803-2807 (1974).

21. e.g.: Wieland, Th.: Naturwissenschaften 64, 303-309 (1977).

22. Faulstich, H., Schäfer, A.J., Weckauf, M: Hoppe-Seyler's Z.Physiol.Chem. 358, 181-184 (1977).

23. Löw, I., Wieland, Th.: FEBS Letters 44, 340-343 (1974).

24. de Vries, J., Wieland, Th.: Biochemistry 17, 1965-1968 (1978).

25. Prentki, M., Chaponnier, Chr., Jeanrenaud, B., Gabbiani, G.: J.Cell Biology 81, 592-606 (1979).

26. de Vries, J., Schäfer, A.J., Faulstich, H., Wieland, Th.: Hoppe-Seyler's Z.Physiol.Chem. 357, 1139-1143 (1976).

27. Jahn, W., Faulstich, H., Deboben, A., Wieland, Th.: submitted for publication (1979).

28. Wulf, E., Deboben, A., Bautz, F.A., Faulstich, H., Wieland, Th.: Proc.Natl.Acad.Sci. USA 76, 4498-4502 (1979).

INSULIN, A CHALLENGE FOR PEPTIDE CHEMISTRY

Rolf Geiger
Hoechst Aktiengesellschaft, D-6230 Frankfurt(Main) 80

After the recognition of its protein character and even before the elucidation of the chemical structure insulin became already a model for chemical modifications at proteins to disclose relationships between functional groups of the amino acid constituents and biological activity of the molecule (1).

While the analytical progress in the determination of blood glucose between 1913 and 1920 was a prerequisite for the later enrichment and isolation of insulin, the new technique of protein sequencing, developed by Sanger, proved its value in the structure elucidation of the hormone.

Similarly, progress in peptide synthesis challenged the preparation of increasingly complicated peptides.

Abbreviations:

Bu^t	tert-butyl
Boc	tert-butyloxycarbonyl
Bpoc	2-(p-biphenyl)-isopropyloxycarbonyl
Msc	methylsulfonyl-ethyloxycarbonyl
Tfa	trifluoroacetyl
Pht	phthaloyl
Trt	trityl
Acm	acetamidomethyl
Ac	acyl
HOBt	1-hydroxybenzotriazole
HOObt	3-hydroxy-4-oxo-3.4-dihydro-1.2.3-benzotriazol
HONSu	N-hydroxysuccinimide
DCC	dicyclohexylcarbodiimide
DMF	dimethylformamide
DBPI	des-B1-phenylalanine-insulin
$Ala(SO_3H)$	cysteic acid

Structure and Activity of Natural Peptides

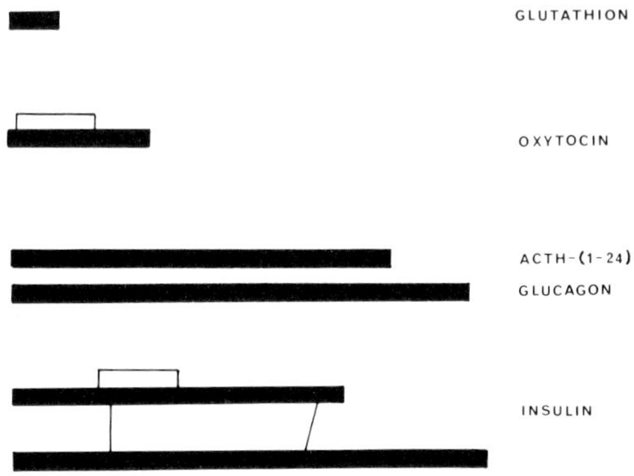

Fig. 1. Some typical peptide hormones.

In figure 1 some peptide hormones are listed to show the extending chain length and complexity of the synthetic peptides prepared. The examples have been choosen because each of them contributed to a distinct progressive step in synthesis strategy and also marks an important stage in our knowledge on hormone action and structure-activity relationship.

In the first half of the fifties synthesis of a peptide like glutathione provided a test for the efficacy of novel synthetic methods while later on oxytocin, vasopressin, angiotensin and similar peptides assumed this role. For instance, the important stepwise peptide synthesis has been elaborated with oxytocin first (2). Even today this hormone as well as vasopressin remain a favorite tool for structure- and conformation-activity studies.

Since about 1960 corticotropin has played a considerable role. Schwyzer developed his strategy for classical segment condensation with corticotropin (ACTH) using tert-butyl-type residues for side-chain protection and benzyloxycarbonyl and other residues for N^{α}-blockade (3).

The importance of corticotropin remains undiminished. Peptides derived of its amino acid sequence turned out to exhibit a significant influence on learning and memory (4) and only recently a biosynthetic precursor containing the amino acid sequences of ACTH, MSH and endorphins has been discovered (5).

In the synthesis of glucagon the Schwyzer-strategy has been developed consistently to maximal protection by Wünsch (6). Another discovery made at this opportunity was the favourable effect of N-hydroxysuccinimide suppressing racemization in the carbodiimide coupling reaction (7).

Already 1963-64 the first syntheses of insulin were announced (8,9). Here most of the problems previously met in the synthesis of the above mentioned hormones were again encountered. Additionally, a new problem arose. The correct combination of the two chains by proper linking of the sulfur atoms was the main goal of every synthesis. Only in the last decade decisive progress has been made in solving this task (9).

Insulin fascinates chemists because with reference to its size it is clearly a peptide. The chains with 21 and 30 amino acids, respectively, can be synthesized with the aid of presently available methods. On the other hand, it behaves like a protein. At the isoelectric point it is sparingly soluble, tending to aggregate. Furthermore, it forms crystals in which the molecules are arranged as hexamers around two zinc ions. Finally, reversible as well as irreversible denaturation is observed.

Thus, synthesis of insulin may pose all problems of peptide synthesis and also come upon every difficulty connected with the protein-like nature of the hormone.

The different biological actions of insulin are detectable by well-elaborated tests. On a cellular level glucose transport into and utilization in muscle and fat cells are determined. The measurement of glucose concentration in the serum mainly represents the liver function of insulin (influence on gluconeogenesis and glycogen depot formation/mobilisation). Description of insulin activity in terms of hormone-receptor interaction is more problematic since such studies have demonstrated that biological activity of hormones can vary with changes induced in the concentration of available hormone receptors (10).

For a chemist two problems are well to the fore:

1. Total synthesis, starting from amino acids
2. Partial synthesis, starting from natural insulin.

The postulate that only an unequivocal synthesis can prove a proposed structure is valid in peptide chemistry, too. In several cases a wrong formula has only been discovered by synthesis of a peptide according to the erroneous structure proposal and comparison of the resulting material with the natural compound.

Thus total synthesis of insulin derives its significance already from this point of view. The question whether chemical synthesis could provide an economical alternative for extraction from pancreatic tissue had to be answered in the negative. Synthetic insulin made by currently available methods proved to be more expensive than natural insulin. Nevertheless, total syntheses of insulin greatly stimulated the development of new methods and synthetic strategies and led to more profound insights into relationships between structure and biological activity.

Figure 2 demonstrates the importance of total synthesis since only by this way the inner amino acids of both A- and B-chains are accessible for amino acid exchange to scan their contribution to the biological effects. Partial synthesis only allows reactions within the terminal regions.

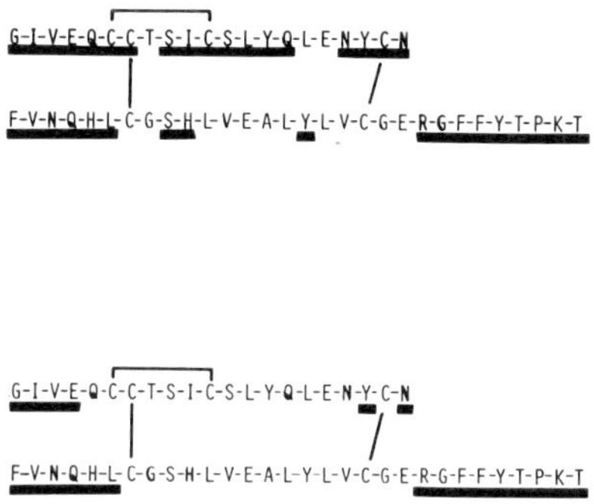

Fig. 2. Exchange and degradation of amino acid residues in total syntheses (above) and partial syntheses (below) (exchanged amino acids underlined).

The total synthesis of insulin is extremely laborious but the principal problems are solved. Three approaches have been successful (9):

1. Random combination of synthetic A- and B-chains
2. Stepwise synthesis of all peptide and disulfide bonds by an unequivocal manner
3. Combination of A- and B-chains with a removable bridge linking the chains.

The random synthesis of disulfide bridges originated in the experiment of Dixon who found 1-2 % biological activity after recombining the previously separated A- and B-chains (9). This observation was the basis for the first attempts to synthesize insulin by the groups of Zahn (Aachen) and Katsoyannis (Pittsburgh) and the chinese team (Shanghai). Du et al. (11) developed an improved technique with an excess of the A-chain and a combination at pH 10.6 to suppress aggregation and polymerization of the B-chain. According to Katsoyannis larger excess of A-chain and use of B-chain as S-sulfonate further increased the yield up to about 20 %, calculated on the amount of B-chain employed (12). Recycling of the excessive material can be considered to improve the yield further. In general, this approach is not highly effective but practicable.

The second route with a predetermined formation of the disulfide bridges was accomplished by Sieber et al. (13) with a sophisticated protecting group and synthesis strategy.

This synthesis followed a principle already proposed by Zervas and further elaborated by Hiskey (9). But only the Ciba-Geigy group succeeded in realizing this approach via the formation of asymmetric disulfides (13).

After the synthesis of a core peptide (scheme 1) only two N^{α}- and two S-protective groups were further required.

Protective groups which had to be eliminated under drastic conditions were out of the question because of the danger of

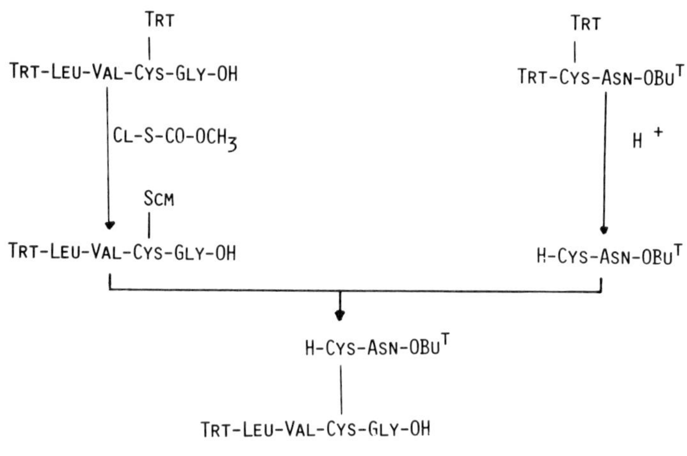

Scheme 1. Synthesis of the asymmetric insulin segment A(20-21)-B(17-20).

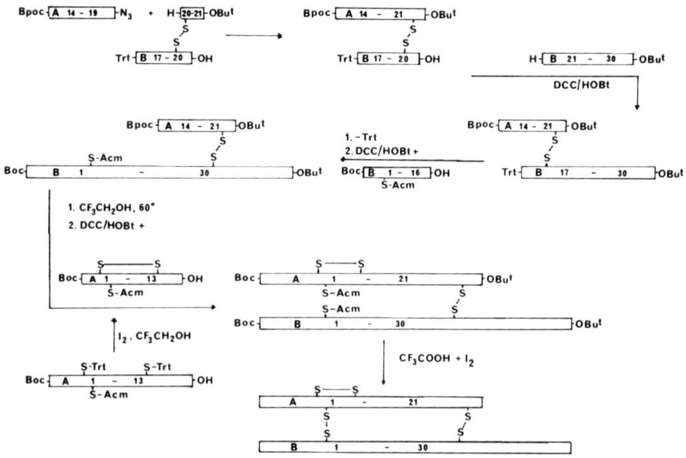

Scheme 2. Synthesis of insulin according to Sieber et al. (13).

disulfide exchange. Thus only tert-butyl-type residues were considered for side-chain protection. For N^{α}-protection two already known groups (Bpoc and Trt) were adapted to the special problem to be cleavable under only slightly different conditions without affecting the tert-butyl residues. The analogous task concerning the S-protective groups was solved by means of acetamidomethyl and trityl residues.

The selective stepwise elimination of these groups with iodine under formation of the corresponding disulfide bonds succeeded. The reaction sequence is outlined in scheme 2.

The high methodical value of this structure-proving synthesis is underlined by the fact that a considerable number of analogues could be prepared in this way without additional difficulties.

The third promising route to insulin comprises the use of a removable bridge linking the chains. This renders the combination of the chains a monomolecular reaction one. The method is derived from semisynthetic work and will be mentioned in that context.

Partial synthesis, although only suitable for derivatisation and amino acid exchange within terminal regions, has been applied intensely during the last decade. It offers a comparatively simple and short way to modified insulins for structure-activity studies.

The very start for a broad working in this field was the observation that insulin reacts with most acylating reagents under certain conditions preferably at the amino groups of A1 and B29, leaving that of B1-phenylalanine free (Fig.3) (9).

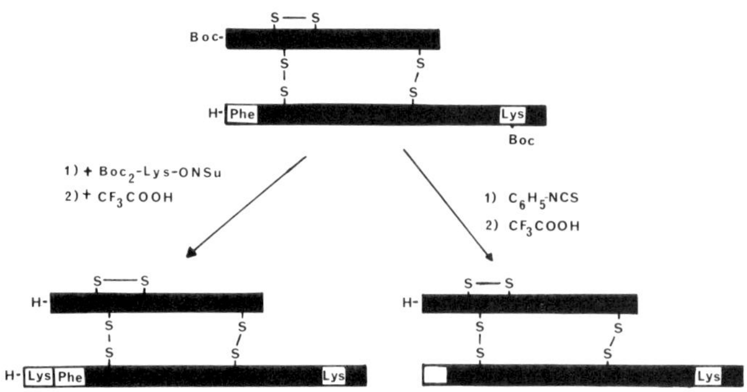

Fig. 3. $N^{\alpha A1}$,$N^{\varepsilon B29}$-Bis-Boc-insulin and its reactions.

With this reaction we are moving on a narrow ridge since under conditions usually applied to peptide bond formation the resulting material encompasses every possible N-acylated compound (Fig. 4).

With the use of easily cleavable protective groups (Table 1) as acyl residues, many techniques of peptide chemistry can be applied to insulin allowing elongation, reduction, or alterations of the individual chains.

Fig. 5 gives an example for stepwise degradation of the B-chain, using the acid-stable Msc residue to protect the N^{α}-

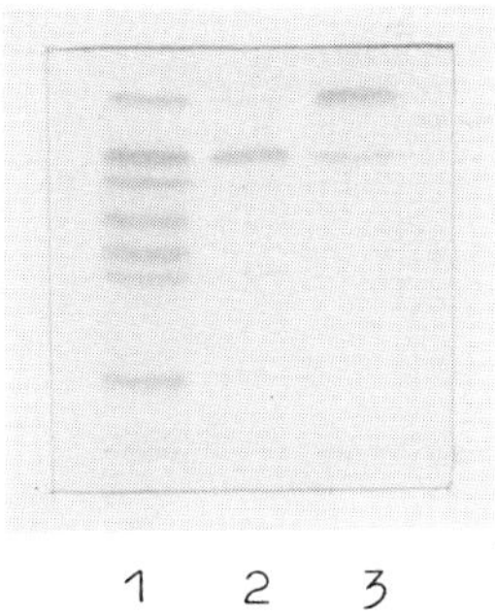

1. pH 7.5 in DMF-Water
2. pH 9.5 in DMF-Water
3. in DMF + Triethylamine (large excess of Boc-N_3)

Fig. 4. Reaction of insulin with Boc-azide under different conditions (tlc).

Residue	Abbrev.	Stability to acid	Stability to alkali	Special Cleavage Conditions
$H_3C-C(CH_3)_2-O-CO-$	Boc	–	+	
$CH_3-SO_2-CH_2-CH_2-O-CO-$	Msc	+	–	
CF_3-CO-	Tfa	+	–	
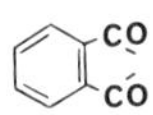	Pht	+	–	Hydrazine in liquified phenol
$Ac-NH-CH(CH_2-CH_2-S-CH_3)-CO-$	Ac-Met	+	+	BrCN

Table 1. Protecting groups for semisyntheses with insulin.

and N^{ε}-amino groups of A1 and B29, respectively, during the Edman degradation cycles. Hybrid insulins can be obtained by exchanging amino acids or peptide segments (Fig. 6).

Such compounds are useful tools for structure-activity and immunological studies as well. For instance, removal of B1-phenylalanine results in an analogue with full biological activity but reduced binding to insulin antibodies (Table 2) (14). Interestingly, this was also true when antibodies

Compound (species: bovine)	Biological activity in vivo	association constants for the binding by guinea pig antibodies raised with bovine insulin
insulin	100 %	$8,3 \times 10^8$ l/m
des-Phe^{B1}- insulin	97	$7,4 \times 10^6$
des-(Phe-Val)$^{B1-2}$- insulin	95	$6,4 \times 10^6$
des-(Phe-Val-Asn)$^{B1-3}$- [pGlu^{B4}]insulin	70	$1,2 \times 10^6$

Table 2. Biological and immunological properties of insulin derivatives.

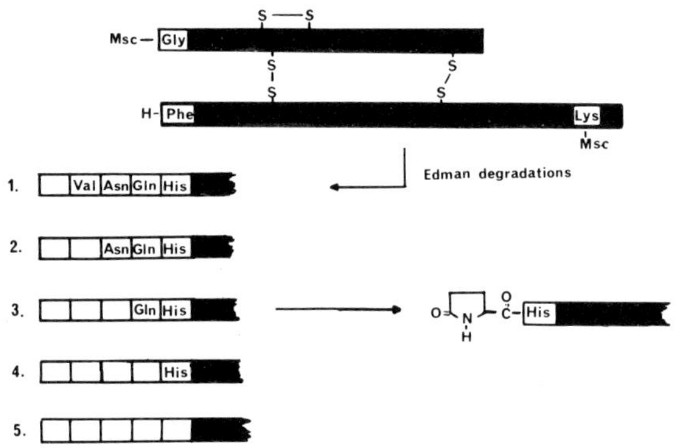

Fig. 5. Stepwise degradation of the B-chain N-terminus of insulin.

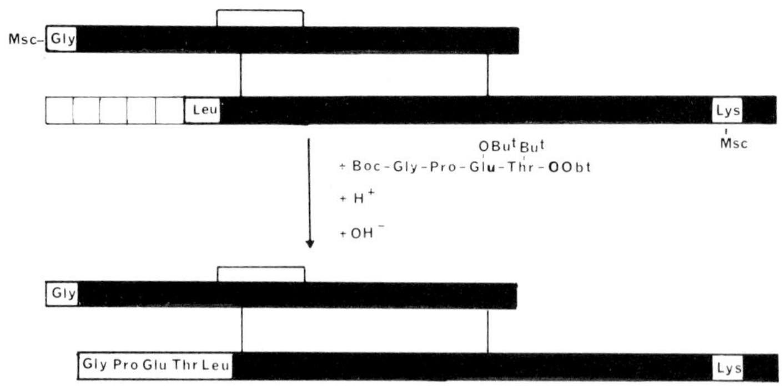

Fig. 6. Semisynthesis of a IGF-I hybrid insulin (IGF = insulin-like growth factor).

against des-B1-phenylalanine-insulin (DBPI) were formed in humans after treatment with the insulin analogue. There is no marked difference between the immunogenicity of DBPI and the parent insulin, porcine or bovine. However, DBPI binds to a lower degree to antibodies,raised against itself, than the parent insulin. Thus the paradoxical situation araises that DBPI produces heterologous antibodies directed more specifically to the parent insulin rather than the homologous antigen. This antigenic behaviour is a valuable advantage of DBPI for the treatment of diabetics with a greatly elevated requirement of insulin caused by a high antibody titre (15).

The position B1 is characterized by having no direct influence on the biological activity in vivo. It is situated on the side of the molecule opposite to the surface binding to the receptor. On the other hand, molecular properties like the tendency to aggregate, solubility, or hydrophilic-hydrophobic behaviour are closely related to structural features within this region. It also seems to influence the structure of areas which determine in part the conformation of the molecule. It is known that the biological activity of insulin derivatives substituted at the amino group of B1 is reduced by enlarging the size of the substituents.

We have recently re-examined the properties of insulin derivatives substituted at the amino group of B1-phenylalanine and found that this statement cannot be generalized (16).

In Table 3 a series of insulin derivatives is listed. These peptides exhibit quite variable biological activities (Table 4). The biological activity drops only with increasing size of a lipophilic substituent. With more hydrophilic polyethylene glycol residues, however, the in vivo activity of the derivatives is retained independently of their size, up to a molecular weight of about 1500. This represents the longest polyethylene glycole (PEG) chain so far attached to the amino group of B1.

R-CO-Phe-Val-Asn-Gln-..... (insulin B-chain)

R =	$CH_3-(CH_2)_4-$	hexanoyl
	$CH_3-(CH_2)_9-$	undecanoyl
	$CH_3-(CH_2)_{16}-$	octadecanoyl
	$HO_3S-CH_2-CH_2-$	3-sulfopropionyl
	$CH_3-O-CH_2-CH_2-O-CO-(CH_2)_2-$	MEG
	$CH_3-(O-CH_2-CH_2)_3-O-CO-(CH_2)_2-$	PEG_{200}
	$CH_3-(O-CH_2-CH_2)_{35}-O-CO-(CH_2)_2-$	PEG_{1500}

Table 3. Insulin derivatives substituted at the amino group of B1-phenylalanine.

Residue at B1	mol. wt.	Biological activity	
		rabbit	fat cell
hexanoyl	100	50	38
undecanoyl	170	40	10 (?)
octadecanoyl	268	0	43
sulfopropionyl	137	100	97
MEG	119	100	26
PEG_{200}	207	100	30
PEG_{1500}	1527	100	36

Table 4. Biological data of insulin derivatives substituted at the amino group of B1-phenylalanine (insulin = 100).

In the fat cell assay both types are nearly equally potent again, independently of the chain length of the individual residue. The only exception is B1-sulfopropionyl insulin which is fully active in both tests.

Insulin derivatives with substitution at B1 generally display a decreased antigenicity compared to insulin. This general observation is also true for these compounds. The affinity of insulin derivatives to insulin antibodies can be tested by displacement of ^{125}I-insulin in vitro (Fig. 7).

The competitive inhibition of ^{125}I-insulin binding to antibodies, raised against porcine or bovine insulin in guinea pigs, by a constant amount of natural insulin ($^*B_i$) or derivative ($^*B_d$) is expressed by the ration $^*B_i/^*B_d$. A low value indicates weak antigenicity (16).

Whereas $[Ala(SO_3H)^{B1}]$insulin (A), an analogue in which B1-phenylalanine is substituted by cysteic acid, binds to antibodies nearly as well as insulin does, the affinity of B1-sulfopropionyl insulin (S) to antibodies is drastically decreased. Thus, a decisive structural feature for the contribution of a sulfo group to antigenicity of insulin derivatives is its position. Exchange in position B1 exhibits a less pronounced effect than substitution at the amino group of B1-phenylalanine. Character and size of such a substituent have only a minor influence.

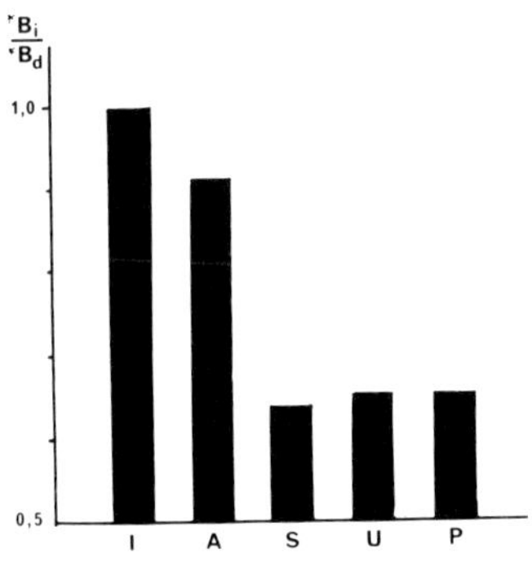

Fig. 7. Immunoreactivity of insulin derivatives.

I Insulin
A $[Ala(SO_3H)^{B1}]$insulin
S B1-Sulfopropionyl-insulin
U B1-Undecanoyl-insulin
P B1-PEG_{1500}-insulin

In contrast to B1-phenylalanine the position of A1-glycine is highly sensitive toward structural variations. Elimination of A1-glycine leads to an almost inactive derivative. Acylation, methylation, or shifting of the amino group in ß-alanine or ε-aminocaproic acid diminish the activity more or less extensively (Table 5) (9, 17).

Residue in A1	biological activity		
	blood sugar lowering (rabbit)	fat cell test	diaphragm test
Sar	80	-	-
ß-Ala	49	58	45
ε-aminocaproyl	33	42	41
L-Ala	20	12	-
D-Ala	96	74	107
L-Leu	-	16	-
D-Leu	69	112	105
L-Glu	-	9	-
D-Glu	81	79	112

Table 5. Biological activities of insulin analogues with different residues in A1 (insulin = 100).

Additional information about this region is obtained by substituting A1-glycine by L- and D-amino acids. Whereas L-amino acids greatly diminish the insulin activity, it is essentially retained with D-amino acids. An enhanced stability to enzymatic degradation has been excluded as a main cause (18), rather is the arrangement of the side-chains considered to be responsible. If one assumes that the amino group of A1 in the analogues always occupies the same distinct position, the side-chain of L-amino acid derivatives would be situated on the surface which is believed to be in contact with the receptor. Side-chains of D-amino acids leave this area accessible and interact only with the region at A4 - A5 which is most likely less important for binding and activity (Fig.8).

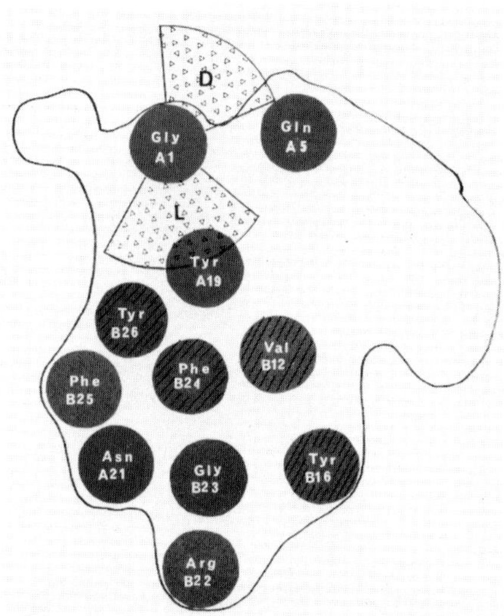

Fig. 8. Possible arrangement of the side-chains of D- and L-amino acids in position A1 of insulin.

The dependence of the different insulin effects on the structure of the side-chains of D-amino acids may give some limited insight into the organization of receptor moieties (19).

The C-terminal regions are not susceptible to similar chemical reactions. However, enzymatic degradation by trypsin yielded desoctapeptide-(B23-30)-insulin. The $N^{\alpha A1}$,$^{\alpha B1}$-bis-Boc-derivative of porcine insulin was reacted with an equimolar amount of partially protected human octapeptide B(23-30) with free N-terminal amino group in the presence of DCC/HOBt to yield 25-40% partially protected human insulin (scheme 3). After deblocking and several purification steps the overall yield was in the range of 5-10% (20). More recently, several approaches have been successful, shifting the equilibrium between cleavage and formation of a peptide bond (21, 22).

As a last reaction sequence deriving from semisynthetic work an approach to total synthesis of insulin made in 1972/73 should be mentioned. The biosynthesis of insulin proceeds via an intermediate single-chain precursor, bearing a peptide chain link between the C-terminus of the B-chain and the N-terminus of the A-chain. Mature insulin is formed after enzymatic removal of the connecting (C) peptide. The vicinity of the amino groups of A1 and B29 made it obvious to substitute the C-peptide by a removable bridge between both amino groups which are fortunately the most reactive ones in insulin. Fig.9 shows several bridges which have been used resulting in about 60-80% yield in the correct formation of the disulfides. The over-all yield was up to 40% (9).

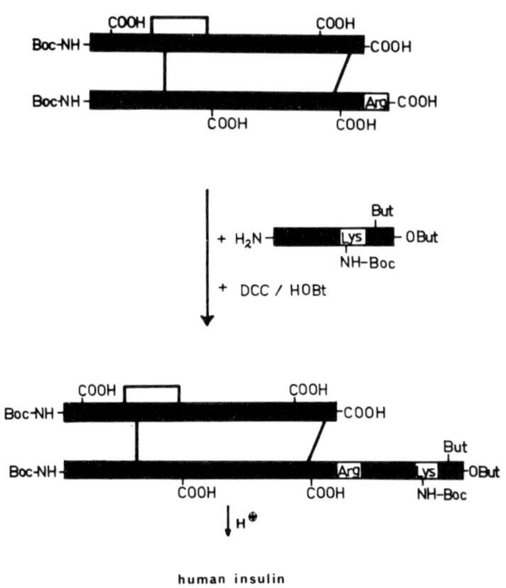

Scheme 3: Human insulin from porcine insulin.

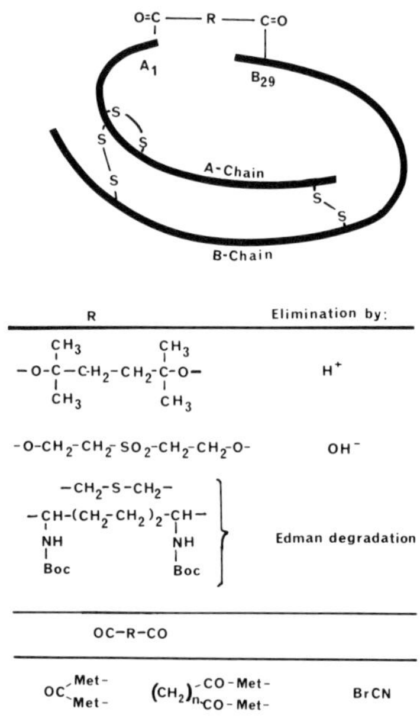

Fig. 9. Crosslinking reagents bridging insulin chains.

The total synthesis of insulin as well as the total and partial syntheses of many derivatives and analogues have improved our knowledge about this complicated molecule which became thus accessible to distinct chemical and enzymatic methods.

Despite the great efforts made mainly during the last 20 years only a few compounds turned out to be of improved value in the treatment of diabetes (15). For the time being, the total synthesis will not provide insulin at a competitive price. There is hope that bacterial synthesis of insulin via genetic modification of plasmides will provide a more economical product.

Another goal, the synthesis of an insulin-like but smaller molecule has not yet been achieved. Attempts to synthesize such molecules have been unsuccessful (17,23). Considering the hitherto existing experience it may be doubtful whether a biologically active insulin of greatly reduced size can be constructed. Our understanding of structural requirements for effective receptor binding is still fragmentary. It is therefore an undiminished challenge to try to fill up all those gaps.

References

1. Fraenkel-Conrat, J., Fraenkel-Conrat, H.: Biochim. Biophys. Acta 5, 89-97 (1950).
2. Bodanszky, M., Du Vigneaud, V.: J. Amer. Chem. Soc. 81, 5688-5691 (1959).
3. Kappeler, H., Schwyzer, R.: Helv. Chim. Acta 44, 1136-1141 (1959).
4. De Wied, D.: Ann. New York Acad. Sci. 297, 263-274 (1977).
5. Nakanishi, S., Inoue, A., Kita, T., Nakamura, M., Chang, A.C.Y., Cohen, S.N., Numa, S.: Nature (London) 278, 423-427 (1979).
6. Wünsch, E.: Z. Naturforsch. 22b, 1269-1276 (1967).
7. Weygand, F., Hoffmann, D., Wünsch, E.: Z. Naturforsch. 21b, 426-428 (1966).

8. Meienhofer, J., Schnabel, E., Bremer, H., Brinkhoff, O., Zabel, R., Sroka, W., Klostermeyer, H., Brandenburg, D., Okuda, T., Zahn, H.: Z. Naturforsch. 18b, 1120-1121(1963).

9. Geiger, R.: Chemiker-Ztg. 100, 111-129 (1976).

10. Roth, J., Kahn, C.R., Lesniak, M.A., Gorder, P., De Meyts, P., Megyesi, K., Neville, Jr., D.M., Gavin, J.R., Soll, A.H., Freychet, P., Goldfine, I.D., Bar, R.S., Archer, J.A.: Rec. Progr. Hormone Research 31, 95-139 (1975).

11. Du, Y.-C., Jiang, R.-Q., Trou, C.-L.: Scientia Sinica 14, 229-236 (1965).

12. Katsoyannis, P.G., Trakatellis, A.C., Johnson, S., Zalut, C., Schwartz, G.: Biochemistry 6, 2642-2655 (1967).

13. Sieber, P., Kamber, B., Hartmann, A., Jöhl, A., Riniker, B., Rittel, W.: Helv. Chim. Acta 57, 2617-2621 (1974).

14. Kerp, L., Steinhilber, S., Kasemir, H., Hahn, J., Henrichs, H.R., Geiger, R.: Diabetes 23, 651-656 (1974).

15. Geiger, R., Enzmann, F.: Excerpta Medica Intntl. Congr. Ser. No. 468. Proinsulin, Insulin, C-Peptide. S.Baba, T. Kaneko, N. Yanaihara Edts., pp. 306-310. Elsevier/North-Holland, Amsterdam · Oxford 1979

16. Geiger, R., Obermeier, R., Teetz, V., Uhmann, R., Summ, H.D., Neubauer, H., Geisen, K., Regitz, G.: Proc. 2nd Internatl. Symposium on Insulin, D. Brandenburg, A. Wollmer Edts., Sept. 4-9, 1979, in press.

17. Geiger, R.: Molecular Endocrinology. I. McIntyre, M. Szelke Edts., pp. 27-41, Elsevier/North-Holland Biomed. Press, Amsterdam 1977

18. Geiger, R., Geisen, K., Summ, H.D., Langner, D.: Hoppe-Seyler's Z. Physiol. Chem. 356, 1635-1649 (1975).

19. Geiger, R., Geisen, K., Regitz, G., Summ, H.D., Langner, D.: Hoppe-Seyler's Z. Physiol. Chem. 361, in press.

20. Obermeier, R., Geiger, R., Hoppe- Seyler's Z. Physiol. Chem. 357, 759-767 (1976).

21. Inouye, K., Watanabe, K., Morihara, K., Tochino, Y., Kanaya, T., Emura, J., Sakakibara, S.: J. Amer. Chem. Soc. 101, 751-752 (1979).

22. Morihara, K., Oka, T., Tsuzuki, H.: Nature (London) 280, 412-413 (1979).

23. Geiger, R., Teetz, V., König, W., Obermeier, R.: Semisynthetic Peptides and Proteins, R.E. Offord, C. di Bello Edts., pp. 141-159, Academic Press, London · New York · San Francisco 1978

RESEARCH ON THYMIC PEPTIDES AND IMMUNOCOMPETENCE

Karl Folkers, Georg Rampold, Naoki Sakura, Johann Leban, Jan Dahmen, Elsa Lundanes, Masahiro Ohta, Ding Chang, Yieh-Ping Wan,

Institute for Biomedical Research, The University of Texas at Austin
Austin, Texas 78712

Wolfgang Voelter, Hubert Kalbacher

Chemisches Institut der Universität Tübingen
74 Tübingen, West Germany

INTRODUCTION

Hormones of the pituitary, adrenal, and pancreas glands have been widely investigated for decades, but it was only in the early 1960's that evidence began to appear which indicated that there may be factors or hormones in the thymus gland which are essential for immunocompetence. There could be just one such hormone. However, the thymus has been investigated for over 150 years in other respects.

Miller and Osoba (1) in 1963, provided direct evidence that the thymus could produce a humoral factor which could be important to the development of immunity. Such evidence was based upon the discovery that deficient immune responses in mice after neonatal thymectomy were prevented by implanting graphs of thymus tissue. In 1965, Osoba (2) conceived the idea that thymus lymphocytes are functionally immature _in situ_, but that after these lymphocytes leave the thymus, they obtain immunological competence through maturation under the influence of a "competence-inducing factor" or hormone produced by the thymus. Osoba also found that this "competence-inducing factor" was in the thymus and not in the spleen or lymph nodes. These and other studies by many investigators were largely physiological, but they set the stage for the entrance of new chemistry of fractionation towards the isolation of one or more such thymic hormones.

As individuals become older, certain normal immune functions are believed to decline. Associated with this decline, is the emergence of diseases which can profoundly affect many tissues.

The immune system is a challenge for both cellular and molecular research, and offers great promise for successful experimental manipulation. Prevention or delay or reversal of the decline in normal immune functions may delay the onset or lessen the severity of diseases associated with aging. Aging is characterized by a declining capacity of an individual to adapt to environmental stress. The immune system is "organismal", because it is in constant contact with most, if not all, of the cells, tissues, and organ systems within the body. Thus, any decline of the immune system could be expected to affect all the other systems.

Many studies today indicate that the process of involution and atrophy of the thymus is the key to the aging of the immune system. Research on the causes and mechanisms of aging of the immune system may be based upon the thymus. Causes of aging may be extrinsic or intrinsic to the thymus. One likely cause of regulatory breakdown may involve neuroendocrine systems.

It is timely to survey the results of those groups of investigators who have been working for several years on the isolation and chemistry of one or more new hormones of the thymus. This survey is primarily an overview to assess importance and significance rather than to provide a detailed account of the data from the experiments. Since the substances which have been isolated in this research are peptides, it is also timely to include some new and largely unpublished information upon the synthesis of peptides with the aid of a new protective group which has advantages over the commonly used protective groups.

It is appropriate to survey separately the chemical data on fractionation and isolation of products toward a hormone or hormones, because of the differences in the chemical data stemming from the several groups of investigators.

THE RESEARCH OF ALLAN GOLDSTEIN ET AL.

Abraham White was one of the early investigators in the research to achieve purification by stepwise fractionation toward a factor or hormone. Allan Goldstein was one of his students in the research which started about 1965 and continued during subsequent years.

In 1975, Goldstein, A., standardized a purification through about five steps and designated the material as "fraction 5". They used this fraction 5 for diversified biological research and provided such fraction to other investigators and also for exploratory clinical studies in patients with immune deficiency diseases.

In 1977, Goldstein, A., et al. (3) reported on the isolation and sequence of thymosin α_1 from fraction 5.

Thymosin α_1 is a peptide having 28 amino acids, and it has AcSer in the N-terminal position. The sequence is as follows.

AcSer-Asp-Ala-Ala-Val^5-Asp-Thr-Ser-Ser-Glu^{10}-Ile-Thr-Thr-Lys-Asp^{15}-Leu-Lys-Glu-Lys-Lys^{20}-Glu-Val-Val-Glu-Glu^{25}-Ala-Glu-Asn^{28}-OH

Synthetic thymosin α_1 is an ongoing subject of diversified biological and clinical investigations.

THE RESEARCH OF GIDEON GOLDSTEIN ET AL.

Gideon Goldstein and his associates began their research upon the thymus about 1969, and have continued this research up to the present time.

In 1974, Goldstein, G., reported on the isolation of thymin from bovine tissue. Thymin was found to consist of two polypeptides which were designated thymin I and thymin II. These two peptides were found to induce marrow cells to develop as intrathymic lymphocytes.

In 1975, Schlesinger and Goldstein (4) reported upon the sequence of thymopoietin II. Goldstein and his associates renamed their two polypeptides as thymopoietin I and II. The expressions thymin I and II were abandoned.

Thymopoietin II has the following sequence:

5 10
NH_2-Ser-Gln-Phe-Leu-Glu-Asp-Pro-Ser-Val-Leu-Thr-Lys-Glu-Lys-
15 20 25 30
Leu-Lys-Ser-Glu-Leu-Val-Ala-Asn-Val-Thr-Leu-Pro-Ala-Gly-Glu-
35 40 45
Gln-Arg-Lys-Asp-Val-Tyr-Val-Gln-Leu-Tyr-Leu-Gln-Thr-Leu-Thr-
49
Ala-Val-Lys-Arg-COOH

Thymopoietin II differs from thymosin α_1 by having more amino acids, but the two peptides do have serine in a position 1 although Ser^1 in thymopoietin II is not acetylated. Both thymopoietin II and thymosin α_1 have C-terminal carboxal groups.

Thymopoietin I differs from thymopoietin II only as follows:

THYMOPOIETIN I: $Gly^{1}Gln^{2}His^{43}$

Weksler, Innes and Goldstein (5) reported in 1978 some immunological studies on aging by utilizing mice. They observed two immune responses which are related to the thymus and which are affected by aging. One response concerned the generation of IgG plaque-forming cells (PFC). The second response concerned the generation of PFC with high affinity for antigen. They found that thymectomy of the mice accelerated both of these two immune responses. They also found that a preparation of thymopoietin increased the immune function of the spleen cells of old mice.

Since thymopoietins I and II are peptides having a relatively large number of amino acids, forty-nine, they designed and synthesized smaller peptides on the basis that such small peptides might simulate the activities of the native peptide. On this basis, Goldstein, G., et al. (6) synthesized a pentapeptide which was found to have a biological activity that was characteristic of the thymic hormone, thymopoietin.

The synthetic pentapeptide has the sequence:

(H) Arg^{32}-Lys^{33}-Asp^{34}-Val^{35}-Tyr^{36}-OH

This pentapeptide induced, in vitro, differentiation of murine prothymocytes to thymocytes. This pentapeptide also inhibited, in vitro, the differential induction of cells which had a B-lineage. These investigators concluded that these two activities of the pentapeptide are unique to thymopoietin.

The pentapeptide was found to reduce, in vivo, autologous Rosette-forming cells from the spleens of athymic mice. This activity was also found to be unique to thymopoietin.

Goldstein, G., et al. concluded that this synthetic pentapeptide may represent an active site of thymopoietin.

Peptides may be synthesized today by the so-called "classical reactions" which may be considered to embody all of those organic reactions which could be used to synthesize peptides before the advent of Merrifield's discovery of the solid-phase method. Since this discovery, the solid-phase method has been increasingly improved so that it can be the method of choice for peptides of an increasing number of amino acids. Over recent years, and because of the increasing knowledge on peptide hormones, continuing advances have been made not only on the "classical methods" but also on the solid-phase methods.

THE RESEARCH OF WOLFGANG VOELTER ET AL.

Kalbacher and Voelter (7), in 1978, published an initial account of a new protecting group for peptide synthesis. The new group is 1-(adamantyl)-1-methylethoxycarbonyl. The starting material and organic reactions used to synthesize Adpoc-OPh are shown in Scheme 1. The generic reaction of Adpoc-OPh with amino acids is also depicted in Scheme 1.

Exemplary amino acid derivatives with the new Adpoc group are as follows:

Adpoc-L-Ala-DCHA
Adpoc-L-Trp-OH
Adpoc-L-Trp-L-Lys(Z)-OH
Adpoc-L-Thr(Bzl)-L-Phe-OH

SCHEME I. 1-(1-ADAMANTYL)-1-METHYLETHOXYCARBONYL:
A NEW PROTECTING GROUP FOR PEPTIDE SYNTHESES

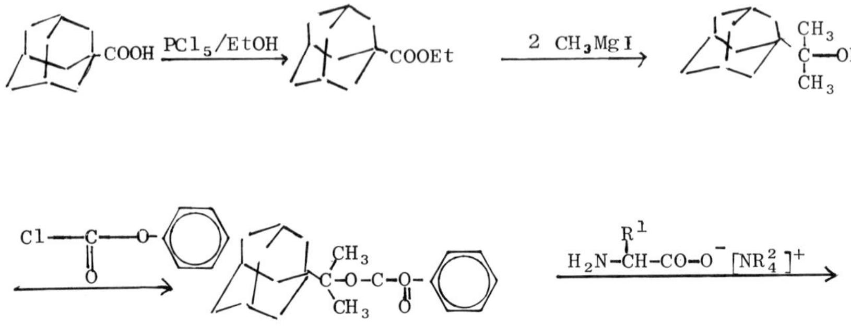

Adpoc-OPh

$$\text{Adpoc-NH-}\underset{}{\overset{R^1}{\text{CH}}}\text{-CO-O}^- [NR^2_4]^+ \xrightarrow{H^+} \text{Adpoc-NH-}\overset{R^1}{\text{CH}}\text{-CO-OH}$$

Kalbacher and Voelter summarized (7) the advantages of the Adpoc group over previously used groups. It was immediately desirable to use amino acids protected by the new Adpoc group in an exemplary synthesis of a peptide by the solid-phase technique. For this purpose, the pentapeptide:

$$\text{(H) Arg}^{32}\text{-Lys}^{33}\text{-Asp}^{34}\text{-Val}^{35}\text{-Thr}^{36}\text{-OH}$$

which represents the amino acids in positions 32-36 of thymopoietin II was selected for synthesis.

The solid-phase synthesis of this pentapeptide with Adpoc-amino acids is diagrammed in Scheme II. The protected peptide on the resin has been liberated by HF, and the purification of the peptide is in progress.

SCHEME II. SOLID-PHASE SYNTHESIS WITH ADPOC-AMINO ACIDS

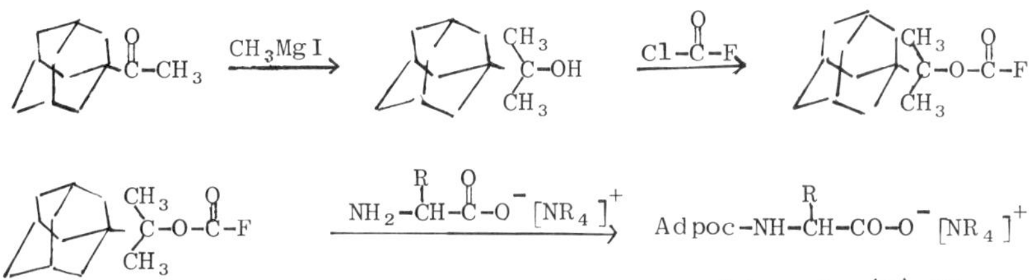

Adpoc-F

Adpoc-Lys(Z)-OH
Adpoc-Asp(OBzl)-OH
Adpoc-Val-OH
Adpoc-Tyr(Bzl)-OH

Aoc-Arg(Tos)-Lys(Z)-Asp(OBzl)-Val-Tyr(Bzl)-(P)

↓

HF, Purification Steps

↓

H-Arg-Lys-Asp-Val-Tyr-OH

Since it was desirable to compare the results of the synthesis of this pentapeptide by the new Adpoc-amino acids with the results of the synthesis of the same pentapeptide by the commonly used t-Boc amino acids, it was necessary to repeat the synthesis of the pentapeptide as depicted in Scheme III.

SCHEME III. SYNTHESIS OF THE PENTAPEPTIDE BY T-BOC AMINO ACIDS

T-Boc-Tyr(Br-Z)-OH
T-Boc-Val-OH
T-Boc-Asp(OBzl)-OH
T-Boc-Lys(Cl-Z)-OH
Aoc-Arg(Tos)-OH

Solid Phase Synthesis ↓ HF, Purification

H-Arg-Lys-Asp-Val-Tyr-OH

Again, the desired pentapeptide was obtained from the resin by treatment with HF and the liberated material is presently being purified. In the near future, it is anticipated that a comparison of the results can be made of the synthesis of this particular pentapeptide by using the new Adpoc-amino acids and the t-Boc amino acids.

THE RESEARCH OF NATHAN TRAININ ET AL.

Nathan Trainin and his coworkers at the Weizmann Institute of Science in Israel initiated their study on a factor in the thymus about 1960, and Trainin and his successive associates have continued their studies to the present time. The basis of their study was the concept of the possible existence of a thymic humoral factor, which they designated THF, in due time.

In 1973, Trainin *et al.* reported upon the hormone-like activity of the THF which induced immune reactivity.

In 1974, Trainin *et al.* described data indicating that THF can induce immunocompeptence through cAMP and DNA syntheses. During the period of 1975-6, Kook, Yakir and Trainin (8) described chemical characterization data indicating that THF can be an acidic peptide having a molecualr weight of *ca.* 3000. and which consists of *ca.* 31 amino acids.

Two amino acid compositions of THF were published in 1975 and 1976 as follows:

1975	1976	
4	3	Asp
1	2	Thr
5	3	Ser
8	5	Glu
2	1	Pro
5	3	Gly
2	2	Ala
-	1	Val
1	2	Leu
-	1	Phe
1	1	Lys
2	2	Arg
-	1	Cys
31	27	

In 1979, Rotter and Trainin (9) described a thymic plasma recirculating factor (TPRF). This factor was partially characterized as a peptide having a molecular weight of about 3000. Their biological data indicated that this factor may be involved in T-cell maturation.

THE RESEARCH OF ALBERTO ASTALDI ET AL.

Alberto Astaldi, of the Central Laboratory of the Netherlands, appeared to have begun his investigations on the thymus about 1970. With his associates, Astaldi has published a series of interesting investigations up to the present time which are important, because the molecular weight of his apparent peptide appears to be lower than that of any other investigator in the field.

In 1978, Kruisbeek *et al.* (10) described a thymic epithelial culture supernatant (TES). They found that TES increased the antibody production to SRBC by spleen cells from athymic mice. Also, TES was found to induce a helper cell function in thymocytes.

In 1979, Wijermans *et al.* (11) described data showing the presence of a thymus-dependent factor in serum (SF). It was found by this group that SF increased cAMP levels and stimulated protein synthesis. Their data also indicated that SF is thymus-dependent.

In 1979, at the Miles International Symposium at Johns Hopkins University in July, Astaldi (12) reported data on the chemical characterization of SF.

SF was apparently pure by the presence of "one spot" in TLC systems. It was estimated to have a molecular weight of less than 500 and probably to consist of four amino acids.

THE RESEARCH OF JEAN FRANCOIS BACH ET AL.

Jean Francois Bach and his associates appear to have initiated their investigations of the thymus about 1969, and have continued to make important contributions in this field up to the present time. During the early years, their investigations focused upon biological aspects of the thymus and immunosuppression. Later, they undertook chemical purification studies on a factor.

In 1973, Bach et al. reported the demonstration of a circulating thymic hormone which they observed to be in the serum of normal mice, but not in the serum of athymic mice. This factor appeared to have a molecular weight of less than 10,000 on the basis of its passage throughout a membrane.

In 1977, Bach et al. described the isolation of a thymic peptide (FTS) from pig serum. They utilized a total of about 1000 liters of serum to achieve isolation of the peptide by steps including ultrafiltration, gel filtration and ion-exchange chromatography.

FTS was found to be a nonapeptide which consisted of 1 Lys, 1 Asp(Asn), 2 Ser, 2 Glu(Gln), 2 Gly, 1 Ala.

Bach et al. (13) further characterized FTS in 1977, and described the sequence as:

Glu-Ala-Lys-Ser-Gln-Gly-Gly-Ser-Asn-OH .

CONVERSION OF PRO-HORMONES TO HORMONES

An increasing number of pro-hormones and their corresponding hormones are becoming known for a diversity of peptide hormones.

Certain structural features which relate the pro-hormones to their corresponding hormones are increasingly evident, at least for certain pro-hormones. It is frequently difficult to make a generalization about structural relationships, and there may be exceptions to a general relationship which may be described as follows.

Lysine and arginine are basic amino acids, and a study of the sequence relationships between pro-hormones and their corresponding hormones reveals that a sequence of two or three or four of one or both of these two basic amino acids, Lys and Arg, may comprise a unit within a pro-hormone which directs or is preferentially susceptible to enzymatic cleavage at one side or the other of the double or triple or quadruple unit consisting of Lys and/or Arg.

Perhaps, a double basic unit is more common than units of three or four of these two amino acids. When there is just a double basic unit, this unit may consist of Lys-Lys or Arg-Arg or Lys-Arg or Arg-Lys. Similar combinations are evident when there are three or four of these two amino acids in some combination in the unit.

One may consider that enzymatic cleavage on either side of one of these double basic units, etc., is facilitated by an extraordinary sensitivity to the enzymatic cleavage in preference to cleavage at single Lys or Arg residues, or that there is a unique enzyme which more specifically recognizes one of these double basic moieties over single Lys or Arg moieties.

Examples of the presence of such double basic moieties, etc., in the conversion of pro-hormones to hormones is exemplified by the conversion of big gastrin to gastrin(Scheme IV) and of proinsulin to insulin(Scheme V), and of proglucagon to glucagon(Scheme VI) and of β-lipotropin to β-melanotropin and to β-endorphin(Scheme VII).

SCHEME IV. CONVERSION OF BIG GASTRIN TO GASTRIN

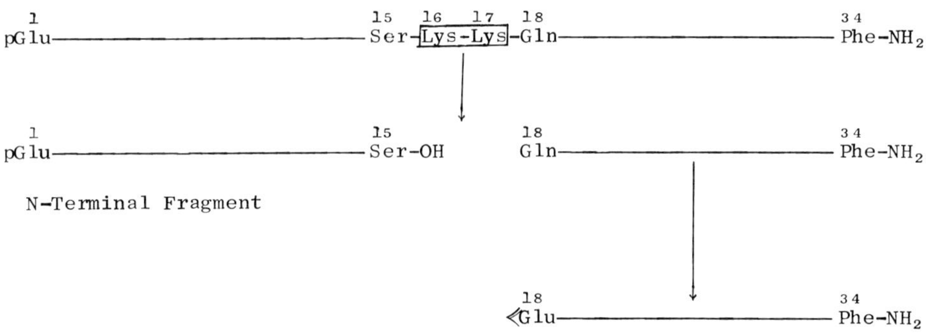

SCHEME V. PROINSULIN TO INSULIN

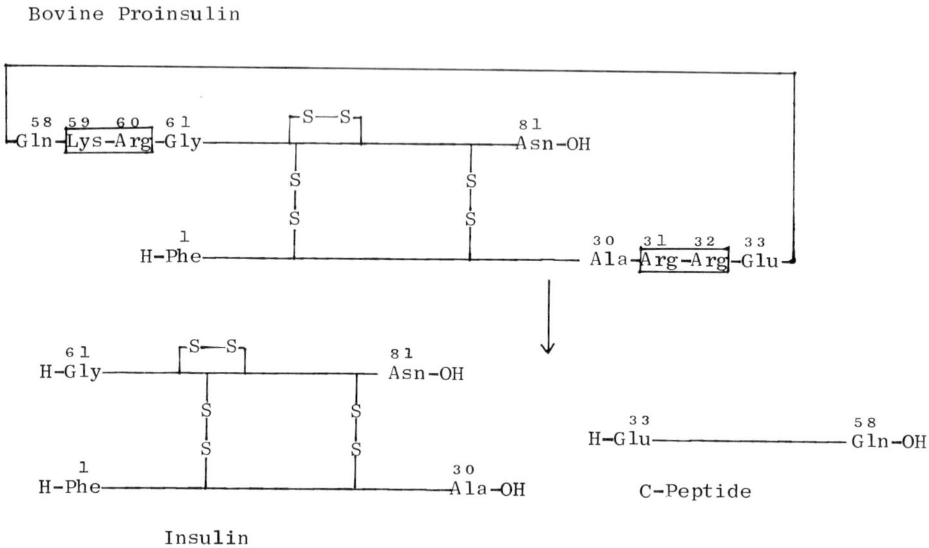

SCHEME VI. PROGLUCAGON TO GLUCAGON

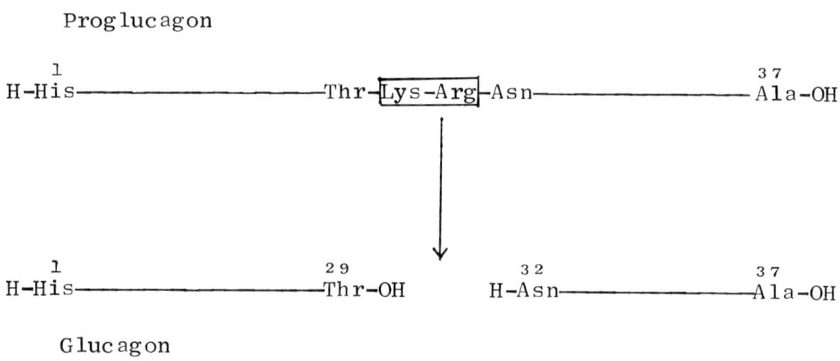

SCHEME VII. β-LIPOTROPIN TO β-MELANOTROPIN AND β-ENDORPHIN

β-Lipotropin

```
1                38  39  40  41          58  59  60  61             91
H-Glu————————————Glu-Lys-Lys-Asp—————————Asp-Lys-Arg-Tyr————————————Glu-OH

1                38              41          58          61             91
H-Glu————————————Glu-OH          H-Asp———————Asp-OH      H-Tyr——————————Glu-OH

    N-Fragment                       β-MSH                   β-Endorphin
```

THE RESEARCH OF KARL FOLKERS ET AL.

We have synthesized thymosin α_1 by an automated solid-phase procedure, and we have isolated a fraction of THF as described by the procedures published by Trainin et al. (14). The behavior of the fraction of THF and the behavior of synthetic thymosin α_1 in HPLC were compared. Under the conditions used, there was no evidence for the presence of thymosin α_1 in the fraction of THF. The comparison was based upon the injection of 1000 µg of the fraction, and no peak corresponding to thymosin α_1 could be seen at a sensitivity which allows the statement that any content of thymosin α_1 in the fraction was less than 0.1% (15). These data and other published evidence on the fraction of THF and thymosin α_1 indicate that THF and thymosin α_1 are different.

Folkers and Wan (16) made a study of a possible structural relationship between thymopoietin and the facteur thymique serique (the FTS of Bach et al.). The known 49-amino acid sequence of thymopoietin and the known 9-amino acid sequence of FTS allowed the concept that Arg^{49} of thymopoietin might be linked to Gln^{1} of FTS in a new 59-amino acid peptide in tissue. Cleavage between Arg^{49} and Gln^{50} adjacent to the unique Lys^{48}-Arg^{49} moiety could liberate thymopoietin and the [H-Gln^{1}]-FTS which could cyclise to FTS by known chemistry, either enzymatic or non-enzymatic. In support of, rather than negating this concept, synthetic FTS and a new dodecapeptide consisting of Val-Lys-Arg linked to the N-terminal of [H-Gln^{1}]-FTS showed comparable immune stimulating activity, *in vivo*.

Wan and Folkers (17) described a synthesis of FTS and the related dodecapeptide as this analog is related to thymopoietin.

Bliznakov (18) described data on the biological assay of synthetic FTS and the related dodecapeptide for reactivation of the impaired immune competence in aged mice. The data unambiguously showed the profound impairment of a humoral, hemolytic, primary, immune response in mice of 22 months in age when compared to this response in young ten-week old mice. The two synthetic peptides showed comparable activity.

These data of Folkers and Wan on thymopoietin and FTS are seemingly compatible with the concept that a new 58-amino acid peptide could be a pro-hormone for both thymopoietin and FTS.

Iwata, Incefy and Good (19) described, in 1979, data on an interaction between thymopoietin and the facteur thymique serique (FTS) in the rosette inhibition assay. They found that there seemed to be an important biologic interaction between thymopoietin and FTS when these two peptides were tested together in the rosette inhibition assay. They concluded that thymopoietin and FTS might interact both *in vitro* and *in vivo*.

THE RESEARCH OF MAURICE HILLEMAN ET AL.

A most important question on all of these diversified investigations on a thymic hormone or hormones concerns the identification of that peptide or those peptides which truly have a hormone role in the complex mechanisms which control immunocompetence. It is evident that in some of the past and diversified chemical and biological investigations that some of the biological activities may reflect only pharmacological activities which are not necessarily physiological or intrinsic to the immune system.

Martinez, Field, Schwam, Tytell and Hilleman (20) evaluated thymopoietin, ubiquitin and the synthetic FTS to restore immunocompetence in T-cell-deficient mice. Specifically, they evaluated these three peptides to restore thymus-dependent immunocompetence in thymectomized C58 mice, because of the previous information that these peptides do induce T-cell markers.

Multiple administrations of these three peptides were made over weeks of time, but no activity to restore resistence to line I_B leukemia and no activity to restore a response to T-cell mitogens was observed. These investigators offered two possible explanations on why these three peptides exhibited no activity.

The most meaningful possible explanation is an irrelevance of these particular substances to cell maturation. Alternatively, on a more encouraging note as considered by other investigators (Allan Goldstein et al.), it could be that a single factor might not show activity since immunologic maturation involves multiple steps.

REFERENCES

1. Miller, J.F.A.P., Osoba, D.: Ciba Found. Study Group 16, 62 (1963).
2. Osoba, D.: J. Exp. Med. 122, 633 (1965).
3. Goldstein, A.L., Law, T.L.K., McAdoo, H., McClure, J., Thurman, G.B., Rossio, J., Lai, C.Y., Chang, D., Wang, S.S., Harvey, A., Ramel, J., Meinenhofer, J.: Proc. Nat. Acad. Sci. 74, 725 (1977).
4. Schlesinger, D.H., Goldstein, G.: Cell. 5(4), 361 (1975).
5. Weksler, M.E., Innes, J.B., Goldstein, G.: J. Exp. Med. 148, 996(1978).
6. Goldstein, G., Scheid, M.P., Boyse, E.A., Schlesinger, D.H., Van Wauwe, J.: Science 204, 1309 (1979).
7. Kalbacher, H., Voelter, W.: Angew. Chem. Int. Ed. Engl. 17, 944 (1978).
8. Kook, A.I., Yakir, Y., Trainin, N.: Cell Immun. 19, 151 (1975), Adv. Exp. Med. Biol. 66, 215 (1976).
9. Rotter, V., Trainin, N.: J. Immun. 122(2), 414 (1979).
10. Kruisbeek, A.M., Astaldi, G.C.B., Blankwater, M.J., Ziljlstra, J.J., Levert, L.A., Astaldi, A.: Cell. Immun. 35, 145 (1978).
11. Wijermans, P., Astaldi, G.C.B., Facchini, A., Schellekens, P.T.A., Astaldi, A.: Biochem. Biophys. Res. Comm. 86, 88 (1979).
12. Astaldi, A.: Miles International Symposium.
13. Bach, J.I., Dardenne, M., Pleau, J.M.: Nature 266, 55 (1977).

14. Kook, A.I., Yakir, Y., Trainin, N.: Cell. Immun. 19, 151 (1975).

15. Folkers, K., Leban, J., Sakura, N., Rampold, G., Lundanes, E., Dahmen, J., Lebek, M., Ohta, M.: Proceedings of the 12th Miles International Symposium on Polypeptide Hormones, Raven Press (in press).

16. Folkers, K., Wan, Y.-P.: Biochem. Biophys. Res. Comm. 80(4), 740 (1978).

17. Folkers, K., Wan, Y.-P.: Bioorganic Chem. 8, 41 (1979).

18. Bliznakov, E., Wan, Y.-P., Chang, D., Folkers, K.: Biochem. Biophys. Res. Comm. 80, 631 (1978).

19. Iwata, T., Incefy, G.S. , Good, R.A.: Biochem. Biophys. Res. Comm. 88, 1419 (1979).

20. Martinez, D., Field, A.K., Schwam, H., Tytell, A.A., Hilleman, M.R.: Proc. Soc. Exp. Biol. Med. 159, 195 (1978).

PROPERTIES OF THE MEMBRANE MODIFYING POLYPEPTIDE ANTIBIOTICS ALAMETHICIN AND TRICHOTOXIN A-40

Günther Jung, Hans Brückner and Heribert Schmitt
Institut für Organische Chemie der Universität Tübingen, Auf der Morgenstelle 18, 7400 Tübingen, Germany

Introduction

An amphiphilic peptide antibiotic with a high content of 2-methylalanine (α-aminoisobutyric acid, Aib) has been isolated 1967 by Meyer and Reusser (1) from the culture medium of a strain of the fungus Trichoderma viride. This polypeptide possesses some bacteriostatic and fungicidic activities. However, after the detection of its unique membrane modifying action on artificial lipid bilayer membranes by Müller and Rudin 1968 (2), it became of high interest. It induces a potential dependent ionic conductance with the characteristics observed on the synaptic transmission of nerve signals (3). On the basis of bilayer experiments an ion conducting pore system has been suggested consisting of oligomers of alamethicin (2,4-6). A single pore can adopt different conductivity states with uptake or release of monomers (7-8). Thus,this unique model system differs in most respects from those of the carrier valinomycin (9) and the channel former gramicidin A (10).

The purpose of this article is to review and summarize the results gained from the various investigations on alamethicin, which span activities in microbiology, biophysics, biochemistry and organic chemistry. Particular emphasis will be given to our studies concerning the elucidation of the relations between structure and membrane modifying properties of natural, modified natural and synthetic analogues.

Structure and Activity of Natural Peptides

Antibiotic properties and structural characteristics of alamathicin and related antibiotics

A relatively large number of lipophilic polypeptide antibiotics has been isolated since 1966 from microorganisms of the class Ascomycetes-Deuteromycetes (Trichoderma spp., Stilbella sp., Cephalosporium sp.and Samarospora sp., table 1). Characteristic residues of the peptide sequences of some of these antibiotics are the α-dialkyl-amino acids α-amino-isobutyric acid (Aib, 2-methylalanine) and D-isovaline (Iva, 2-ethylalanine), the C-terminal aminoalcohols L-phenylalaninol (Pheol or Phol) and L-valinol (Valol or Vol) and a N-terminal acetyl group:

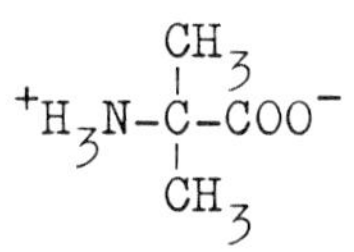

α-Aminoisobutyric acid

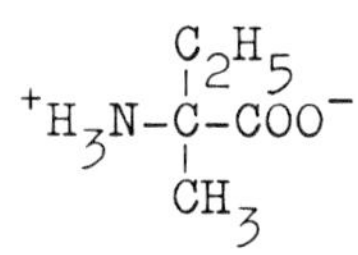

Isovaline

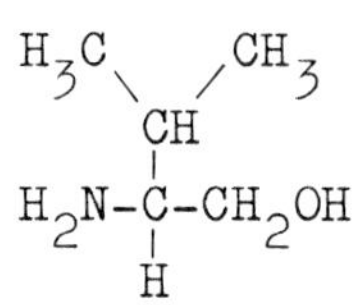

Valinol

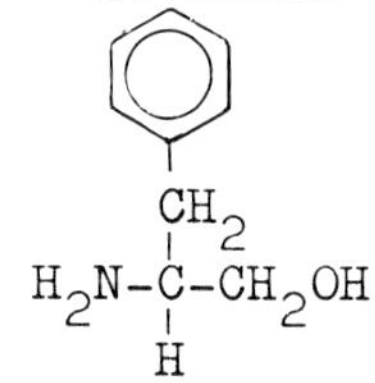

Phenylalaninol

Besides the dominating Aib residues the common protein amino acids Ala, Val, Leu , Phe, Gly, Gln, Glu, Hyp (all L-configuration) constitute only about half of the sequences of these antibiotics (Table 1).

The antibiotic activities of this class of polypeptide antibiotics can be summarized as follows: Suzukacillin exhibits antibacterial properties against Bacillus subtilis, Xantho-

Table 1. Amphiphilic polypeptide antibiotics containing α-dialkyl-α-amino acids.

Name	Microorganism and Producer	Antibiotic in culture filtrate (c) resp.mycelium (my)	Occurence	Ref.
Suzukacillin A and B	Trichoderma viride PERS EX.FR. Strain 63 Cl (Asahi Chemical Industrial Co ., Ltd., Tokyo)	c	earth	11-14
Alamethicin	Trichoderma viride PERS EX.FR. Antibioticum U-22 324 (The Upjohn Company, Kalamazoo, Michigan, USA) NRRL 3199	c	-.-.-	1,15-25
Antiamoebin I and II	Emericellopsis poonensi THIRUM. E.synnematicola MATHUR u.THIRUM. Cephalosporium pimprina THIRUM. (Hindustan Antibiotics Ltd.,Pimpri)	c (20 %) my(80 %)	earth	26-34
Stilbellin	Stilbella sp. (Shionogi u.Co.,Ltd., Osaka, Japan)	my	(deer dung)	34, 35
Trichotoxin A	Trichoderma viride PERS EX.FR. NRRL 5242	my	Texas (T) male sterile corn	36
Trichotoxin B	Trichoderma viride PERS EX.FR. NRRL 5243	my	see trichotoxin A	36
Trichotoxin A-40 and A-50	Trichoderma viride PERS EX.FR. NRRL 5242	my	corn,see trichotoxin A,B	24,25, 34,37-39

Table 1. (Continued).

Name	Microorganism and Producer	Antibiotic in culture filtrate (c) resp. mycelium (my)	Occurence	Ref.
Emerimicin II-IV also known as antibiotics EM-2(U-40588) EM-3(U-40589) EM-4(U-40590)	Emericellopsis microspora (addition of trans-4-n-propyl-L-proline to the culture medium) strain 333 (The Upjohn Company, Kalamazoo, Mich., USA)	my	-.-.-	34,40, 41
Zervamicin I and II (also known as Zervacin I and II)	Emericellopsis salmosynnemata strain 336 (The Upjohn Company, Kalamazoo, Mich., USA)	my	-.-.-	42
Samarosporin	Samarospora sp. No. F-7762 (Daiichi Seiyaku Co., Ltd., Tokyo) FERM-P 224	my	soil	34,43
Tü 165	Em.salmosynnemata (CBS 382.62)+			44

NRRL = U.S. Department of Agriculture, Agricultural Research Service, Northern Utilization Research and Development Division, Peoria, Ill. 61604, IMI = Commonwealth Mycological Institute, FERM = Fermentation Research Institute, Chiba (Japan), CBS = Centraalbureau voor Schimmelcultures, Baarn, Holland

+CBS 382.62 was cultivated in the Institut für Mikrobiologie II der Universität Tübingen under the strain number Tü 165. The Antibiotic isolated from Tü 165 is identical with antiamoebin.

monas oryzae, Sarcina lutea, Streptococcus pyrogenes, Staphylococcus aureus, Corynebacterium diphteriae and Mycobacterium phlei, and it shows antifungal activity against Aspergillus niger, Trichophyton fumigatus and Saccharomyces sake (c=10-100 µg/ml) (11). It can suppress the intestinal disease coccidiose on animals (45). Antiamoebin has been shown to exhibit protozoidal and antihelminthic properties and activities against Dracunculus medinensis, Trypanosoma evansii, Trichomonas vaginalis (28). Stilbellin is active against Bacillus subtilis, Staphylococcus aureus, Escherichia coli, Klebsiella pneumoniae, Trichomonas vaginalis (35). Emerimicin II-IV inhibits the growth of protozoae such as Ochromonas danica, Crythidia fasciculata and Tetrahymena pyriformis (40). Zervamycins I, II are active against gram positive bacteria and protozoae (42). Samarosporin inhibits the growth of bacteria, fungi, protozoae and yeast (43) and has an antiphlogistic activity (46). Alamethicin is active against pathogenic fungi such as Blastomyces dermatitis, Nocardia asteroides, Hormodendrum compactum (1). The alamethicins (F-30, F-50), trichotoxins (A-40, A-50) and suzukacillin A are similarly active against Bacillus subtilis, Staphylococcus aureus and Saccharomyces cerevisiae tested under the same conditions. Interestingly these latter polypeptide antibiotics are not able to inhibit the growth of Mycoplasma laidlawii, a bacterium without protecting cell wall (47).

Action on mammalian cells

The cytolytic effects of alamethicin were first detected on Ehrlich ascites tumor cells. At micromolar concentrations the peptide antibiotics cause hemolysis in human erythrocytes.

A detailed study of the hemolytic action with variations of concentration, temperature and incubation time revealed characteristic differences in the mode of hemolysis between the peptide antibiotics which can be correlated with structural differences between these natural analogs. The addition of bovine serum albumin reduces the hemolytic activity. The lytic action increases considerably from hypertonic to hypotonic media. The bee venom constituent melittin is more active and sodium dodecylsulfate is less active under the same conditions. The lipophilic peptide antibiotics constitute a new class of membrane disrupting agents with particular properties not known from other lytic compounds.

Assuming that the lytic activity is proportional to the concentration of the peptide antibiotic, the temperature dependence of the concentration/activity curves allows the calculation of activation energies of 62 kJ/mol for alamethicin, 67 kJ/mol for suzukacillin A, and 79 kJ/mol for trichotoxin A-40. However, melittin revealed the very low activation energy of 49 kJ/mol. These values may be correlated with results from circular dichroism and ^{13}C nuclear magnetic resonance (see below) experiments which indicate conformational changes of the peptides aggregating at the membrane/water interphase.

Hemolytic activity on erythrocytes (22) may be used as screening test within this class of peptide antibiotics, since all analogues exhibiting pore forming activities in lipid membranes were found to be lytically active, too. Differences in the hemolytic activity of the natural, modified and synthetic polypeptide analogues are observed (48). Thus, the lytic activity of the trichotoxin A-40 series increases in the following order of Glu^{6} side chain modifications (Fig.2): Trichotoxin ($-COO^{-}$) $<$ hydrazide ($-CONHNH_3^{+}$) $<$ amide ($-CONH_2$) $<$ trichotoxin ester ($-COOCH_3$) . The same behaviour is found upon modification of alamethicin at Glu^{18} : Alamethicin F-30 ($-COO^{-}$) $<$ alamethicin F-50 ($-CONH_2$) (22) $<$ alamethicin F-30 ester ($-COOCH_3$). The Glu^{18}-benzyl ester of the synthetic nonadecapeptide (49) is more active, too, than the peptide with a free Glu^{18}-γ-carboxyl group. Thus all these analogues ex-

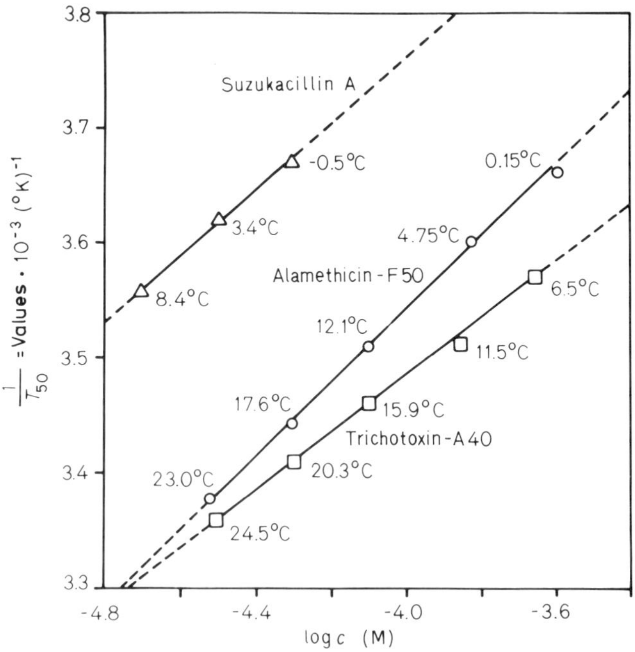

Fig. 1. Temperature concentration diagram of the hemolysis of human erythrocytes by alamethicin, trichotoxin and suzukacillin A. T_{50}^{-1} values are the temperatures which cause 50 % hemolysis. c: Concentration of the peptide antibiotics.

hibit hemolytic properties including the synthetic nonadecapeptide (fully protected and with free Glu^{18}-γ-COOH) and the following peptides with unknown sequences: Trichotoxin A-50 analogues (50), trichotoxin A-55 (50) with about 50 % isovaline residues and the suzukacillin analogues. Similar results are obtained at bilayer membranes using the method of current voltage characteristics (51,52).

However, the following partial sequences were found to be inactive in, both, lysis and bilayer experiments: N-terminal dodecapeptide of trichotoxin A-40 Ac-Aib-Gly-Aib-Leu-Aib-Glu-Aib-Aib-Aib-Ala-Aib-Aib-OH (38), the C-terminal hexapeptides $^{+}H_2$-Pro-Leu-Aib-Iva(Aib)-Gln-Vol (39), the synthetic deca-

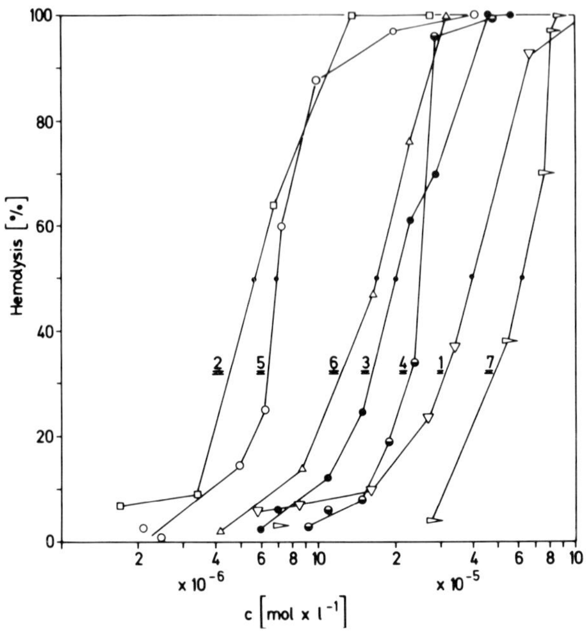

Fig. 2. Concentrations dependence of lytic activities of membrane modifying peptides on erythrocytes: Trichotoxin A-40 [1], its methyl ester [2] and hydrazide [3], trichotoxin A-50 [4], alamethicin F-50 [5], synthetic nonadecapeptide [6] and sodium dodecylsulfate [7], (experimental conditions, see ref. 22).

peptide Boc-(Aib-Ala)$_5$-OH(-OMe) (48) and the synthetic hexadecapeptide Boc-(Aib-Ala)$_5$-Pro-Ala-Aib-Aib-Glu-Gln-OMe (49). These results suggest a minimal length of 18 residues with a large variability in particular regions of defined conformation. The lytic action is diminished in the case of the more polar analogues with charged glutamic acid side chain.

Effects of alamethicin on the activity of adenyl and guanyl cyclase were reported (53). Bovine, rat and mouse leukocytes were exposed to alamethicin, suzukacillin and trichotoxin (54); the antibiotics caused cell lysis at 5 x 10^{-5} mol/l. In rat mast cells histamine release was found simultaneously.

Thymidine incorporation into DNA of bovine and mouse lymphocytes was inhibited at sublytic concentrations around 2 x 10^{-6} mol/l. Lymphocyte stimulation by mitogens was inhibited below 10^{-6} mol/l (55).

Potential dependent pore formation in bilayer membranes

Several groups were involved in the earlier experiments on lipid bilayers using alamethicin as pore forming agent (2,56, 57). The formation of voltage dependent ion conducting pores in bilayer lipid membranes by alamethicin involves conformational changes of the molecule. The peptide molecules which were added to the aqueous solution change their conformation by lengthening their helix (58) during contact with the hydrophobic membrane region. It is concluded that a formation of preaggregates occurs and then pore induction starts (59). Single pore experiments suggest that the pores adopt different conductance states by uptake or release of monomers (57). For details of the bilayer experiments we refer to recent publications (60,61).

In order to elucidate the relations between strucure and membrane modifying properties for this type of antibiotics our group studied the conformational change of alamethicin and its natural analogues suzukacillin A and trichotoxin A-40 which is caused by the transference from aqueous to lipophilic media (58). Voltage-jump current-relaxation experiments were carried out by G.Boheim at membranes modified by alamethicin and its analogues (52). On the basis of the results of these structural and biophysical investigations an artificial ion conducting and lytically active nonadecapeptide analogue was synthesized which possesses a conformationally designed, modified sequence (49). In 1979 in our group the complete sequences of the trichotoxin A-40 analogues were elucidated using selective cleavage conditions and preparative isolations of pure N-terminal dodecapeptide and C-terminal hexapeptide partial sequences (38,39,62).

The results of the bilayer experiments at low concentrations of these antibiotics are rather complex, since two modes of membrane modifying activities are observed: 1. Activity in form of resolvable, stable ($\tau > 1$ ms) pores and 2. activity in form of voltage dependent pore formation (not resolvable fluctuations) which finally leads to membrane lysis.

Stable pores are formed by alamethicin and suzukacillin (mode 1), whereas trichotoxin A-40 and the synthetic nonadecapeptide induce a voltage dependent conductance in form of non-resolvable bursts at low conductance level (mode 2). Our trichotoxin probes of highest purity, isolated by an optimized procedure and purified by counter-current distribution, differ in their pore forming properties from the sample reported earlier (51); resolvable pores are no longer observed. Spikes and bursts are recorded, however, and still all these trichotoxin probes exhibit the same lytic activities (22) on, both, natural and artificial membranes. Surprisingly the pores are stabilized by addition of fragments of trichotoxin A-40 (24) which are inactive themselves. This stabilization effect may be explained by reduction of the number of collisions and/or by slower movement of pore forming peptide molecules in the bilayer. Experiments with mixtures containing alamethicin and the synthetic nonadecapeptide or the mixture of partial hydrolysates of trichotoxin A-40 showed a similar increase of the mean life time of the alamethicin pores.

On the other hand, the addition of helical decapeptide Boc-(Aib-Ala)$_5$-OH increases the lytic activities in, both, hemolysis and bilayer experiments. Supposing resolvable pores are formed only by specific intermolecular arrangements, the Gln residue within the helical moiety of the alamethicin resp. trichotoxin molecules may be important for the kind of helix interactions within the bilayer. Thus, the requirements for stable single pore formation may be fulfilled by introduction of Gln^6 into the synthetic nonadecdapeptide (24).

Detergent-like properties

The detergent-like properties of alamethicin were studied with respect to its surface tension at the air/water interphase (63). These experiments revealed some information on the aggregation of alamethicin. The pH dependence of the partition coefficients of trichotoxin A-40, alamethicin F-50 and suzukacillin were determined in the system 1-octanol/water (37). The concentration in octanol decreases sharply from pH 4 to 5,5 for suzukacillin A and trichotoxin A-40. Alamethicin F-50, supposed to have no free carboxyl group,shows a similar but more steady decrease from pH 4 to 8.

Culture of Trichoderma viride strains and isolation of the peptide antibiotics

Alamethicin F-50 is isolated from the culture medium of the strain T. viride NRRL 3199 according to scheme I (22).
The membrane modifying peptide antibiotics trichotoxin A-40 and A-50 can be isolated from the dry mycelium of the trichoderma viride strain NRRL 5242 (37). An improved isolation procedure includes extraction by dichloromethane, Sephadex LH-20 chromatography in methanol and counter-current distribution in 2-butanol/0.015 M ammonium acetate (pH = 8.9) for the separation of A-40 and A-50 components. Crystalline trichotoxin A-40 can be obtained from saturated methanolic solutions (Scheme II) (38).

Scheme I. Isolation and purification of alamethicin F-50.

Culture filtrate (20 l, pH 8.5)

↓ HCl conc.

Culture filtrate (pH 3.0)

↓

Crude alamethicin I (7 g, after fivefold extraction with 4 l of $CHCl_3$)

↓

Crude alamethicin II (5.5 g, after threefold extraction with 30 ml portions of ether and petrolether)

↓

Alamethicin III, lytic fraction (800 mg, after Sephadex LH-20 gel filtration, CH_3OH)

↓

Alamethicin (550 mg, after treatment with Darco G 60)

↓

Alamethicin F-50 (after reversed phase column chromatography)

Purification and microheterogeneity

The peptide antibiotics are produced as a mixture of closely related polypeptides with free and amidated carboxyl groups which can be separated in gram amounts by counter-current distribution. The Craig system used for the separation of trichotoxin A-40 and A-50 allows also the enrichment of pure natural analogs A-40 arising from amino acid exchange (e.g. Ala/Aib or Aib/Iva). Thin-layer chromatographic systems (e.g. on silicagel, $CHCl_3/CH_3OH/CH_3COOH/H_2O$, 65:25:3:4 (v/v), separate mixtures of analogs. 100 mg portions can be purified

Scheme II. Isolation and purification of trichotoxin from Trichoderma viride NRRL 5242.

Mycelium of T. viride NRRL 5242 (500 Fernbach flasks à 500 ml)

↓

Mycelium powder (2700 g, after washing and lyophilization)

↓

Brown, oily extract (400 g, after Soxhlet extraction)

↓

Brown powder (70 g, after extraction with methanol)

↓

Crude trichotoxin (4 g, after Sephadex LH-20 gel filtration, CH_3OH)

↙ ↘

Trichotoxin A-40 (1.6 g) Trichotoxin A-50 (1.5 g)

(After counter current distribution, 650 transfers, 2-butanol/ 0.015 M ammonium acetate (pH 8.9; 1:1 (v/v), lyophilization).

on the reversed phase LIchrosorb RP-8. Furthermore HPLC on the reversed phases silicalgel RP-8 and µBondapak C_{18} were used in our groups for very efficient separations of trichotoxins and alamethicins (21).

Structure elucidation of alamethicin

The almost correct amino acid sequence of alamethicin was determined by Payne et al. (16). These findings were confirmed later by a MS study on a partial hydrolysate (17). Already in 1974 we reported our finding of a so far unknown phenyl component (20), which turned out later to be L-phenylalaninol

3.b)

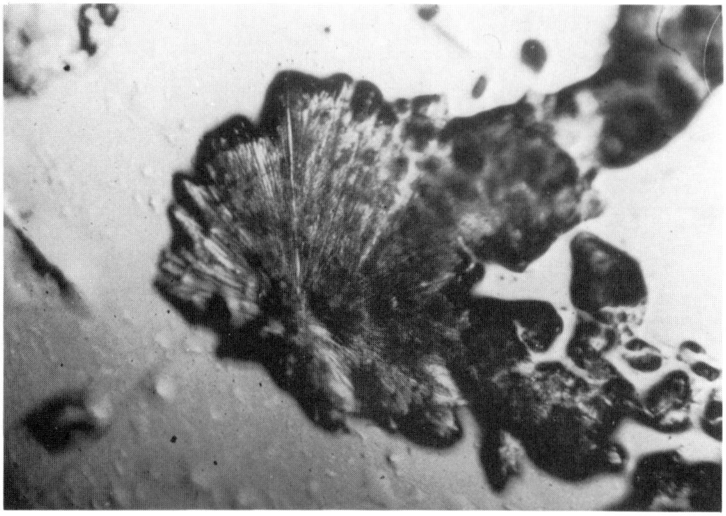

3.a)

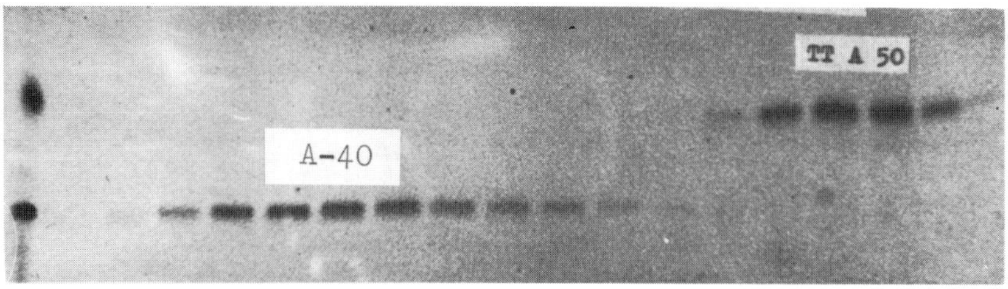

Fig. 3. a) Thin-layer chromatogram of trichotoxin A-40 and A-50 after counter-current distribution (215 transfers, 2-butanol/ ammonium acetate (pH 8.7), 1:1 (v/v)) of the natural antibiotic mixture. b) Crystals of trichotoxin A-40 obtained from a methanolic solution.

(58). At the same time Martin and Williams (19) found N-terminal Ac-Aib and also L-phenylalaninol using ^{1}H NMR spectroscopy and hydrolysates obtained by acidic cleavage. Rinehart et al. (23) confirmed the linear structure found by Martin and Williams using HRFDMS (23). However, still additional evidence using other methods is required to give final proof, since α,γ-transpeptidation and further rearrangements on the C-terminal Glu,Gln,Pheol residues may occur readily. Table 2 summarizes all sequences known so far in the field of alamethicin-like and related peptides.

Sequence determination of trichotoxin A-40

Amino acid analysis and gas chromatography revealed the amino acid composition of table 2. Selective cleavage conditions for trichotoxin A-40 using anhydrous trifluoroacetic acid yielded three N-acetylated dodecapeptides and two N-prolylhexapeptides ending with valinol (Fig. 4). These fragments were uniform after fractionation on silica gel column (chloroform/methanol/water/acetic acid, 65:25:4:3 (v/v)) followed by counter-current distribution (dodecapeptides: 2-butanol/ethyl acetate/methanol/0.3 % ammonium acetate, 3:5:2:3 (v/v), hexapeptides: 1-butanol/ethyl acetate/ammonium acetate (pH = 4.6), 2:4.5:5 (v/v)). The sequences of the fragments were determined via GLC-MS of trifluoracetylated methyl esters of partial hydrolysates and solid phase Edman degradation (64), furthermore FDMS, ^{13}C NMR and CD were recorded. Chiral phase gas chromatography and GLC of diastereomeric derivatives established the L-configuration of Ala, Leu, Glu, Pro, Val and the D-enantiomer of Iva (34).

Table 2. Amino acid sequences of polypeptide antibiotics with α-dialkyl-α-amino acids and amino alcohols.

```
Alamethicin I (II)
    1   2   3   4   5   6   7   8   9   10  11  12  13  14  15  16  17  18  19  20
Ac-Aib-Pro-Aib-Ala-Aib-Ala-Gln-Aib-Val-Aib-Gly-Leu-Aib-Pro-Val-Aib-Aib-Glu-Gln-Phol
                      (Aib)

Suzukacillin A
    1   2   3   4   5   6   7   8   9   10  11  12  13  14  15  16  17  18  19  20  21
Ac-Aib-Pro-Val-Aib-Val-Ala-Aib-Ala-Aib-Aib-Gln-Aib-Leu-Aib-Gly-Leu-Aib-Pro-Val-Aib-Aib-
                                                                                22  23  24
                                                                               Glu-Gln-Phol

Trichotoxin A-40
    1   2   3   4   5   6   7   8   9   10  11  12  13  14  15  16  17  18
Ac-Aib-Gly-Aib-Leu-Aib-Gln-Aib-Aib-Aib-Ala-Aib-Aib-Pro-Leu-Aib-Iva-Glu-Vol
      (Ala)                       (Ala)                       (Aib)

Emerimicin III (IV)
    1   2   3   4   5   6   7   8   9   10  11  12  13  14  15
Ac-Phe-Aib-Aib-Aib-Val-Gly-Leu-Aib-Aib-Hyp-Gln-Iva-Hyp-Ala-Phol
                                                      (Aib)

Antiamoebin I (II)
    1   2   3   4   5   6   7   8   9   10  11  12  13  14  15  16
Ac-Phe-Aib-Aib-Aib-Iva-Gly-Leu-Aib-Aib-Hyp-Gln-Iva-Hyp-Aib-Pro-Phol
                                      (Pro)
```

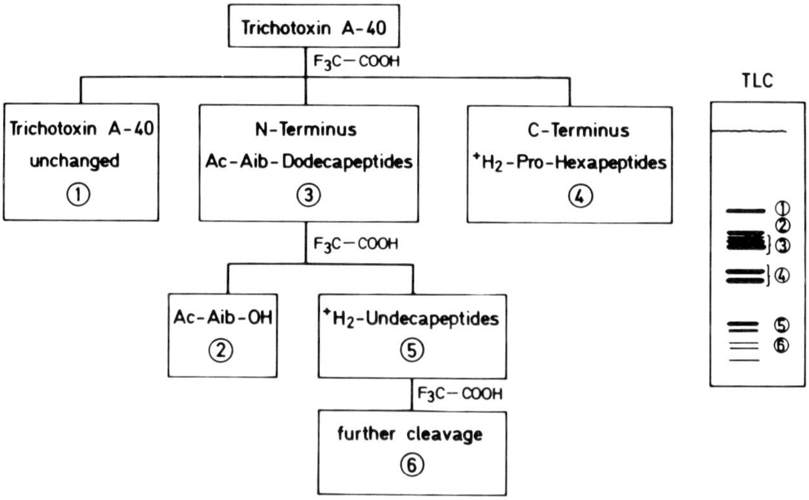

Fig.4. Scheme of the trifluoroacetolysis of trichotoxin A-40 (3 h 15 min, 37°C) and TLC of cleavage products in chloroform/ methanol/water/acetic acid,65:25:4:3,on silica gel.

Comparison of the mass spectra of low temperature-derivatized Gln-Valol with that of trichotoxin A-40 indicates the α-linkage. Rearrangements can be ruled out under these conditions, but esterification at 100°C leads to transpeptidation (39). Additional experiments are carried out at present in order to proof finally the Glu, Gln residues:There is strong evidente that the natural sequence of trichotoxin A-40 has a Gln residue in position 6 and a Glu residue in position 17 (65). Particular emphasis was given to the determination of the configuration of the isovaline residues found in trichotoxins (22,37,50), suzukacillins (14) and several other peptide antibiotics such as antiamoebin, stilbellin and samarosporin (emerimicin IV) (34). The configuration of isovaline in antiamoebins and emerimicins was erroneously assigned as L (23, 31-33, 41) by Rinehart et al. We used the chiral phase Chirasil-Val (66,67) and chromatography of diastereomeric derivatives (68). These methods indicated unequivocally D-Iva in antiamoebin, stilbellin, samarosporin (emerimicin IV) (34).

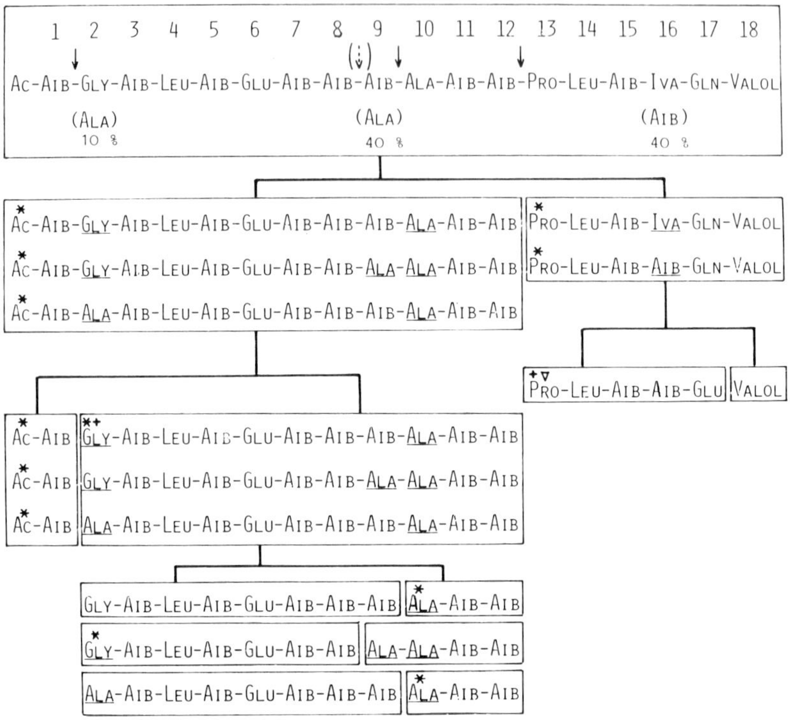

Fig.5. Sequences obtained after trifluoroacetolysis of trichotoxin A-40 and separation according to Fig. 3 (* GLC-MS of trifluoroacetylated peptide methyl esters; + solid-phase Edman degradation via pentafluorophenyl and phenylhydantoin derivatives; ∇ selective cleavage of valinol with dioxane/6 N HCl). Order of cleavage: 12-13 > 1-2 > 9-10 resp. 8-9; % amino acid exchange based on amount of isolated peptides and on TLC).

Acid labile peptide bonds

In sequencing trichotoxin A-40 (38) various hints for particularly labile peptide bonds such as Aib-Pro were found. Model peptides such as Ac-Aib-Pro-NH_2, Ac-Aib-Gly-OH, Ac-Aib-Ala-OH, Ac-Ala-Aib-Ala-OMe (62) and others were investigated kinetically using various techniques such as DC, polarimetry, NMR and amino acid analysis (69).

Due to the labile Aib-Pro bond N-Acetyl-α-aminoisobutyric acid (Ac-Aib-OH) can be split off by selective trifluoroacetolysis from the amphiphilic polypeptide antibiotics of the alamethicin type and detected in thin-layer chromatograms by chlorine/o-tolidine or chlorine/4,4'-tetramethyldiamino-diphenylmethane (TDM). Furthermore Ac-Aib-OH is easily identified directly from the partial hydrolysate after esterification with diazomethane by GC-MS using glass capillary columns coated with Carbowax 20 M. These two methods allowed the unequivocal determination of Ac-Aib-OH in the antibiotics alamethicin F-30 and F-50, trichotoxin A-40 and A-50 and suzukacillin A and B. In addition to Ac-Aib-OH the GC-MS method allowed the detection of about 20 % N-acetyl-isovaline in suzukacillin B (25).

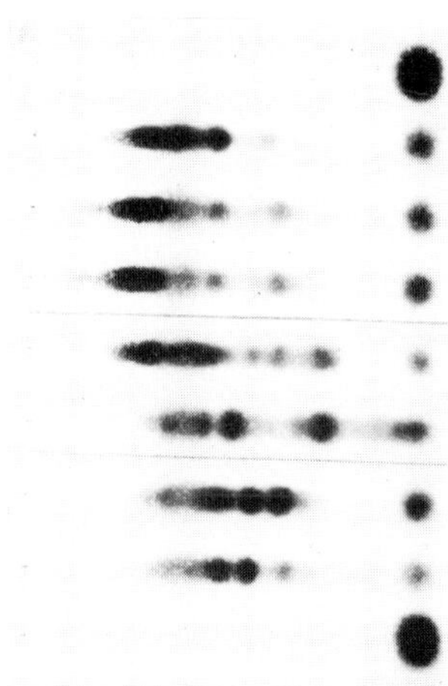

Fig.6 . Detection of Ac-Aib-OH by TLC (silica gel, 1-butanol/ CH_3COOH/H_2O, 4:1:1, (v/v)) after trifluoroacetolysis (TFA, 19h, 37°) of samples (1 %), [4] alamethicin F-30 (Upjohn company), [5] alamethicin F-50, [6] alamethicin F-50 (crude extract), [7] suzukacillin A, [8] suzukacillin B (isolated from crude extract), [2] trichotoxin A-40, [3] trichotoxin A-50; distance from start (left side) to solvent front (right side) 10 cm (25).

Syntheses of alamethicin sequences and analogues

Several groups including our own synthesized a number of partial sequences of alamethicin and recognized soon the special problems involved in Aib peptide synthesis. For example, no active ester coupling of Aib residues can be used and the saponification of Aib esters proceeds considerably slower compared to the usual esters. Two syntheses of alamethicin resp. alamethicin-like peptides have been reported (70-72). However, probably because of very low yields, insufficient data have been collected for a complete proof of the identity of these synthetic with the natural products. Particularly difficulties seemed to be involved if one uses the normal Merrifield procedure, first, because of the low coupling yields and second, because of the acidic conditions used for deprotection of the Boc groups. Trifluoroacetic acid cleaves the acid labile Aib peptide bonds (see Fig. 6). The following synthetic analogues of the membrane modifying peptide antibiotic alamethicin were obtained by fragment condensation (49): The nonadecapeptide Boc-(Aib-Ala)$_5$-Gly-Ala-Aib-Pro-Ala-Aib-Aib-Glu-(OBzl)-Gln-OMe, its derivative with a free carboxyl group of glutamic acid in position 18 and the hexadecapeptide Boc-(Aib-Ala$_5$)-Pro-Ala-Aib-Aib-Glu(OBzl)-Gln-OMe. Mixed anhydride coupling with isobutyl chloroformate was most useful for the preparation of di- and tripeptide fragments (table 3), whereas larger fragments were coupled by the dicyclohexylcarbodiimide method with 1-hydroxybenzotriazole. The nonadecapeptide was synthesized from the crystalline, analytically pure fragments A to D, investigated by ^{13}C NMR and CD, by a strategy based on considerations concerning the mode of action of alamethicin and its conformation. Boc-Aib^1-Ala-Aib-Ala-Aib^5-OH (A) was condensed with H-Ala^6-Aib-Ala-Aib-Ala^{10}-OMe (B) and Boc-Gly^{11}-Ala-Aib^{13}-OH (C) with H-Pro^{14}-Ala-Aib-Aib-Glu(OBzl)-Gln^{19}-OMe (D). Finally the decapeptide 1-10 was coupled with the nonapeptide 11-19. The hexadecapeptide, synthesized from the decapeptide and fragment D, is inactive. The nonadecapeptide possesses the same hemolytic activity as alamethicin, and in bilayer experiments it exhibits a potential dependent pore formation in the form of bursts.

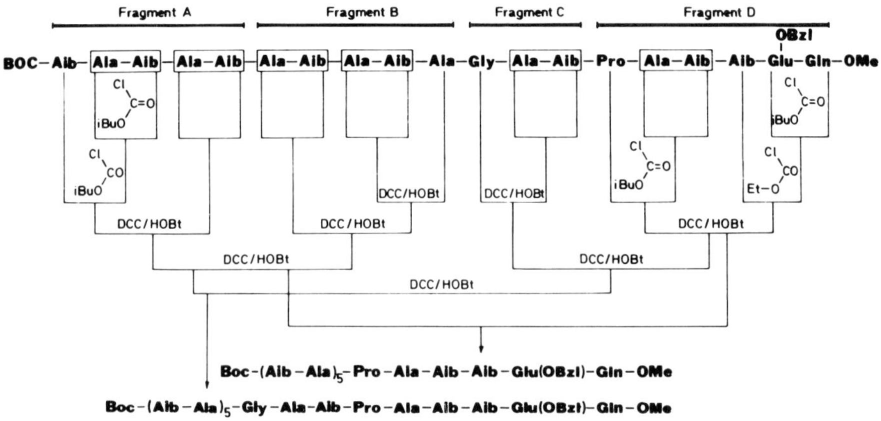

Fig. 7. Syntheses of the membrane modifying nonadecapeptide and the inactive hexadecapeptide.

Table 3. Coupling yields for the mixed anhydride and the DCC/HOBt method (49).

Coupling components	Yield [%] Mix. anhydride	Yield [%] DCC/HOBt
Boc-Ala-OH + H-Aib-OMe	97	52
Boc-Aib-OH + H-Ala-Aib-OMe	80	25
Boc-Gly-OH + H-Ala-Aib-OMe	30	50
Boc-Pro-OH + H-Ala-Aib-OMe	68	70
Boc-Glu(OBzl)-OH + H-Gln-OMe	91	-
Boc-Aib-OH + H-Glu(OBzl)-Gln-OMe	93	-

A variaty of natural and modified sequences was built up for a systematic spectroscopic investigation of the helical and non-helical regions of the alamethicin molecule. The liquid phase method with polyoxyethylene as solubilizing support was found to be very useful for this purpose (74).

The N-terminal helical part of the sequence of the polypeptide antibiotic alamethicin was synthesized on polyoxyethylene as solubilizing support (69,75). The N-acylated undecapeptide was split off by hydrazinolysis as the crystalline hydrazide (Ac-Aib-Pro-Aib-Ala-Aib-Ala-Gln-Aib-Val-Aib-Gly-$NHNH_2$). The results of elemental and amino acid analysis, gas chromatography/mass spectrometry of the esterified and trifluoracetylated partial hydrolysate and gas chromatography of the N-pentafluoropropionyl-amino acid isopropyl esters on N-propionyl-L-valine-tert-butylamidepolysiloxane as a chiral phase were in full agreement with the synthesized sequence. Furthermore, selective removal of the N-terminal acetyl-2-methylalanine was accomplished by trifluoroacetolysis of the particularly acid-labile Aib-Pro peptide bond in the model Ac-Aib-Pro-NH_2, in the undecapeptide and in alamethicin. The helical conformation of the N-acetyl-undecapeptide hydrazide was proven by circular dichroism, using three different evalution methods, and by ^{13}C NMR spectroscopy. By conventional fragment condensation the two undecapeptides Boc-(Ala-Aib)$_2$-Ala-Glu(Bzl)-(Ala-Aib)$_2$-Ala-OMe and Boc-(Ala-Aib)$_2$-Ala-Gly-(Ala-Aib)$_2$-Ala-OMe were synthesized as models for the N-terminal helical part of alamethicin using again the mixed anhydride and DCC/HOBt methods. The undecapeptides were obtained in crystalline state suitable for X-ray crystallography. Recently the synthesis of the natural sequence of the C-terminus Boc-Pro-Val-Aib-Aib-Glu-Gln-Pheol was accomplished,too (73).

Circular dichroism and conformational analysis

In the solvents trifluoroethanol and hexafluoroacetone trihydrate alamethicin shows three Cotton effects typical of a right-handed α-helix: The n-π^* transition at 221 nm (-), and the two π-π^*exciton bands at 207 nm (-) and 190 nm (+) (Fig. 8 , (58)).

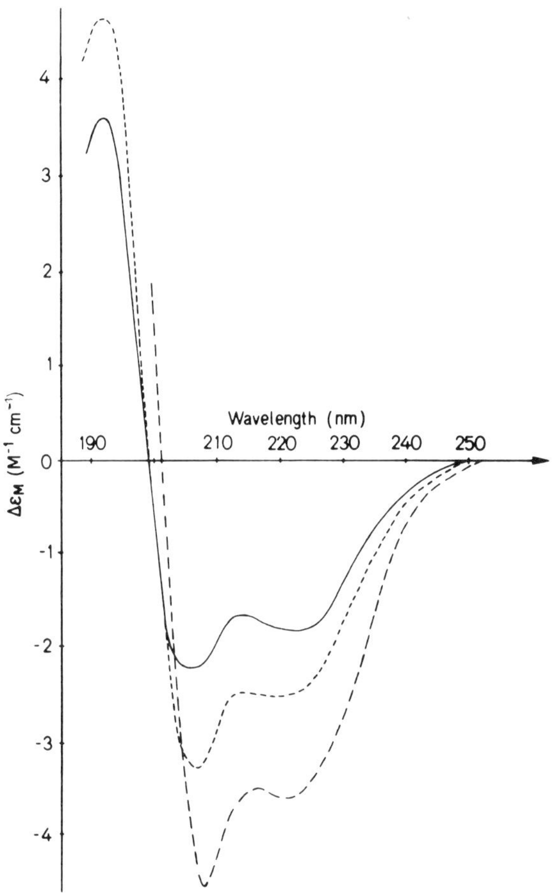

Fig.8. Circular dichroism spectra of alamethicin in 1-octanol (---), trifluoroethanol (...) and hexafluoroacetone trihydrate (——);temp. 20°C, c = 0.2 mM.

The circular dichroism of alamethicin is independent of the concentration in the solvents. Thus it is beyond doubt that the polypeptide which is monomeric in these solvents has indeed a rigid, helical, intrachenar portion as has already been assumed on the basis of the solvent dependence of the two negative Cotton effects (76) and of a calculation of conformation (77). There is, however, strong dependence on the solvent. The intensities of both Cotton effects are increased on going from aqueous solvents to lipophilic solvents and are decreased on going to the strongly polar, hydrogen-bond-forming, hexafluoroacetone trihydrate (Fig. 9). While the ratio of intensities $\Delta\varepsilon_{221nm}$ / $\Delta\varepsilon_{207nm}$ is approximately 0.80 $\pm$ 0.03 in all organic solvents, the predominantly aqueous solutions show a reversal of the intensities of the Cotton effects, i.e. a ratio of above 1.0. If the content of α-helix is calculated with reference to the CD spectra of poly(L-glutamic acid) then a reasonable relation is obtained between helix content and lipophilic character of the solvent, although the applicability of this standard is under limitations in this particular case. Unfortunately detailed CD investigations of the solvent and temperature dependence with oligo-(amino acids) and poly(amino acids) which could serve as appropriate models cannot be found in the literature. Aqueous solutions show the smallest helix content (18-23 %), in organic solvents the helix content increases to 40 %. As shown earlier (76) , aqueous solutions with less than 10 % alcohol or even those with 6 M guanidine and 10 M urea show a small helix content. Alamethicin was also measured by temperature dependent CD and in membrane simulating environments such as nematic phases and in lipid vesicles (58). A helical conformation was established also for suzukacillin A. These results on conformational analysis could be refined by the availability of many partial sequences of natural and synthetic origin, both, free of analogues. As an example the CD spectra of the trichotoxin series (Fig. 10) clearly indicate the location of the helix turns within the N-terminal undecapeptide

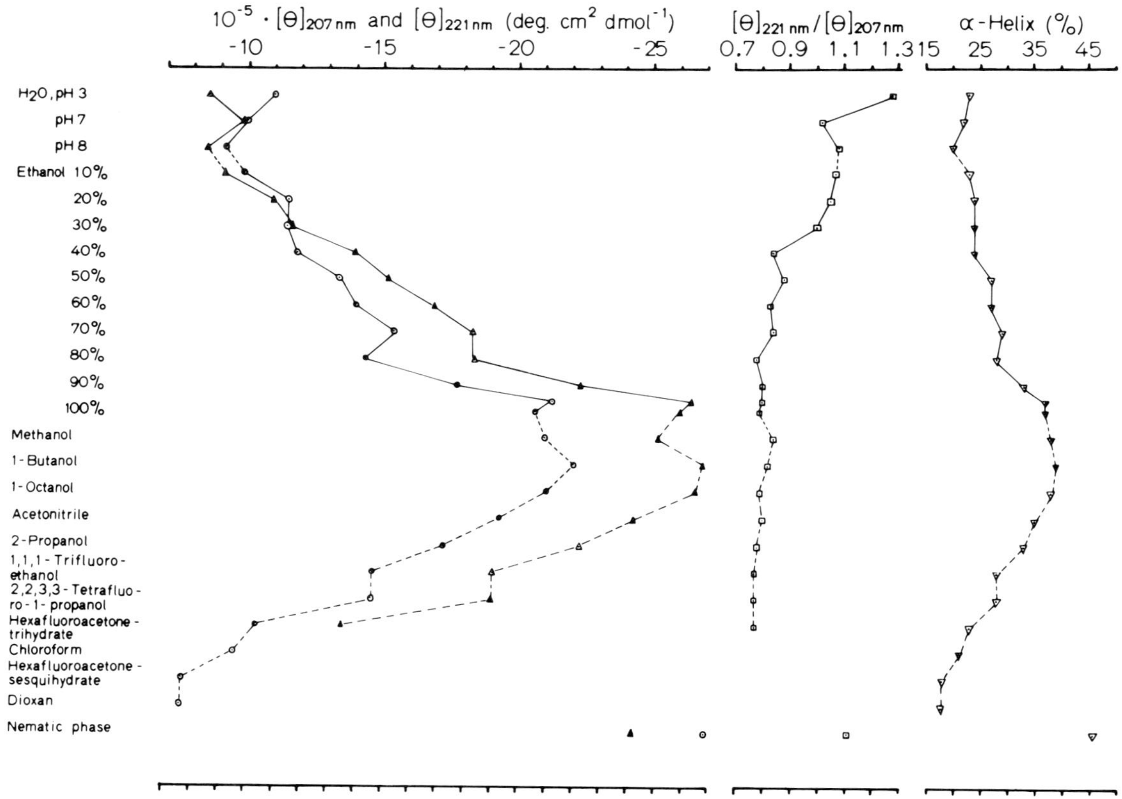

Fig.9. Solvent dependence of the circular dichroism spectra of alamethicin. Relations between the circular dichroism $\Delta\varepsilon$ (and $\Delta\varepsilon_M$) of the negative 207-nm (Δ) and 220 nm ($\circ$) Cotton effects, their ratio $\Delta\varepsilon_{221}/\Delta\varepsilon_{207}$ ($\square$) and α-helix content (∇). Temp. 20°C, concn. 0.2 mM.

region. The CD spectrum of the helical synthetic decapeptide Boc-(Aib-Ala)$_5$-OMe (Fig. 11) is similar to that of the trichotoxin undecapeptide (Fig. 10). The C-termini of trochotoxin A-40 $^+H_2$-Pro-Leu-Aib-Iva(Aib)-Gln-Valol (Fig. 10) exhibit almost exactly the same CD curves as the corresponding synthetic C-terminal hexapeptide Boc-Pro-Ala-Aib-Aib-Glu(OBzl)-Gln-OMe (Fig. 11). Conformational analysis which is based on three different evaluation methods of the CD spectra revealed the following α-helix contents: 28 % for Boc-(Aib-Ala)$_5$-OMe is comparable to that of the hexadecapeptide Boc-(Aib-Ala)$_5$-Pro-Ala-Aib-Aib-Glu(OBzl)-Gln-OMe (24 %), the α-helix content of the nonadecapeptide (44 %) is higher than that of alamethicin (35 %). The alamethicin value is similar to that of the dodecapeptide Boc-(Aib-Ala)$_6$-OPOE (33 %) what corresponds to about 6-7 residues in helical turns. The N-terminus of the helix can be shielded from polar interactions by formation of a ß-turn involving hydrogen bridges of Ac-Aib-Pro (resp. Ac-Aib-Gly). The C-terminal hexapeptides have predominantly unordered conformations with no evidence of ß-turns, which, on the basis of CD, seem to occur only in the short Aib containing tri -to pentapeptides.

We pointed out earlier (21,22) that the Gly^{11}-Ala-Aib^{13} region may be responsible for lengthening the helix when the peptide enters the hydrocarbon region. It is noteworthy that the glycine residue is found in trichotoxin A-40 at the N-terminal end of the helix replacing the proline in alamethicin. The synthetic hexadecapeptide lacking these three residues is inactive. The C-terminal aminoalcohols L-phenylalaninol resp. L-valinol and the N-acetyl group are obviously nonessential residues. Our cleavage and purification methods (38,50) allow the preparation of partial sequences of natural origin in amounts and states of purity for spectroscopic and synthetic purposes. A conformational analysis of the CD spectrum for the synthetic and membrane modifying nonadecapeptide analog of alamethicin Boc-(Aib-Ala)$_5$-Gly-Ala-Aib-Pro-Ala-Aib-

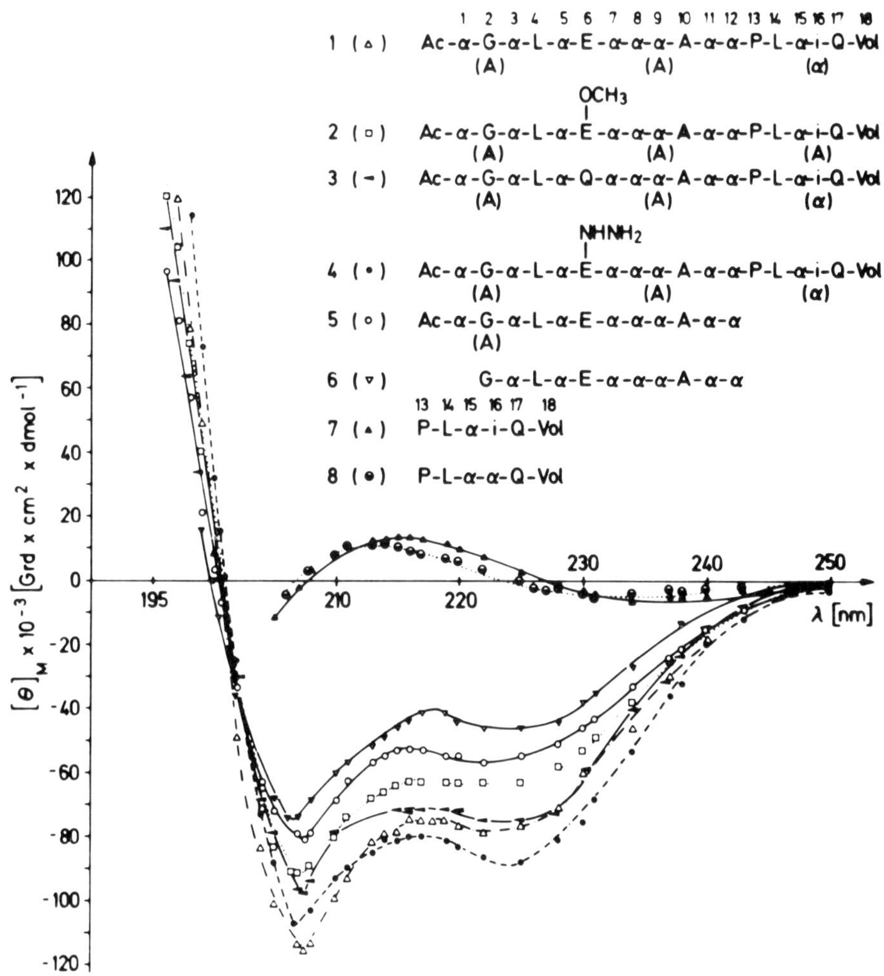

Fig. 10. Circular dichroism spectra of trichotoxin A-40 [1], its methyl ester [2], amide [3], hydrazide[4] and of the partial sequences N-acetyl-dodecapeptide [5], undecapeptide [6], and two hexapeptides [7,8]. Conditions: $c = 5 \times 10^{-4}$ mol/l in methanol; T = 25°C (α = Aib; others see (78)).

Aib-Glu(OBzl)-Gln-OMe was carried out (79). The CD data were evaluated according to three different methods and discussed with respect to those obtained from natural alamethicin and suitable models such as Boc-$(\text{Aib-Ala})_7$-OPOE, fragments of the

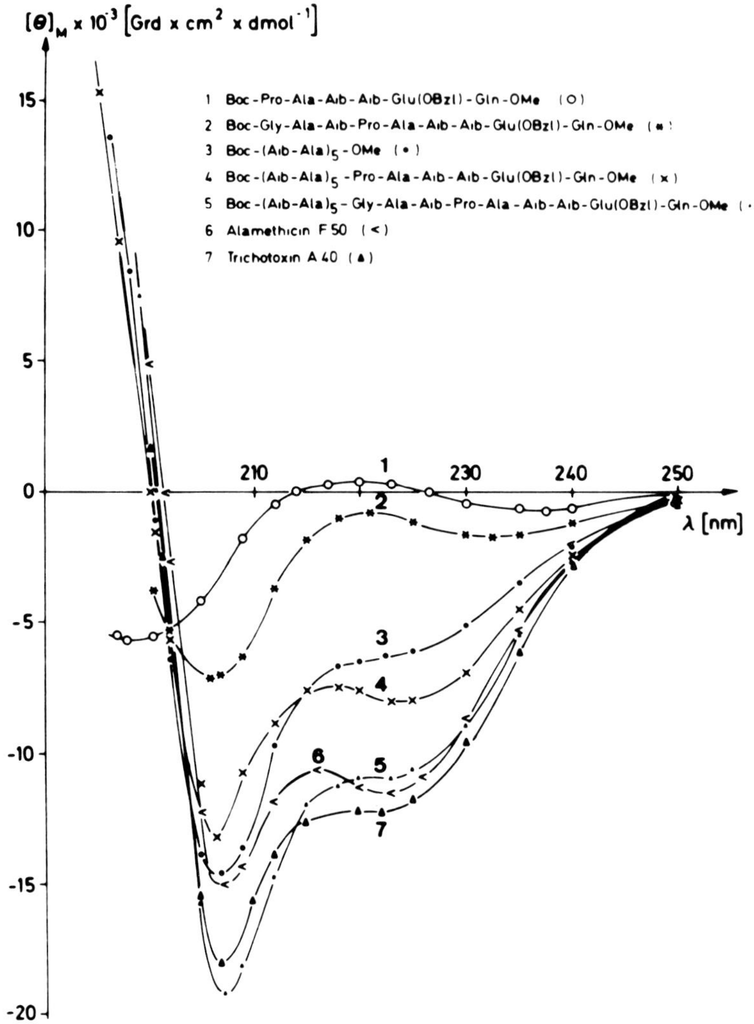

Fig.11. CD-spectra of synthetic peptides and some peptide fragments in comparison to alamethicin F-50 and trichotoxin A-40. Conditions: c = 5 x 10^{-4} mol/l(ethanol/H_2O (95:5)), T = 25°C (OPOE = polyoxyethylene).

synthetic nonadecapeptide and the hexadecapeptide N-Boc-(Aib-Ala)$_5$-Pro-Ala-Aib-Aib-Glu(OBzl)-Gln-OMe. The synthetic nonadecapeptide with the longer helical region exhibits membrane activities comparable to those of alamethicin, whereas the hexadecapeptide with the shorter helix is inactive (Fig. 11).

The stepwise synthesis and conformational studies of the N-terminal helical partial sequence of the membrane modifying polypeptide antibiotic alamethicin were described by our group (69). The polyoxyethylene esters of the fragments Boc-Pro-Aib-Ala-Aib-Ala-Gln-Aib-Val-Aib-Gly-OH and Ac-Aib-Pro-Aib-Ala-Aib-Ala-Gln-Aib-Val-Aib-Gly-OH were synthesized using polyoxyethylene (molecular mass 10, 000) as solubilizing support (80). CD spectra of each intermediate in ethanol show α-helix formation of the N-protected peptide polymers beginning with the nonapeptide and of the N-protonated sequences beginning with the decapeptide (Fig. 12). Compared to the helix of alamethicin, temperature- and solvent-dependent CD measurements indicate analogous conformational behavior. The results suggest that in lipophilic media the alamethicin helix can extend the full length of the partial sequence between the two proline residues and that aqueous media favor an increase of random-coil conformation (75).

For model studies of the particular lipid interaction of alamethicin, the stepwise synthesis of peptides with the alternating (Aib-Ala) sequence (n = 1-7) was carried out on a polyoxyethylene support (molecular mass 6000 (75)). CD and ORD studies in ethanol showed a change from the random coil to a right-handed α-helix with increasing peptide length. This change is observed for the N-protected peptides at a chain length of 8 residues and for the N-protonated peptides at a length of 9 residues. The comparison of the CD data of free and polyoxyethylene-bound peptides revealed that the solubilizing polymeric support cannot induce conformational changes. The intensities of the CD bands of Boc-(Aib-Ala)$_n$-

OPOE (n≥6) are higher than those of alamethicin, and these model peptides show similar temperature and solvent inducible changes of their helix contents.

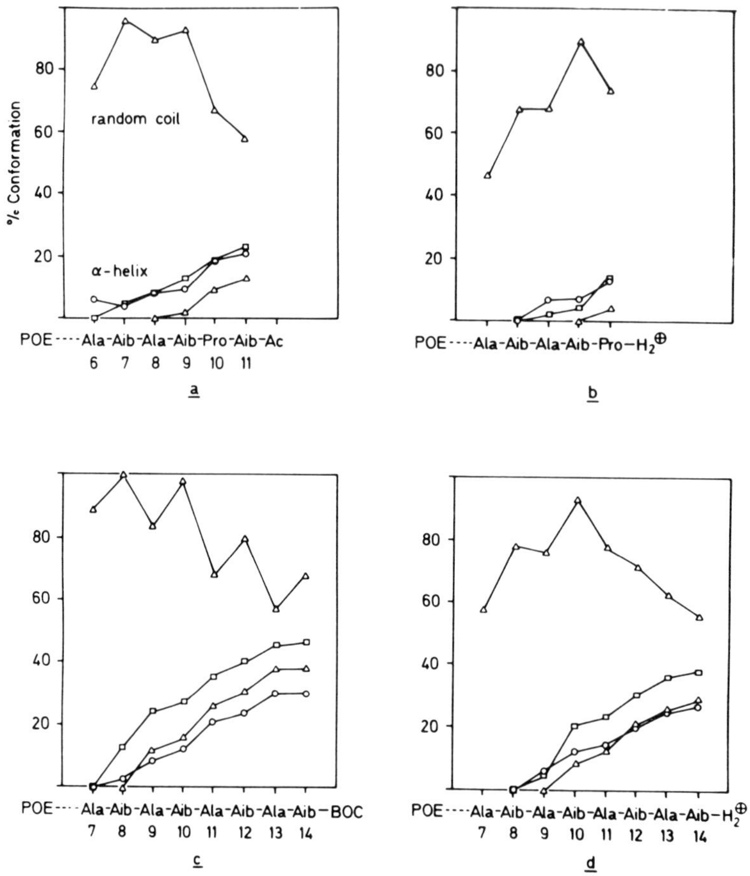

Fig. 12. Conformational changes of the growing peptide chains Ac-Aib-Pro-Aib-Ala-Aib-Ala-Gln-Aib-Val-Aib-Gly-OPOE (a), $^{+}H_2$-Pro-Aib-Ala-Aib-Ala-Gln-Aib-Val-Aib-Gly-OPOE (b), Boc-(Aib-Ala)$_7$-OPOE (c) and $^{+}H_2$-(Aib-Ala)$_7$-OPOE (d) determined according to Rosenkranz and Scholtan (△)(81,82), Greenfield and Fasman (□)(83) and Holzwarth and Doty (○)(84).

Furthermore, helical conformation was found for the conventionally synthesized undecapeptides Boc-(Ala-Aib)$_2$-Ala-Glu-(Bzl)-(Ala-Aib)$_2$-Ala-OMe and Boc-(Ala-Aib)$_2$-Ala-Gly-(Ala-Aib)$_2$-Ala-OMe (49). The latter peptide demonstrates that the helix is not destabilized by a glycine residue even in the middle of the sequence. However, Aib peptides of the same length but of higher Aib content show somewhat smaller Cotton effects. Thus, there is a pronounced helix-stabilizing influence of Aib residues.

^{13}C NMR and conformation

^{13}C nuclear magnetic resonance (NMR) has been used with success for studies on the conformation of alamethicin. The ^{13}C-NMR spectrum was assigned with the aid of signals of synthetic partial sequences and selective proton decoupling. The solvent and temperature dependence of the ^{13}C NMR spectra, T_1 measurements and the use of lanthanide-shift reagents allow the differentiation between the amino acids belonging to a rigid helical portion of the alamethicin sequence and those belonging to a more flexible part (58). The ^{13}C NMR results are in agreement with results obtained from extended solvent and temperature-dependent CD studies which indicate a highly stabilized nonpolar and intrachenar helical part.

The ^{13}C NMR spectra of suzukacillin A (14) and trichotoxin A-40 (37) revealed strong similarities with respect to their conformational evolution. Details of these studies will be described elsewhere. ^{13}C NMR revealed also the structural components Ac-Aib (14), phenylalaninol (58) and isovaline (62) which could not be detected by the conventional sequencing techniques. The signal assignements and the conclusions drawn from the NMR experiments on natural polypeptides are supported by detailed investigations on a series of synthetic peptides. Thus, pH-dependent spectra were recorded on va-

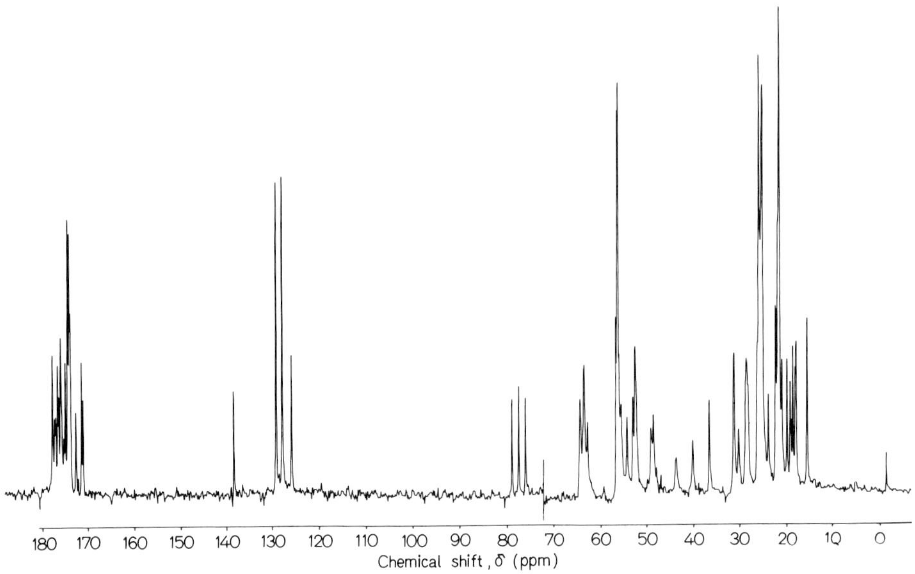

Fig. 13. Partially relaxed Fourier transform ^{13}C NMR spectrum of alamethicin, 150 mg alamethicin in 1.5 ml ^{12}C-enriched $C^2H_3O^2H/C^2HCl_3$, 22.628 MHz, 64 000 scans, 30°C (58).

rious model peptides with alternating Aib-Ala sequences in order to find out the maximal pH-induced shift differences for the Aib residues in different sequential positions (24, 85). Conformational effects on the ^{13}C chemical shifts of Aib residues are illustrated on the helical decapeptide Boc-(Aib-Ala)$_5$-OH. The spectrum confirms very clearly our earlier assignements of the alamethicin, trichotoxin A-40 and suzukacillin A spectra with respect to helical Aib residues exhibiting a magnetic nonequivalence of up to 4 ppm. Nonhelical Aib methyl groups differ by less than 1.5 ppm (Fig. 14). Temperature-dependent ^{13}C NMR studies on trichotoxin A-40 and the isolated N-acetyl-dodecapeptides suggest that one of the two Ala residues (-Aib^9-Ala^{10} - resp. -Ala^9-Ala^{10} -) is helical and the other not (62). In alamethicin both alanines (Ala^4 and Ala^6) have values typical for helical

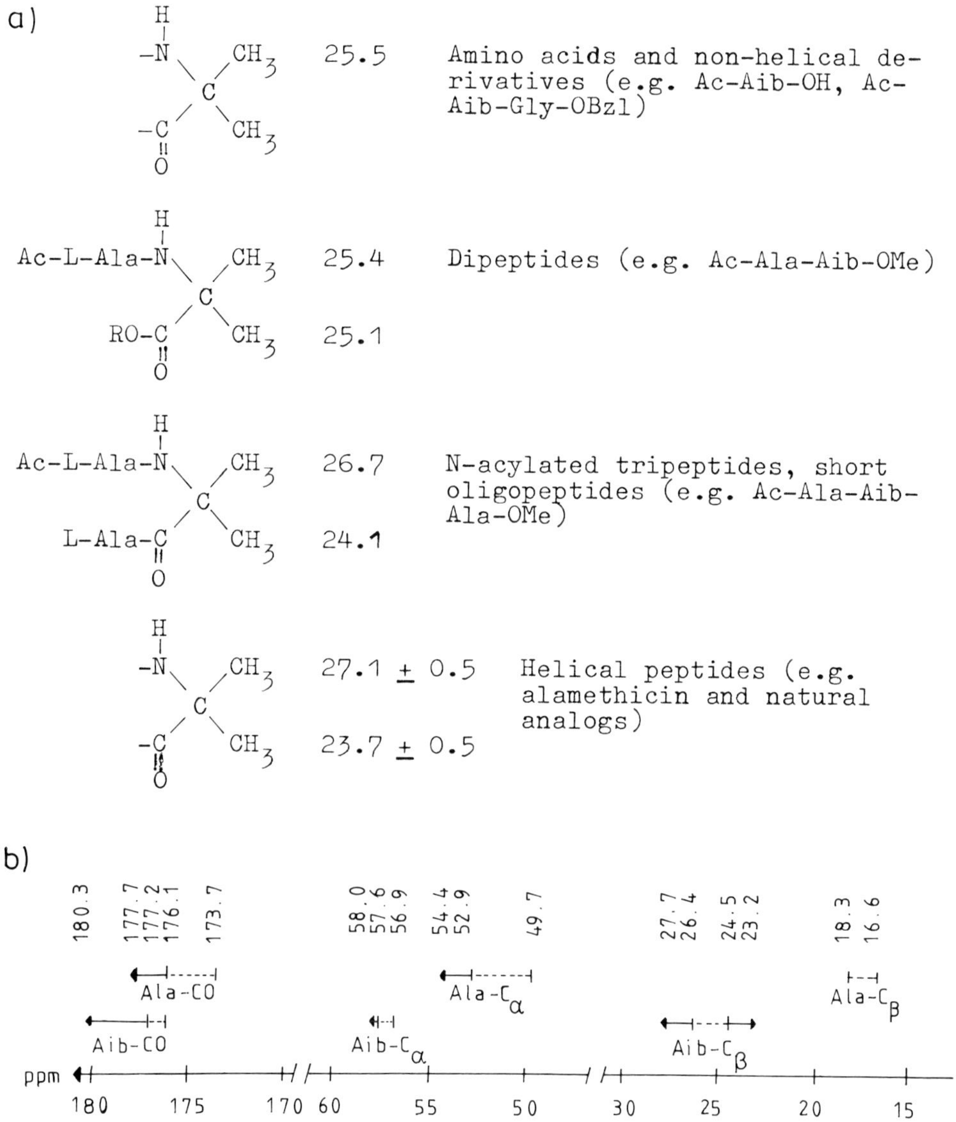

Fig. 14 . ^{13}C chemical shifts of ß-carbons of α-aminoisobutyric acid in different peptides (a). ^{13}C chemical shift changes of Aib and Ala carbons in oligo (⊢--⊣ up to pentapeptides) and polypeptides (b).

conformation. Synthetic Aib peptides such as the two undecapeptides Boc-(Ala-Aib)$_2$-Ala-X-(Ala-Aib)$_2$-Ala-OMe (X = Gly and Glu(Bzl)) and the undecapeptide with N-terminal alamethicin sequence, the decapeptide Boc-(Aib-Ala)$_5$-OH and various protected and free oligopeptides revealed basic information on the NMR chemical shifts of the Ala and Aib residues. Some of these results are summarized in Fig. 14. The magnetic nonequivalence of the geminal methyl groups of Aib is of particular value for conformational assignements. However, the uncritical use of ^{13}C NMR data without support from CD can lead to wrong interpretations. For example, the tripeptide Ac-Ala-Aib-Ala-OMe exhibits almost the same magnetic nonequivalence of Aib-C_β as the helical polypeptides. This is probably due to a strong stabilization by a ß-turn.

Crystallographic analysis of Aib peptides and conformational aspects

The achiral α-aminoisobutyric acid belongs to the rare amino acids. However, it has been found in antibiotics: alamethicin (1), suzukacillin (11), trichotoxin (37), samarosporin (43), stilbellin (35), emerimicin (40), chlamydocin (86) and also in casein (87). Furthermore it is of interest for the modification of peptide hormones in order to prevent enzymatic attack or to induce specific conformations (72,88). Theoretical considerations show that the additional methyl group reduces drastically the conformational space around the Aib residue. Several groups (72,88,89) calculated Ramachandran conformation diagrams which indicate the possibility either for an α-helix or a 3_{10}-helix for Aib polypeptides. Burgess and Leach (77) considered Aib as an "obligatory α-helical amino acid residue". This is in contradiction to IR studies on Poly-Aib indicating a 3_{10} conformation (90). Several X-ray structural analyses on shorter peptides such as Boc-Pro-Aib-Ala-Aib-

OBzl (91), Z-Aib-Pro-Aib-Ala-OMe (92), Tos-$(Aib)_5$-OMe (93) and Z-Aib-Pro-NH-CH_3 (94) show φ, Φ angle combinations which could also fit the values of an α-helix. However, ß-turns are assumed by these shorter peptides indicating a 3_{10} helical type. Since the values of, both, the 3_{10} and the α-helix are very close to each other the final proof for the type of alamethicin will not be found until the crystal structure of alamethicin or our synthetic crystalline undecapeptide will be known. Recently the crystal structure of Boc-Aib-OH was published (95). Boc-Aib-OH is of some interest because it is an important chain-lengthening residue for all synthetic intermediates studied by spectroscopic methods. N-tert-butyloxycarbonyl-2-methylalanine crystallizes from ethylacetate/light petroleum in monoclinic crystals of the space group $P2_1/c$, with a = 1085, b = 1099, c = 1172 pm, ß = 122.0° and Z = 4. Intermolecular hydrogen bonding between urethane carbonyl and carboxy groups is observed between two antiparallel molecules in form of 14-membered rings. These rings are somewhat inclined each to the next in the plane (010), and they are linked again by intermolecular hydrogen bonds between the 2-methylalanine-CO and-NH groups in the direction [010]. The resulting layers are held together only via hydrophobic interaction between opposite tert-butyl groups.

References

1. Meyer, P., Reusser, F.: Experientia 23, 85-86 (1967).
2. Mueller, P., Rudin, D.O.: Nature 217, 713-719 (1968).
3. Katz, B.: Nerv, Muskel und Synapse, Thieme, Stuttgart, 1971
4. Eisenberg, M., Hall, J.E., Mead, C.A.: J.Membrane Biol. 14, 143-176 (1973).
5. Gordon, L.G.M., Haydon, D.A.: Phil.Trans.R.Soc.Lond.Ser B 270, 433-447 (1975).
6. Boheim, G., Hall, J.E.: Biochim.Biophys.Acta 389, 436-443 (1975).

7. Gordon, L.G.M., Haydon, D.A.: Biochim.Biophys.Acta 436, 541-556 (1976).

8. Boheim, G., Kolb, H.A.: J.Membrane Biol.38, 99-150 (1978).

9. Pinkerton, M., Steinrauf, L., Dawkins, P.: Biochem.Biophys. Res.Commun.35, 512 (1969).

10. Urry, D.W.: Proc.Natl.Acad.Sci. 68, 672 (1971).

11. Ooka, T., Shimojima, Y., Akimoto, T., Takeda, I., Senoh, S., Abe, J.: Agr.Biol.Chem.30, 700-702 (1966).

12. Ooka, T., Takeda, I.: Agr.Biol.Chem. 36, 112-119 (1972).

13. Ooka, T., Takeda, I.: Agr.Biol.Chem. 38, 19-27 (1974).

14. Jung, G., König, W.A., Leibfritz, D., Ooka, T., Janko, K., Boheim, G.: Biochim.Biophys.Acta 433, 164-181 (1976).

15. McMullen, A.I.: Biochem.J. 119, 10 P - 11 P (1970).

16. Payne, J.W., Jakes, R., Hartley, B.S.: Biochem.J.117, 757-766 (1970).

17. Ovchinnikov, Y.A., Kiryushkin, A.A., Kozhevnikova, I.V.: Zhurnal Obshchei Khimii 41, 2085-2099; J.Gen.Chem.USSR 41, 2105-2116 (1971).

18. Martin, D.R., Williams, R.J.P.: Biochem.Soc.Trans.Proc. 533 rd Meeting, Vol.3, 166-167 (1975).

19. Martin, D.R., Williams, R.J.P.: Biochem.J.153, 181-190 (1976).

20. Jung, G., Dubischar, N., Leibfritz, D., Ottnad, M., Probst, H., Stumpf, C.: in: Peptides 1974, Wolmar, Y. (ed.), Wiley, New York, pp. 345-354 (1975).

21. Jung, G., Dubischar, N., Irmscher, G., Mayr, W., Oekonomopulos, R.: Chemiker-Ztg. 101, 196-201 (1077).

22. Irmscher, G., Jung, G.: Eur.J.Biochem.80, 165-174 (1977).

23. Pandey, R.C., Cook, J.C., Rinehart, K.L.: J.Amer.Chem. Soc.99, 8469-8483 (1977).

24. Jung, G., Brückner, H., Oekonomopulos, R., Boheim, G., Breitmaier, E., König, W.A.: in: Peptides: Structure and Biological Function, Gross, E. and Meienhofer, J.(eds.), Pierce Chemical Company, Rockford, Ill., pp. 647-654 (1979).

25. Brückner, H., Jung, G.: Chromatographia 13, 170-184 (1980).

26. Meena, D., Laddu, A.R., Saryal, R.K.: Current Sci. 36, 347-348 (1967).

27. Thirumalachar, M.J., Rhadhakrishnan, C.V., Vores, A.P.B: Hindustan Antibiot.Bull.8, 201-204 (1966).

28. Thirumalachar, M.J.: Hindustan Antibiot.Bull. 10, 287-289 (1968).

29. Deshmukh, P.V., Vaidya, M.G.: Nature 217, 849 (1968).

30. Vaidya, M.G., Deshmukh, P.V., Chari, S.N.: Hindustan Antibiot.Bull.11, 81-89 (1968).

31. Rinehart, K.L., Meng, H., Olson, K.L., Cook, J.C., Pandey, R.C.: Nature 269, 832-833 (1977).

32. Pandey, R.C., Meng, H., Cook, J.C., Rinehart, K.L.: J.Amer.Chem.Soc.99, 5203-5205 (1977).

33. Pandey, R.C., Cook. J.C., Rinehart, K.L.: J.Antibiot.31, 241-243 (1978).

34. Brückner, H., Nicholson, G.J., Jung, G., Kruse, K., König, W.A.: Chromatographia 13, 209-214 (1980).

35. Sasaki, K., Minato, H., Katagiri, K., Hayakawa, S., Matsushima, T.: J.Antibiot.24, 67-68 (1971).

36. Hou, C.T., Ciegler, A., Hesseltine, C.W.: Appl.Microbiol. 23, 183-185 (1972).

37. Irmscher, G., Bovermann, G., Boheim, G., Jung, G.: Biochim.Biophys.Acta 507, 470-484 (1978).

38. Brückner, H., König, W.A., Greiner, M., Jung, G.: Angew. Chem.91, 508-509 (1979); Angew.Chem.Internat.Ed.Engl.18, 476-477 (1979).

39. Brückner, H., Jung, G., König, W.A., Wachter, E.: in: Peptides: Structure and Biological Function, Gross, E. and Meienhofer, J. (eds.), Pierce Chemical Company, Rockford, Ill., pp. 723-726 (1979).

40. Argoudelis, A.D., Johnson, L.E.: J.Antibiot. 27, 274-282 (1974).

41. Pandey, R.C., Cook, J.C., Rinehart, K.L.: J.Amer.Chem.Soc. 99, 5205-5206 (1977).

42. Argoudelis, A.D., Dietz, A., Johnson, L.E.: J.Antibiot. 27, 321-328 (1974).

43. Inoue, N., Inoue, A., Furukawa, M., Kanda, N.: J.Antibiot. 29, 618-622 (1976).

44. Werner, G.: Diplomarbeit, University of Tübingen, 1977

45. Japanese patent no. 750517 (1975), Chem.Abstr. 83, 37917 (1975).

46. Japanese patent no. 7336200 (1973), Chem.Abstr. 79, 35134 (1973).

47. Irmscher, G.: Thesis, University of Tübingen, 1977

48. Jung, G., Oekonomopulos, R.: in: Peptides 1978, Siemion, I.Z., Kupryszewski (eds.), Wrocław University Press, Wrocław, pp. 689-691 (1979).

49. Oekonomopulos, R., Jung, G.: Liebigs Ann.Chem. 1979, 1151-1172.

50. Jung, G., Brückner, H., Oertel, B.: Symp.Chem.of Peptides

and Proteins,, Grainau (Germany), Abstr.pp. 165-166 (1978).

51. Boheim, G., Irmscher, G., Jung, G.: Biochim.Biophys.Acta 507, 485-506 (1978).

52. Boheim, G., Janko, K., Leibfritz, D., Ooka, T., König, W.A., Jung, G.: Biochim.Biophys.Acta 433, 182-199 (1976).

53. Besch, H.R., Jones, L.R., Fleming, J.W., Watanabe, A.M.: J.Biol.Chem. 252, 7905-7908 (1977).

54. Jung, G., Leibfritz, D., Ottnad, M., Dubischar, N., Probst, H.: Hoppe-Seyler's Z.Physiol. Chem. 355, 1213-1214 (1974).

55. Bessler, W.G., Ottenbreit, B., Irmscher, G., Jung, G.: Biochim.Biophys.Res.Commun. 87, 99-105 (1979).

56. Finer, E.G., Hauser, H., Chapman, D.: Chem.Phys.Lipids 3, 386-392 (1969).

57. Boheim, G.: J.Membrane Biol. 19, 277-303 (1974).

58. Jung, G., Dubischar, N., Leibfritz, D.: Eur.J.Biochem. 54, 395-409 (1975).

59. Baumann, G., Mueller, P.: J.Supramol.Structure 2, 538-557 (1974).

60. Hanke, W., Boheim, G.: Biochim.Biophys.Acta 596, 456-462 (1980).

61. Boheim, G., Hanke, W., Eibl, H.J.: Proc.Nat. Acad.Sci. 77, 3403-3407 (1980).

62. Brückner, H.: Thesis, University of Tübingen,1979

63. McMullen, A.I., Stirrup, J.A.: Biochim.Biophys.Acta 241, 807-814 (1971).

64. Laursen, R.: Solid Phase Methods in Protein Sequence Analysis, Pierce Chemical Company, Rockford, Ill., 1975

65. Brückner, H., König, W.A., Greiner, M., Jung, G.: in preparation.

66. Frank. H., Nicholson, G.J., Bayer, E.: J.Chrom.Sci. 15, 174-176 (1977).

67. Frank, H., Nicholson, G.J., Bayer, E.: J.Chromatogr. 146, 197-206 (1978).

68. König, W.A., Rahn, W., Eyem, J.: J.Chromatogr. 133, 141 (1977).

69. Mayr, W., Jung, G.: Liebigs Ann.Chem., submitted (1980).

70. Kendrick, N.C.E.: Thesis, University of Washington,1973

71. Gisin, B.F., Kobayashi, S., Hall, J.E.: Proc.Natl.Acad. Sci. 74, 115-119 (1977).

72. Marshall, G.R., Bosshard, H.E., Kendrick, N.C.E., Turk, J., Balasubramanian, T.M., Cobb, S.M.H., Moore, M., Le-

duc, L., Needleman, P.: in: Peptides 1976, Loffet, A. (ed.), Editions de l'Université de Bruxelles, Brussels, pp. 361-368 (1976).

73. Schmitt, H., Jung, G.: In preparation.

74. Leibfritz, D., Mayr , W., Oekonomopulos, R., Jung, G.: Tetrahedron 34, 2045-2050 (1978).

75. Mayr , W., Oekonomopulos, R., Jung, G.: Biopolymers 18, 425-450 (1979).

76. McMullen, A.I., Marlborough, D.I., Bayley, P.M.: FEBS Lett. 16, 278-280 (1971).

77. Burgess, A.W., Leach, S.J.: Biopolymers 12, 2599-2605 (1973).

78. IUPAC-IUB Commission on Biochemical Nomenclature. A One-Letter Notation for Amino Acid Sequences: Biochemistry 7, 2703-2705 (1968).

79. Oekonomopulos, R., Jung, G.: Biopolymers 19, 203-214 (1980).

80. Bayer, E., Mutter, M.: Nature 237, 512-513 (1972).

81. Rosenkranz, H., Scholtan, W.: Hoppe-Seyler's Z. Physiol. Chem. 352, 896-904 (1971).

82. Rosenkranz, H.: Z. Klin. Chem. Klin. Biochem. 12, 415-422 (1974).

83. Greenfield, N., Fasman, D.: Biochemistry 8, 4108-4126 (1969).

84. Holzwarth, G., Doty, P.: J. Amer. Chem. Soc. 87, 218-228 (1965).

85. Severge, A., Jüttner, F., Breitmaier, E., Jung, G.: Biochim. Biophys. Acta 437, 289-300 (1976).

86. Closse, A., Huguenin, R.: Helv. Chim. Acta 57, 533-545 (1974).

87. Kandatsu, M., Kikuno, K.: Bull. Agr. Chem. Soc. Jap. 22, 339 (1958).

88. Marshall, G.R., Bosshard, H.E.: Circ. Res. Suppl. II, 43 (1972).

89. Nagaraj, R., Shamala, N., Balaram, P.: J. Amer. Chem. Soc 101, 16-20 (1979).

90. Malcolm, B.R.: Biopolymers 16, 2591-2592 (1977).

91. Smith, G.D., Duax, W.L., Kendrick, N.C.E., Marshall, G.R., Mathews, F.S.: in: Peptides, Goodman, M. and Meienhofer, J. (eds.), Wiley, New York, pp. 277-279 (1977).

92. Shamala, N., Nagaraj, R., Balaram, P.: Biochem. Biophys. Res. Commun. 79, 292-298 (1977).

93. Shamala, N., Nagaraj, R., Balaram, P.: J.C.S. Chem. Commun.

996-997 (1978).

94. Venkataram Prasad, B.V., Shamala, N., Nagaraj, R., Chandrasekaran, R., Balaram, P.: Biopolymers 18, 1635-1646 (1979).

95. Mayr, W., Jung, G., Strähle, J.: Liebigs Ann. Chem. 1980, 715-724.

NEUROPEPTIDES: SUBCELLULAR LOCALIZATION OF ACTH AND RELATED PEPTIDES

Karl H. Voigt, Eckard Weber*
Labor für Neuroendokrinologie, Abteilung Physiologie

Rainer Martin
Sektion Elektronenmikroskopie, Universität Ulm, Oberer Eselsberg, D-7900 Ulm, Germany

Introduction

The corticotroph cells of the pituitary contain in addition to corticotropin (ACTH) also other biologically active peptides such as ß-endorphin and MSH-analogues (1-7). These peptides are synthesized concomitantly in a 31 000 dalton prohormone (8-11) and secreted simultaneously in vivo (12, 13) and also in vitro (14-16, 18, 19). In extracts of anterior pituitaries from several different species the peptide fragments listed in Fig. 1 were identified and shown to be cleaved from the common precursor molecule during posttranslational processing (8-11, 16, 17). In addition to these peptides, the cells of the intermediate lobe contain α-MSH and the corticotropin-like intermediate lobe peptide (CLIP, e.g. $ACTH^{18-39}$) (16, 20-24). Although the Met-enkephalin sequence is contained within ß-endorphin, there is now evidence to indicate that the latter is not the precursor of the pentapeptide (16, 18, 24, 25). It has been found only in small amounts in extracts of the anterior pituitary but in significant quantities in those of the neurohypophysis (26-28). Besides the presence in the

* Present address: Stanford University, Medical Center, Dept. of Psychiatry, Nancy Pritzker Lab., Stanford, Calif. U.S.A.

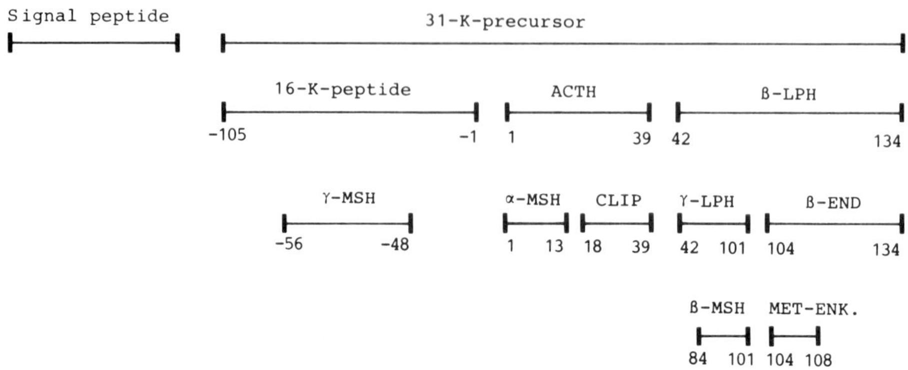

Fig. 1. Schematic representation of the ACTH/LPH-precursor molecule and its fragments. Important, Met-enkephalin is not derived from ß-endorphin. ACTH seems to be the major fragment in the anterior pituitary, whereas in the intermediate lobe it is cleaved to generate α-MSH and CLIP. ß-Endorphin is equally distributed between the two lobes. The MSH-peptides share the common amino acid sequence Met-Glu/Gly-His-Phe-Arg-Trp (sequences: minus 53-48, 4-9, 47-52 of the prohormone) (17).

pituitary gland, most of the peptides derived from the 31-k-precursor are localized obviously also in central and peripheral nervous tissue and in the gut (review, 29).

By applying specific antibodies against ACTH, ß-endorphin and the 16-k-fragment in this report we will demonstrate that the immunoreactivities to all these peptides are stored together within the same secretory granules in the corticotroph cell of the rat anterior pituitary and the additional occurrence of α-MSH and CLIP immunoreactivities in the intermediate lobe cells. The localization of Met-enkephalin was different, we found it in somatotrophs of the anterior pituitary and in nerves of the neurohypophysis.

The information about biosynthesis and localization of ACTH/

LPH-related peptides is restricted to different species of vertebrates. In this article we describe the occurrence of immunoreactivity to Met-enkephalin and α-MSH also in a cephalopod neurohemal organ.

Material and methods

Immunocytochemistry: Pituitaries from normal male rats (strain ChBB, Thomae, Biberach), sacrificed by decapitation, were quickly removed and fixed by immersion in either picric formaldehyde (30) (light microscopy) or in 2 % glutaraldehyde in 0.1 M phosphate buffer, pH 7.3 (electron microscopy). The tissue was embedded in Epon 812. Semithin (1μm) sections were cut from the polymerized blocks and mounted on slides by heat. After removal of the resin with sodium methoxide (31), the sections were stained by the unlabeled antibody enzyme method involving the soluble peroxidase-antiperoxidase complex (PAP) (32) or by a double antibody method using horseradish peroxidase (HRPO)-labeled anti rabbit IgG. The sections were dehydrated, embedded in Eukitt and observed with phase contrast optics. For subcellular localization the immunocytochemistry was applied on ultrathin sections(about 50 nm thick). The adjacent sections were collected, each on a single-hole nickel grid,and stained for demonstration of the different peptides. For immunocytochemical studies on human pituitaries the same methods were applied on microadenomas and the surrounding tissue. The tissue was removed by neurosurgery (Dr. Fahlbusch, Neurochirurgische Abteilung, Universität München, Germany) by the parapharyngeal route for treatment of patients suffering from ACTH hypersecretion (Cushing's disease). In another experimental series we used the extensive neuropil layer in the vena cava of Octopus vulgaris for phylogenetic studies on the occurrence of the ACTH/LPH peptides. The tissue of the vena cava was prepared according to the description for pituitary material. We used consecutive ultrathin-semithin sections for the localization of immunostaining. The ultrathin section without immunostaining was contrasted with 2 % aqueous uranyl acetate and lead citrate for electron microscopy. The adjacent 1 µm or 0.5 µm thick section was immunostained as described. (For details of the immunocytochemical procedures see 33-37).

Peptides: The following peptides were used for specificity controls: Natural human ß-LPH ($ß_h$-LPH) (C.H. Li, San Francisco, USA). Natural sheep ß-LPH ($ß_s$-LPH), porcine ß-MSH ($ß_p$-MSH) and synthetic sheep ß-endorphin (M. Chrétien, Montreal, Canada). Synthetic human ß-endorphin, α-endorphin, α-MSH and γ-MSH (Bachem, Basel, Switzerland; Peninsula, San Carlos, USA). Met-enkephalin and synthetic analogues (E. Wünsch, Munich, Germany). Synthetic preparations of ACTH ($^{1-39, 1-24, 1-10, 1-13, 25-39, 4-10, 4-16, 18-39}$) (Ciba-Geigy and Bachem, Basel, Switzerland; Ferring AB, Malmö, Sweden). Purified mouse tumor cell 16-k-fragment (B.A. Eipper and R. Mains, Denver, USA; see 19). These peptides were added at a final concentration of 20µg/ml to the working dilutions of the antisera for 24h in preabsorption tests (Table 1, 2, 3).

Antisera: 1. ACTH-antisera: Affinity purified antiserum Bertha against the middle portion of ACTH ($ACTH^{11-24}$), affinity purified antiserum Freddie against the C-terminal sequence $^{34-39}$ (B.A. Eipper and R. Mains, Denver, USA; see 38, 39), and antiserum K3 against $ACTH^{1-39}$ and $ACTH^{1-24}$ (40). 2. α-MSH-antiserum: Antisera R36 and R33 against α-MSH, crossreacting with the N-terminal portion $^{1-13}$ of ACTH (36, 37). 3. ß-LPH antiserum: This antiserum (C.H. Li, San Francisco, USA) was preabsorbed for 24h prior to use with an excess (20µg/ml) of $ß_h$-endorphin in order to block any potential binding sites for ß-endorphin; the remaining immunoreactivity will be directed against γ-LPH. 4. 16-k-fragment antiserum: It is directed against the non-ACTH/ß-LPH-portion of the common precursor, prepared and described by Eipper and Mains (19, 39). 5. Met-enkephalin antiserum: The antiserum R26 is specific for Met-enkephalin, the crossreaction to Leu-enkephalin was about 2 % in a radioimmunoassay (RIA) and not detectable in cross-absorption tests using affinity purified antiserum (35, 41). 6. ß-Endorphin antiserum: The affinity purified antiserum Melinda (B.A. Eipper and R. Mains, Denver, USA) recognized α-endorphin, ß-endorphin and ß-LPH (39). The unpurified antiserum R56 was also used (14). All antisera were raised in rabbits and were diluted with saline, 0.01 M phosphate buffer + 0.15 M NaCl; working dilutions 1:1 000 to 1: 20 000. The antisera R26 (Met-enkephalin), K3 (ACTH), R36 (α-MSH), and R56 (ß-endorphin) were used in radioimmunoassays and tested for specificity under RIA conditions and in cross-absorption experiments (Table 2, 3).

Table 1. Specificity control of the antisera. Cross absorption experiments were performed with the working dilutions of the antisera and an excess of the antigens (20µg/ml) for 24h prior to use for immunostaining of anterior pituitary slices (rat and human origin). +, immunostaining not altered; (+), weakened; -, completely blocked .(*Antiserum K3, ** antiserum Melinda).

Antigen	Antiserum against			
	$ß_h$-LPH	ACTH*	ß-ENDORPHIN**	16-k
$ß_h$-LPH	-	+	-	+
$ß_s$-LPH	-	+	-	+
$ß_p$-MSH	-	+	+	+
ß-ENDORPHIN	+	+	-	+
α-ENDORPHIN	+	+	-	+
MET-ENKEPHALIN	+	+	+	+
$ACTH^{1-39}$	+	-	+	+
$ACTH^{1-24}$	+	-	+	+
$ACTH^{25-39}$	+	+	+	+
α-MSH	+	+	+	+
16-k-FRAGMENT	(+)	+	+	-

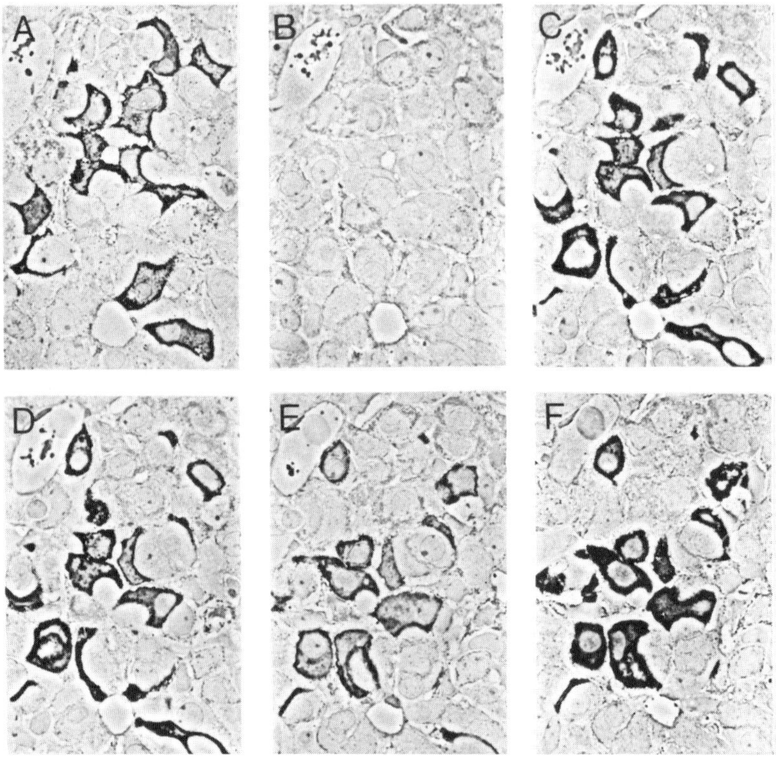

Fig. 2. Serial semithin (1μm) sections through a rat anterior pituitary stained for: (A) 16-k-fragment; (C) ACTH; (D) ß-endorphin and (E) γ-LPH (ß-LPH antiserum was preabsorbed with 20μg ß-endorphin; thus γ-LPH immunoreactivity is shown by the treated ß-LPH antiserum). All fragments are present in the same cells. To demonstrate the specificity of the antiserum against 16-k-fragment, the section in B was treated with 16-k-fragment antiserum saturated with an excess of purified 16-k-fragment, and the section in F was stained with 16-k-fragment antiserum preincubated with purified ß-LPH, ß-endorphin, and ACTH. (From Weber, Martin, Voigt, ref. 7).

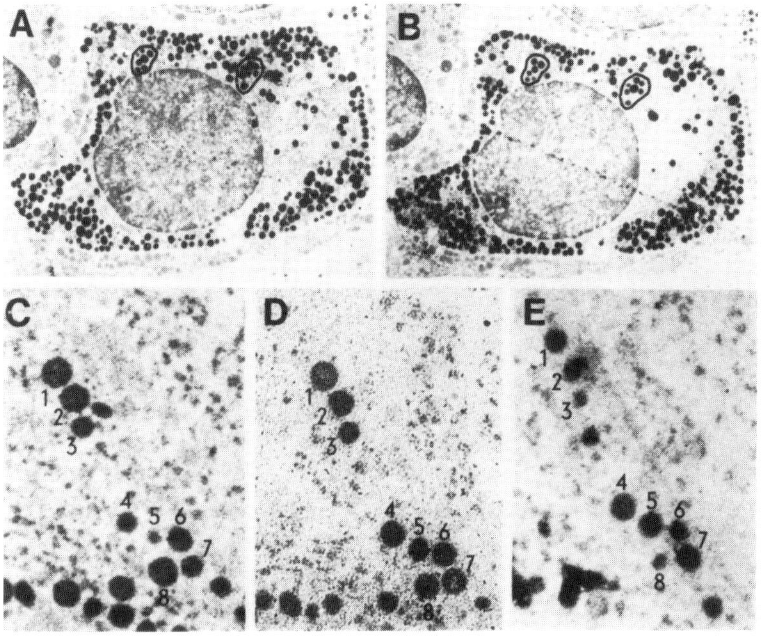

Fig. 3. Serial ultrathin sections through secretory granules of ACTH/endorphin cells from a rat anterior pituitary. The granules were labeled for (A) ß-LPH and (B) ß-endorphin or for (C) ß-endorphin, (D) ACTH and (E) 16-k-fragment. This proves conclusively that immunoreactivities to all segments of the ACTH/endorphin precursor are contained within the same secretory granules. In A and B some groups of granules which can easily be identified in both sections have been encircled (From Weber, Voigt, Martin, ref.7).

Results

Immunocytochemistry of the ACTH/endorphin cell in the anterior pituitary

The corticotroph cells of the rat anterior pituitary contain more than only ACTH: When serial semithin sections were stained sequentially with antibodies against ß-endorphin, ß-LPH, ACTH or 16-k-fragment it became evident that immunoreactivities to these fragments of pro-opiocortin were present in the same cells (Fig. 2). Cross-absorption controls (Table 1) indicated that the immunostaining was specific for each fragment. The antisera against ACTH, ß-endorphin and 16-k-fragment also recognized the common precursor (39, 42). The ß-LPH antiserum has not been tested, but we assume that it also will bind to the prohormone. The ß-endorphin immunostaining was blocked by ß-endorphin and by ß-LPH, indicating that this antiserum cannot distinguish the ß-endorphin sequence contained within ß-LPH from the isolated peptide. The antiserum against ß-LPH was blocked by ß-LPH and by ß-MSH indicating that the antibody is directed against the MSH-portion of LPH. However, it is questionable if ß-MSH is a naturally occurring peptide hormone (43). When the anti-ß-LPH was preabsorbed with 16-k-fragment the immunostaining was weakened, probably due to a contamination of the 16-k-peptide with LPH or ß-MSH. The immunostaining of the antiserum against 16-k-fragment was not affected by preabsorption with the other fragments, but was completely blocked by purified 16-k-fragment.

To determine the subcellular localization of the pro-opiocortin-fragments in the anterior pituitary cells serial ultrathin sections were stained with the different antibodies (Fig. 3). The reaction product was strongly confined to the secretory granules which have a mean diameter of about 200 nm. Therefore it was possible to recognize the same granules in up to three adjacent sections if they were cut about 50 nm thin. When adjacent sections through the same secretory granules were labeled separately with antibodies against ACTH, ß-endorphin/ß-LPH or 16-k-fragment, the immunoreactivity to these portions of pro-opiocortin occurred in the same granules.

By applying the immunocytochemical methods in the same way to human pituitary tissue very similar results were observed (Fig. 4). In microadenomas of pituitaries from patients suffering from Cushing's disease, the majority of cells

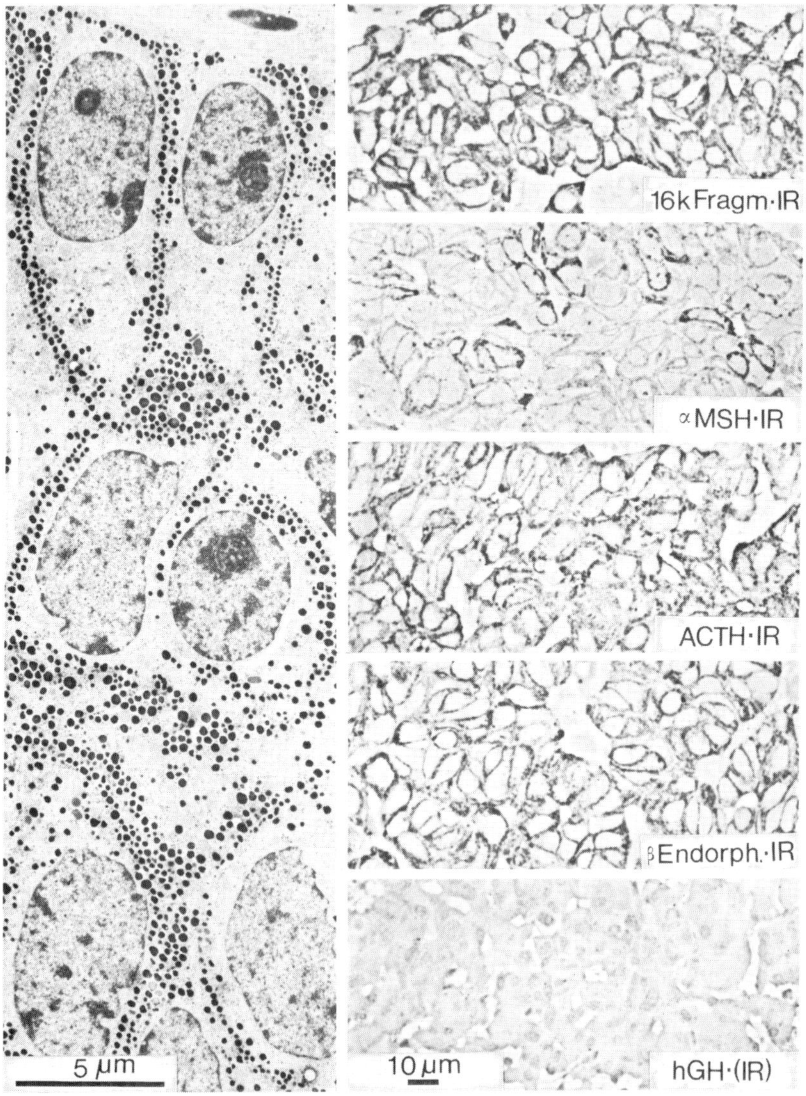

Fig. 4. ACTH/ß-endorphin cells of a human pituitary adenoma (Nelson tumor). The cells are homogeneous and display immunoreactivity to 16-k-fragment, ACTH, ß-endorphin/ß-LPH, but not to human growth hormone (hGH) and to Met-enkephalin (not included); some cells exhibit α-MSH immunoreactivity. On the left the ultrastructure of several not immunostained tumor cells is shown.

exerted ACTH-immunoreactivity. When serial semithin sections (0.5µm) of the adenoma were incubated with the antisera against pro-opiocortin fragments, it was evident that all these peptides occurred in the same cell. However, in some microadenomas different immunostaining was observed: a subpopulation of cells was not immunostained by ACTH antiserum, but by antisera against 16-k-fragment, ß-endorphin/ß-LPH and α-MSH; in some cells α-MSH-immunoreactivity was detected in addition to immunoreactivities to all the other fragments. The adenoma in Fig. 4 consists of homogeneous cells exhibiting immunoreactivity to ACTH, ß-endorphin/ß-LPH and to 16-k-fragment. Additionally, a subpopulation of cells was immunostained by α-MSH antiserum. (In this case the adenoma was a Nelson tumor that had developed years after bilateral adrenalectomy as a treatment of Cushing's disease.)

Immunocytochemistry of the intermediate lobe of the rat pituitary

Intense immunostaining of all cells of the intermediate lobe (PI) was found by applying antibodies against α-MSH, ACTH, ß-endorphin/ß-LPH and the 16-k-fragment. α-MSH appeared to be the dominant peptide of the rat PI, but present only in subpopulations of the corticotrophs in the anterior pituitary (Fig. 5). The specificity of the used antiserum against α-MSH is shown in Table 2, tested in a radioimmunoassay (RIA) and by preabsorption experiments. The antiserum is directed mainly against the C-terminal part of α-MSH. A clear cross-reactivity with $ACTH^{1-13}$ but not with $ACTH^{1-10}$ may indicate, that N-acetylation as present in α-MSH is not important for the binding of this antiserum. After immunostaining of two consecutive ultrathin sections with either ACTH- and α-MSH antibodies or ACTH- and ß-endorphin antibodies it was evident that single secretory granules are immunoreactive with all

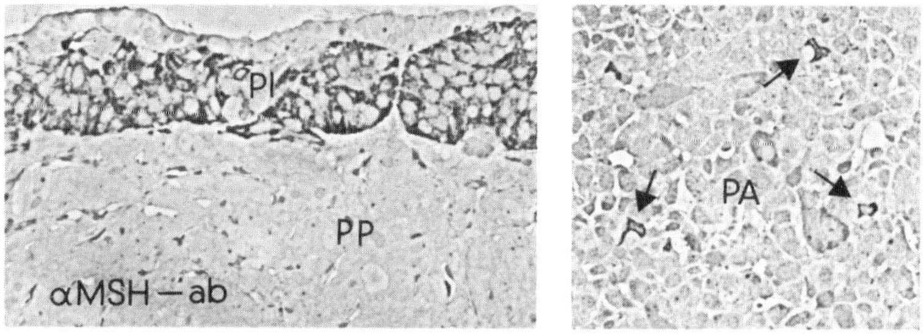

Fig. 5. Immunostaining with antiserum R36 against α-MSH on semithin sections (1μm) of the pars anterior (PA), pars intermedia (PI) and pars posterior (PP) of a rat pituitary. All secretory cells of the PI, none of the PP and very rare cells of the PA were recognized by the antibody. (From ref. 36).

three antibodies (Fig. 6). In addition to the fragments of pro-opiocortin found in the anterior pituitary (Fig. 3), the granules of PI cells additionally contained α-MSH and CLIP immunoreactivity (Fig. 6, 8).

Occurrence of Met-enkephalin-like immunoreactivity in the anterior pituitary

Met-enkephalin immunoreactivity was not detected in, both, the corticotrophs of the anterior pituitary and the intermediate lobe cells. When semithin sections from picric acid formaldehyde- or glutaraldehyde-fixed rat anterior pituitaries were stained with affinity purified antibodies against Met-enkephalin, a large number of cells were immunoreactive. Staining of consecutive sections with antibodies against Met-enkephalin, ß-endorphin and growth hormone revealed that all somatotroph cells were immunostained with Met-enkephalin antibodies and antibodies against growth hormone, respectively, but not with anti-ß-endorphin (Fig. 7A-C). There was no cross-reaction of the Met-enkephalin antibody with different preparations of growth hormone (human GH, hGH, and porcine GH, pGH, prepared by M. Schleyer, Ulm, Germany; rat GH, rGH, obtained from NIH,

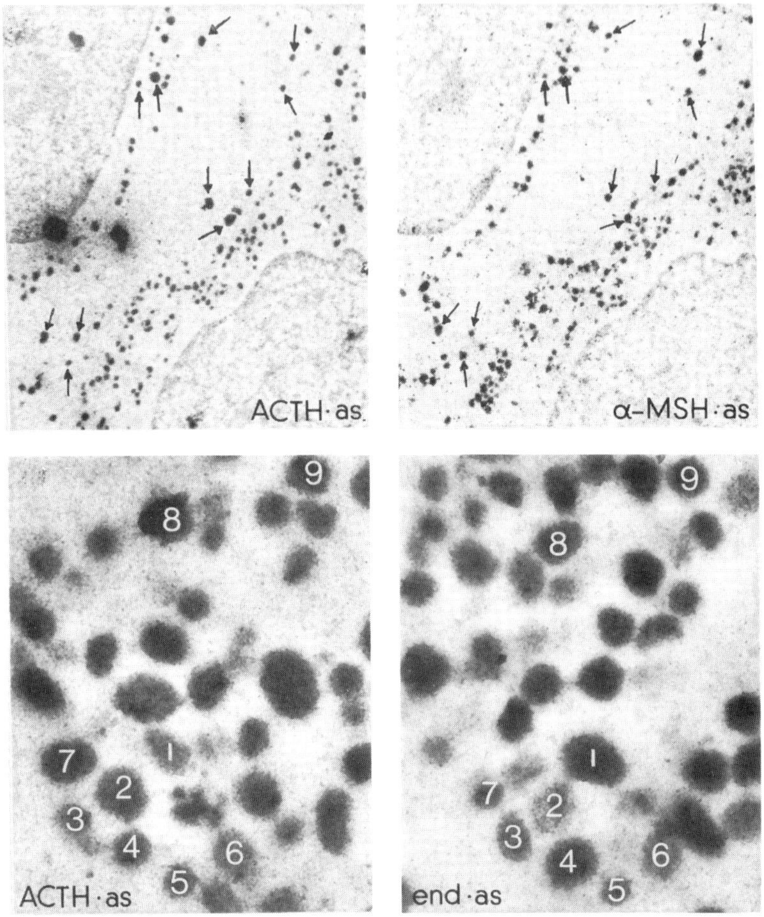

Fig. 6. Electron micrographs of PI cells incubated with antibodies against ACTH, α-MSH, and ß-endorphin. In adjacent sections identical granules were stained for ACTH and α-MSH (arrows, top) and for ACTH and ß-endorphin (compare the granules with numbers, bottom). (From ref. 36).

Bethesda, USA) in radioimmunoassays as well as in preabsorption experiments. Antiserum against hGH was chosen because it did not produce background staining and the immunoreaction was completely abolished after preabsorption with rGH (and with hGH; only weakened with pGH). Since Met-enkephalin is identical with the NH_2-terminal pentapeptide sequence of ß-endorphin, corticotrophs were carefully examined for occurrence

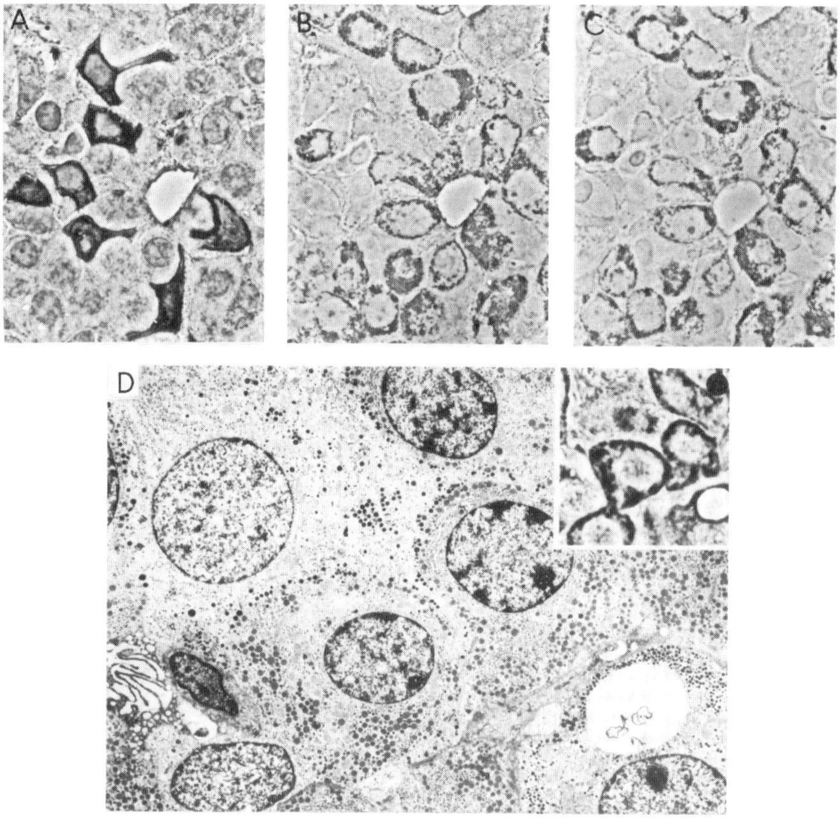

Fig. 7. Phase contrast light micrographs of immunocytochemically stained serial semithin sections (0.5μm) of rat anterior pituitaries. The rats were adrenalectomized two weeks before fixation in order to enhance the number of corticotroph cells. Three adjacent sections sequentially immunostained for ß-endorphin (A), Met-enkephalin (B) and growth hormone (C). None of the cells containing ß-endorphin immunoreactivity also contain Met-enkephalin-like reactivity; the latter is strongly confined to the somatotrophs.

To localize the Met-enkephalin immunoreactivity, a semithin section through a glutaraldehyde-fixed rat anterior pituitary was stained for Met-enkephalin (D, inset) and the ultrastructure of these cells was examined in immediately adjacent ultrathin sections (D) contrasted with uranyl acetate and lead citrate. The stained cells display the ultrastructure typical of somatotrophs. (From ref. 35).

Table 2. Cross-reactivity of antiserum R36 (anti α-MSH) measured under RIA-conditions* (displacement of α-MSH-^{125}I antiserum dilution 1:40 000) and in preabsorption tests** on semithin sections of pituitary intermediate lobe and on vena cava neuropil of Octopus vulgaris (see also Table 1).

Antigen	Crossreactivity	
	RIA*	Immunoreactivity**
α-MSH		
Peninsula	100 %	-
Bachem	100 %	-
Bioscience Prod.	100 %	-
$ACTH^{1-13}$	10.0 %	-
$ACTH^{1-10}$	2.0 %	+
$ACTH^{1-16}$	6.0 %	
$ACTH^{4-10}$	<0.1 %	+
$ACTH^{1-24}$	1.0 %	+
$ACTH^{1-39}$	1.0 %	+
$ACTH^{18-39}$	<0.1 %	+
D-Ser($^{4-16}$)Lys-NH_2	0.2 %	(+)
$ß_h$-MSH	<0.1 %	+
$ß_p$-MSH	<0.1 %	+
γ-MSH	<0.1 %	+

of Met-enkephalin immunoreactivity, but Met-enkephalin immunostaining was distinctly confined to somatotroph cells (Fig. 7A-C). On the other hand, no endorphin-like immunoreactivity was observed in GH-cells. Because the somatotrophs can easily be identified by their characteristic ultrastructure, the localization of Met-enkephalin immunoreactivity was ascertained on adjacent ultrathin sections which were routinely contrasted with heavy metal salts. As to be expected from the immunostaining experiments all cells immunostained for Met-enkephalin displayed the characteristic ultrastructural features of somatotrophs (Fig. 7D).

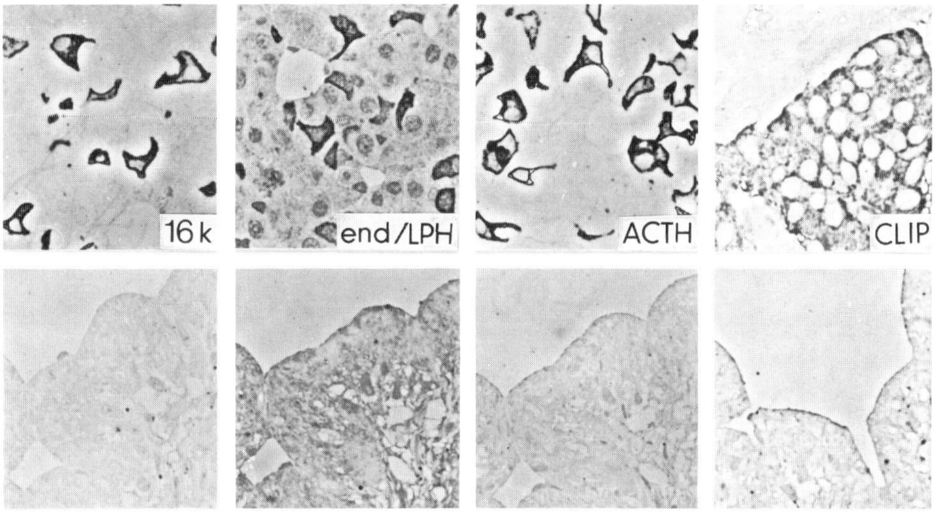

Fig. 8. Semithin sections through the vena cava neuropil of Octopus vulgaris (bottom) and of PA and PI of the rat pituitary (top). Two sections of the different tissues were incubated simultaneously in the same drop of antiserum to demonstrate the immunoreactivity. No immunostaining was observed in the neuropil, but an intense reaction was seen in the rat anterior pituitary for 16-k-fragment, ß-endorphin/ß-LPH and ACTH, and in the intermediate lobe for CLIP.

Extensive specificity controls of the antiserum against Met-enkephalin were performed to corroborate the unexpected finding of enkephalin immunoreactivity in GH-cells (Table 3). The results indicate that the C-terminal part of Met-enkephalin is important for the antibody-antigen binding; cross-reactivity with Leu-enkephalin was only 0.2 %, alterations at the NH_2-terminal part did not result in an evident loss of binding, cross-reactivity with different endorphin preparations and ß-LPH was less than 0.01 %.

Table 3. Cross-reactivity of antiserum R26 against Met-enkephalin in a radioimmunoassay* (displacement of ^3H(Tyr)-Met-enkephalin, antiserum dilution 1:16 000). **Preabsorption controls of affinity purified antiserum R26, performed on semithin sections of the three pituitary lobes and with most antigens also on the vena cava neuropil of the octopus (see also Table 1).

Antigen	Crossreactivity	
	RIA*	Immunostaining**
Tyr-Gly-Gly-Phe-Met	100 %	-
Tyr-Gly-Gly-Phe-Leu	0.2 %	+
Tyr-Gly-Gly-Phe-NLeu	11.0 %	(-)
Gly-Gly-Phe-Met	-	+
Phe-Met	-	+
Phe-Leu	-	+
Phe-NLeu	-	+
Phe-Gly-Gly-Phe-Met	56 %	-
Tyr-Gly-Gly	-	+
Tyr-DAla-Gly-Phe-Met	22 %	-

Immunocytochemical reactions in a cephalopod neurohemal organ

The nerve terminals of the neurosecretory system of the vena cava from Octopus vulgaris (51) contain masses of dense granules (52) and between the nerve endings lie large, granule-loaded cells. Since the content of the granules and the function of the neurosecretory system is obcure, a collection of antibodies against vertebrate peptides were tested for their reaction in sections of the vena cava neuropil applying the same immunocytochemical methods as for the pituitary. Using antisera against ACTH, ß-endorphin/ß-LPH, 16-k-fragment and CLIP, none of the antisera produced immunoreaction (Fig. 8). When pituitary slices were incubated in the same drop of antiserum for demonstration of a successful incubation an intense immunostaining was evident (Fig. 8). In contrast, applying antisera against α-MSH, Met-enkephalin and arginine-vasopressin (AVP), distinct immunocytochemical reactions were observed in the vena cava neuropil (Fig. 9). Met-enkephalin immunoreactivity was found in large granules of the cells that were embedded in the nerve terminals (Fig. 9D, 10). These cells

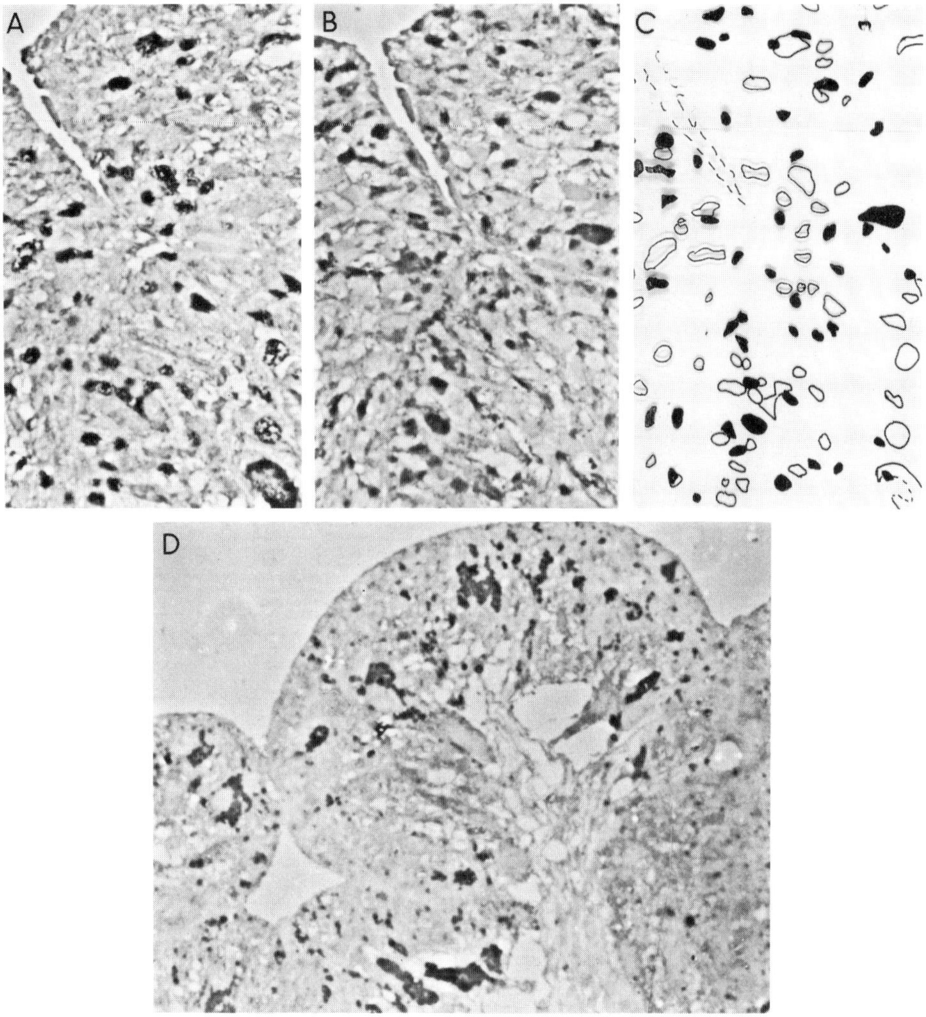

Fig. 9. Light micrographs of the vena cava neuropil. α-MSH-like immunoreactivity is shown in (A). In (B) AVP-like immunoreactivity is present in the section following that of (A). In (C) α-MSH-like immunoreactive terminals (A) with diameters larger than 1µm have been drawn as open contours and endings exhibiting AVP-like immunoreactivity (B) as black spots. In no case immunostainings reside in identical terminals. (D) shows immunostaining for Met-enkephalin contained in cells with very large dense granules that can be clearly identified on electron micrographs (see Fig. 10).

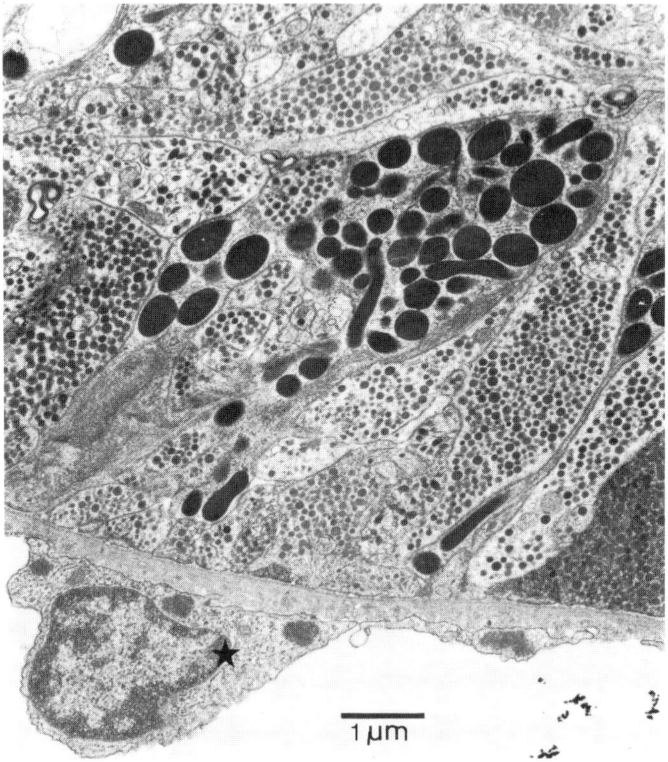

Fig. 10. Electron micrograph of the neuropil of the octopus vena cava, no immunostaining. In the center of the picture a typical 'Met-enkephalin-cell' is shown with huge dense granules (300 nm - 1.0 µm diameter) and with short processes. Numerous nerve terminals filled with smaller granules (100 nm - 150 nm diameter). Some of these nerve terminals display immunoreactivity for α-MSH or AVP, but these specific endings do not account for the total number of terminals in the neuropil. The neuropil is surrounded by a 1 - 2 µm thick basal lamina, one of the scarce endothelial cells (*) is also shown.

are purely peripheral, without obvious connection to the brain. The structural difference to the nerve terminals is clearly visible in the electron micrograph (Fig. 10). Immunostaining with anti-Met-enkephalin was specific as shown by preabsorption experiments (Table 3). None of the other portions of proopiocortin were detectable in the 'enkephalin-cell'. Two different populations of the large number of nerve endings were

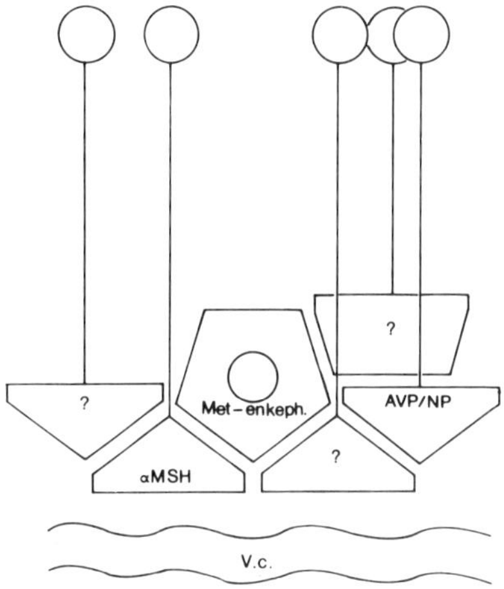

Fig. 11. Schematic drawing of the vena cava neuropil, showing a Met-enkephalin immunoreactive cell and different nerve terminals, containing α-MSH-like and AVP-like, concomitantly with neurophysin-like (NP) immunoreactivity. The nature of the granule content of the other terminals is unknown.

immunostained by anti- α-MSH and anti-AVP (Fig. 9A, B, D). The antiserum against AVP was raised in rabbits (R. Lang, Heidelberg, Germany). It recognized also arginine-vasotocin, isotocin, lysine-vasopressin and in a lesser degree oxytocin. The nerve terminals exhibiting α-MSH-like immunoreactivity outnumbered the nerve endings showing a reaction with the AVP-antiserum. On electron microscopical examination of the immediately preceding section larger immunoreactive terminals could be identified by their position and shape (Fig. 10). The immunoreactivity was restricted to the nerve terminals, nerve bundles and neuronal perikarya did not exhibit immunostaining.

Discussion

Immunocytochemical reactions on tissue samples produced by 'specific' antisera or affinity-purified antibodies need critical interpretation. The localization of pituitary peptides reported here should be considered as 'immunoreactivity raised by binding to a defined antiserum', the true occurrence of the peptides has to be shown by biochemical analysis. Nevertheless, the demonstration of concomitant subcellular storage of ACTH, ß-LPH/ß-endorphin and 16-k-fragment immunoreactivities in the corticotroph of the anterior pituitary (7, 34) and in the intermediate lobe cell (36) confirms the biosynthesis studies which have evidenced that all these peptides are formed from a common precursor protein (8-11, 16, 39). Some of the antibodies used in the present study did not only recognize the processed fragments but also the precursor itself. In these cases the method used cannot reveal whether the precursor molecule is intact or has been cleaved into the different fragments. But the antibody against α-MSH did not crossreact significantly with ACTH or other larger precursor sequences. Thus, the occurrence of α-MSH immunoreactivity in all cells of the intermediate lobe (36), in a subpopulation of the corticotrophs in the anterior pituitary (36, 44) and in greater extent in some human pituitary microadenomas (45) may demonstrate the cleavage of the precursor by intragranular enzymes. In vitro secretion studies with anterior and intermediate lobe cells have shown that different fragments of pro-opiocortin (ACTH, ß-endorphin/ß-LPH, α-MSH, 16-k-fragment) can be released simultaneously by granule extrusion (14-16, 18, 19, 39). The two types of ACTH/ß-endorphin containing cells, the corticotroph of the anterior pituitary and the intermediate lobe cell, are not only different in their immunocytochemical reaction, but also in the regulation of their hormone secretion (15, 39), in the content of peptides (21, 22) and in the processing of their final products (20, 23, 39). Some cells of the pituitary microadenomas seem to resemble the intermediate lobe cells with respect to the appearance of α-MSH immunoreactivity. It is not known, whether or not α-MSH is released into the bloodstream in patients with Cushing's disease.

Met-enkephalin, although identical with the NH_2-terminal sequence of ß-endorphin was not detectable by immunocytochemical methods in the corticotrophs of the anterior pituitary, in the intermediate lobe and in human pituitary microadenomas. The differential distribution of these endogenous peptides within the brain (26), their affinity with different opiate receptor populations (46) and the recent finding of circulating Met-enkephalin independent of the fluctuations of ACTH and ß-endorphin (47) give further support for the argument that they belong to separate neuropeptide systems. Our finding that somatotrophs contain Met-enkephalin immunoreactivity is difficult to interpret. Met-enkephalin and Leu-enkephalin have been observed in pituitary extracts from different species (26-28) but the nature of this enkephalin activity is unknown. Significant amounts of growth hormone (48) as well as of Met-enkephalin (49, 26) have been found in the rat brain amygdaloid nucleus. But it is unknown whether growth hormone and Met-enkephalin are stored together in the same cells as in the pituitary. Possibly the immunostaining of somatotrophs by the Met-enkephalin antiserum is simply caused by a cross-reaction with a part of the prohormone for growth hormone. (Recently, we could confirm Met-enkephalin immunostaining in somatotrophs using other antisera against Met-enkephalin).

Three different antibodies - anti-α-MSH, anti-Met-enkephalin and anti-AVP - recognized material in the neurosecretory system of the octopus (37, 41). It has to be proven that the immunoreaction in the octopus is due to peptides closely related to the'typical vertebrate' hormones. (Preliminary attempts to detect the peptides by radioimmunoassay in extracts of the octopus vena cava have not yet been conclusive). However, the preabsorption tests may indicate that the used antibodies recognize peptides with sequences not very different from the pituitary hormones against which they have been produced. The immunocytochemical data did not indicate that α-MSH-like peptides are associated with other pro-opiocortin fragments in the nerve endings of the neuropil. The octopus α-MSH-like material may be synthesized in a way different from that in the rat intermediate lobe cells, not including ACTH as an immediate precursor (16, 20, 23). Using the same antiserum as in this study also in the rat brain, α-MSH immunoreactivity was observed outside the ACTH/endorphin neurons (50).

The octopus nerve endings which reacted with the AVP antiserum were also stained by a neurophysin antibody (37). It is not known against which part of neurophysin the antibody is directed, but it seems conceivable that in the octopus as in the vertebrate posterior pituitary a common precursor contains AVP/neurophysin sequences.

Whereas AVP- and α-MSH-like immunoreactivities occurred in two different populations of nerve endings in the vena cava neuropil, the antibody against Met-enkephalin recognized material in large granules within soma and processes of cells that are situated in close contact with the neurosecretory endings (Fig. 11). This system in the octopus may represent a favourable object for the examination of a functional role of the Met-enkephalin, α-MSH and AVP-like material that has been detected by the antibodies applying immunocytochemical methods.

Acknowledgment

We are very grateful for the gift of antibodies and/or peptides to Drs. M. Chrètien, C.H. Li, B.A. Eipper, R.E. Mains, R. Lang, P. Schwandt, E. Wünsch, M. Schleyer.
We are also greatly indepted to the Laboratoire Arago, Banyuls, France, and the Naples Zoological Station, Napoli, Italy, for supply of animals and laboratory facilities.

This work has been supported by Deutsche Forschungsgemeinschaft: SFB 87/B2; K.H.V. is recipient of a Heisenberg Stipendium.

References

1. Dubois, P., Vargues-Regairaz, H., Dubois, M.P.: Z. Zellforsch. 145, 131-143 (1973).
2. Moon, H.D., Li, C.H., Jennings, B.M.: Anat. Rec. 175, 529-537 (1973).
3. Phifer, R.F., Orth, D.N., Spicer, S.S.: J. Clin. Endocrinol. Metab. 39, 684-693 (1974).
4. Pelletier, G., Leclerc, R., Labrie, F., Cote, J., Chrètien, M., Lis, M.: Endocrinology 100, 770-776 (1977).
5. Bloom, F., Battenberg, E., Rossier, J., Ling, N., Leppaluoto, J., Vargo, T.M., Guillemin, R.: Life Sci. 20, 43-48 (1977).
6. Voigt, K.H., Weber, E.: In: Way, E.L., (ed.): Endogenous and exogenous opiate agonists and antagonists, Pergamon Press, New York 1980
7. Weber, E., Martin, R., Voigt, K.H.: Life Sci. 25, 1111-1118 (1979).
8. Mains, R.E., Eipper, B.A., Ling, N.: Proc. Natl. Acad. Sci. USA 74, 3014-3018 (1977).
9. Roberts, J.L., Herbert, E.: Proc. Natl. Acad. Sci. USA 74, 4826-4830 (1977).
10. Nakanishi, S., Inoue, A., Taii, S., Numa, S.: FEBS Lett. 84, 105-109 (1977).
11. Crine, P., Gianoulakis, C., Seidah, N.G., Gossard, F., Pezalla, P.D., Lis, M., Chrètien, M.: Proc. Natl. Acad. Sci. USA 75, 4719-4723 (1978).
12. Rossier, J., French, E.D., Rivier, C., Ling, N., Guillemin, R., Bloom, F.E.: Nature 270, 618-620 (1977).
13. Krieger, D.T., Liotta, A.S., Suda, F., Yamaguchi, H.: In: Wuttke, W., Voigt, K.H., Weindl, A., Dries, R.R., (ed.): Brain and Pituitary Peptides, Karger, Basel, pp. 35-45, 1980
14. Voigt, K.H., Weber, E., Fehm, H.L., Martin, R.: In: Wuttke, W., Weindl, A., Voigt, K.H., Dries, R.R. (ed.): Brain and Pituitary Peptides, Karger, Basel, pp. 54-64, 1980
15. Vale, W., Rivier, C., Mitnik, S., Guillemin, R.: Endocrinology 103, 1910-1915 (1978).
16. Mains, R.E., Eipper, B.A.: J. Biol. Chem. 254, 7885-7894 (1979).
17. Nakanishi, S., Inoue, A., Kita, T., Nakamura, M., Chang, A.C.Y. Cohen, S.N., Numa, S.: Nature 278, 423-427 (1979).

18. Mains, R.E., Eipper, B.A.: J. Biol. Chem. 253, 651-655 (1978).

19. Eipper, B.A., Mains, R.E.: J. Supramol. Struct. 8, 247-262 (1978).

20. Crine, P., Benjannet, S., Seidah, N.G., Lis, M., Chrètien, M.: Proc. Natl. Acad. Sci. USA 74, 4276-4280 (1977).

21. Scott, A.P., Lowry, P.J., Bennett, H.P.J., McMartin, C., Ratcliffe, J.G.: J. Endocrinol. 61, 369-380 (1974).

22. Scott, A.P., Lowry, P.J., Ratcliffe, J.G., Rees, L.H., Landon, J.: J. Endocrinol. 61, 335-367 (1974).

23. Seidah, N.G., Gianoulakis, C., Crine, P., Lis, M., Benjannet, S., Routhier, R., Chrètien, M.: Proc. Natl. Acad. Sci. USA 75, 3153-3157 (1978).

24. Giagnoni, G., Sabol, S.L., Nirenberg, M.: Proc. Natl. Acad. Sci. USA 74, 2259-2263 (1977).

25. Kimura, S., Lewis, R., Stern, A.S., Rossier, J., Stein, S., Udenfriend, S.: Proc. Natl. Acad. Sci. USA 77, 1681-1685 (1980).

26. Rossier, J., Vargo, T.M., Minick, S., Ling, N., Bloom, R.E., Guillemin, R.: Proc. Natl. Acad. Sci. USA 74, 5162-5165 (1977).

27. Simantov, R., Childers, S.R., Snyder, S.H.: Brain Res. 135, 358-367 (1977).

28. Miller, R.J., Kwen-Jen Chang, Cooper, B., Cuatrecasas, P.: J. Biol. Chem. 253, 531-538 (1978).

29. Hökfelt, T., Johansson, O., Ljungdahl, A., Lundberg, J. M., Schultzberg, M.: Nature 284, 515-521 (1980).

30. Zamboni, L., De Martino, C.: J. Cell Biol. 35, 148A (1967).

31. Mayor, H.D., Hampton, J.C., Rosario, B.: J. Cell Biol. 9, 909-910 (1961).

32. Sternberger, L.A.: Immunocytochemistry ,Englewood Cliffs, N.J., Prentice-Hall, Inc., pp 159, 1974

33. Weber, E., Voigt, K.H., Martin, R.: Endocrinology 102, 1466-1474 (1978).

34. Weber, E., Voigt, K.H., Martin, R.: Brain Res. 157, 385-390 (1978).

35. Weber, E., Voigt, K.H., Martin, R.: Proc. Natl. Acad. Sci. USA 75, 6134-6138 (1978).

36. Martin, R., Weber, E., Voigt, K.H.: Cell Tissue Res. 196, 307-319 (1979).

37. Martin, R., Frösch, D., Voigt, K.H.: Gen. Comp. Endocr. (1980) (in press).

38. Mains, R.E., Eipper, B.A.. J. Biol. Chem. 251, 4115-4120 (1976).

39. Mains, R.E., Eipper, B.A.: Symp. Soc. Exp. Biol. 33, 37-55 (1979).

40. Voigt, K.H., Fehm, H.L., Pfeiffer, E.F.: Horm. Metab. Res. 3, 313-317 (1971).

41. Martin, R., Frösch, D., Weber, E., Voigt, K.H.: Neurosci. Lett. 15, 253-257 (1979).

42. Eipper, B.A., Mains, R.E.: J. Biol. Chem. 253, 5732-5744 (1978).

43. Scott, A.P., Lowry, P.J.: Biochem. J. 139, 593-602 (1974).

44. Weber, E., Voigt, K.H.: In: Endogenous and exogenous opiate agonists and antagonists, Way, E.L., (ed.), Pergamon Press, New York 1980

45. Voigt, K.H., Martin, R., Fahlbusch, R., Fehm, H.L.: Acta Endocr. 234, 53-54 (1980).

46. Chang, K.J., Cooper, B.R., Hazum, E., Cuatracasas, P.: Molecular Pharmacol. 16, 91-104 (1979).

47. Clement-Jones, V., Lowry, P.J., Rees, L.H., Besser, G.M.: Nature 283, 295-297 (1980).

48. Pacold, S.T., Kirsteins, L., Hojvat, S., Lawrence, A.M., Hagen, T.C.: Science 199, 804-806 (1978).

49. Simantov, R., Kuhar, M.J., Uhl, G.R., Snyder, S.H.: Proc. Natl. Acad. Sci. USA 74, 2167-2171 (1977).

50. Watson, S.J., Akil, H.: Brain Res. 182, 217-223 (1980).

51. Alexandrowicz, J.S.: J. Mar. Biol. Ass. (U.K.) 44, 111-132 (1964).

52. Berry, C.F., Cottrell, G.A.: Z. Zellforsch. 104, 107-115 (1970).

MOLECULAR WEIGHT AND STRUCTURE ANALYSIS OF OLIGOPEPTIDE HORMONES AND PEPTIDE CONJUGATES BY FIELD DESORPTION MASS SPECTROMETRY

Michael Przybylski
Institut für Organische Chemie, Universität Mainz, Johann-Joachim-Becher-Weg 18-20, D-6500 Mainz, West Germany

Wolfgang Voelter
Institut für Organische Chemie, Universität Tübingen, Auf der Morgenstelle 18, D-7400 Tübingen, West Germany

Introduction

Although oligopeptides are, in principle, an attractive class of compounds for mass spectrometry because of their regular structure and thermal stability, their mass spectrometric analysis has been greatly hampered by their involatility. Hence, a variety of methods developed and tested in the last years have dealt with derivatization reactions to enhance volatilization and enable the analysis of peptides by conventional, electron impact (e.i.) mass spectral techniques (1-3). An effective multiple-step derivatization procedure for oligopeptides has been developed by Biemann et al. (4) that allows the sequence determination of polypeptides, after hydrolysis into small fragments, by a gas chromatograph- e.i. mass spectrometer- computer system. However, only di- up to tetra/pentapeptide derivatives can be analysed by this approach. More recent studies have been concerned with chemical ionization (c.i.) mass spectrometry, as a "soft" ionization technique, for higher oligopeptides (5, 6) but still some degree of N- and C- terminal derivatization is necessary to attain vapori-

Structure and Activity of Natural Peptides

zation of the peptide in the mass spectrometer, and molecular weight information also seems to be limited.

Since its introduction some years ago (7), field desorption (f.d.) mass spectrometry has already become a powerful tool especially for the analysis of highly polar and extremely involatile organic and even inorganic compounds. The basic work on the field-induced ionization of organic molecules has been pioneered for years by Beckey and coworkers who showed that the thermal energy normally required for evaporating a neutral molecule is minimized when the molecule is ionized and "desorbed", by a high external electric field, from the surface of a field emitter anode into the gas phase (8, 9). Since the initial development and advances in f.d. mass spectrometric techniques, a relatively large number of successful analyses of organic compounds not amenable to e.i. or c.i. mass spectrometry has been already reported (10 - 12). Thus, abundant molecular ions in f.d. mass spectra of oligopeptides with up to 15 amino acid residues were obtained without any derivatization (13, 14). Certainly, a direct molecular weight determination would be most valuable in the case of biological peptides, such as many peptide hormones, with special end groups or side chain structures that will be lost during derivatization procedures required for conventional mass spectral methods. In addition to the molecular weight analysis, f.d. mass spectral studies carried out with some oligopeptide hormones and model compounds so far indicate that structural information might be obtained from fragmentation reactions occurring at increased temperatures applied to the field ion emitter. However, little systematic work has yet been done to investigate this possibility for directly obtaining sequence information of peptides.

In this paper, no attempt will be made to completely cover the currently available f.d. mass spectral data of peptides which have been included in a recent review on basic methodology and biomedical applications of f.d. mass spectrometry (15). Instead, some examples from recent studies have been selected to show the utility and particular role this relatively new mass spectrometric method may play in problems of molecular weight and structure identification of biologically important peptides and peptide derivatives.

Experimental Procedures Pertinent to F.D. Mass Spectrometric Analysis of Peptides

Several papers have appeared which summarize the particular features of f.d. mass spectral techniques (10), and many advances in the various experimental techniques have been made since the initial development of commercially available f.d. ion sources (16). Only the most important experimental conditions as applied to f.d. mass spectrometric studies in the author's laboratory will therefore briefly be described.

For the preparation of "activated" field ion emitters as the crucial precondition for successful f.d. mass spectral analysis, the high temperature activation of tungsten wires of 10 µm diameter with benzonitrile, first described by Beckey and Schulten (17), has proved the preferred method in most laboratories. This procedure usually generates a sufficient number of microneedles (carbon dendrites) of 20 - 30 µm length on the emitter surface. Both,the dipping technique (10) and techniques using microliter syringes (15, 18) have been applied for the deposition of the sample solution onto the emitter, depending on the amount of material present and the solubility of the peptide. With the dipping technique, typical concentrations of 50 - 100 µg/20 µl were used, while 10-20 µg were loaded with a syringe for repetitive magnetic scans at various emitter temperatures. The selection of a suitable solvent for sample loading seems to be particularly important for peptide analysis by f.d. mass spectrometry. In most cases, solutions of DMSO and water at slightly acidic pH were found quite appropriate to minimize alkali ion attachment phenomena (15), especially for large peptides with many functional groups. However, alkali ion addition has been purposely used in several studies of oligopeptides (14) and may be quite helpful for molecular weight analysis. DMSO and H_2O solutions are somewhat more difficult to handle for emitter loading by a syringe than more volatile solvents, and are most conveniently applied by a freeze-loading procedure (18).

All f.d. mass spectrometric studies were carried out with the combined e.i./f.d./f.i. (field ionization) ion source of a Varian MAT 711 double focussing spectrometer coupled to a Varian SS-100 data system. The field emitter was loaded with the sample, introduced in the ion source by a direct insertion lock and heated gradually in typically 3 - 4 min from ambient up to a maximum heating current of approximately 25 mA, for peptides of higher molecular weight. The "best anode temperature" (BAT) (10) for maximum molecular ion abundance, and f.d. fragmentation reactions above BAT was determined in two consecutive experiments with the same emitter, using electric ion detection with an electron multiplier and acquisition of spectra on a magnetic tape of the computer. Typically, the following experimental conditions were used: 1000 (10 %) resolution, 11 kV total potential difference between field emitter anode and cathode (at 8 kV accelerating voltage), 180 - 220 ^{o}C ion source temperature, 2.5 kV electron multiplier voltage. In experiments with reduced ion accelerating voltage (6 or 4 kV), the total potential difference was 10 - 10.5 kV. Mass calibration was carried out in the e.i. mode with high-boiling perfluorokerosine (Merck, Darmstadt, West Germany) or with e.i. and f.i. ions of several perfluoroalkyl triazine derivatives (PCR, Gainesville, USA). The latter compounds have proved particularly useful as f.i. reference ions for high resolution f.d. peak matching studies (19).

A. Molecular Weight Determination of Oligopeptide Hormones by F.D. Mass Spectrometry

F.d. mass spectrometric studies by Winkler and Beckey of some oligopeptides, either unprotected or only slightly modified at the N- and C- terminus, were among the first examples to show the potential of this new ionization technique for exhibiting highly abundant molecular ions (13). The molecular ions $M^{+\cdot}$ or "quasimolecular" ions MH^+ were found as base peaks in f.d. spectra of up to nonapeptides even with Arg and His containing sequences which are thermally and chemically labile and usually present particular difficulties in peptide analysis by classical means. Since the initial experiments, several model studies have been carried out with oligopeptide hormones as a class of biologically important peptides which was thought to particularly profit from a direct molecular weight determination method because of a) their biologically essential end or side groups that are taken off by cleavage or derivatization reactions, b) the often very small amount of material available which makes a direct mass spectral method highly desirable. F.d. mass spectra of the thyrotropin-releasing tripeptide hormone (TRH, Pyr-His-Pro-NH_2) and several biologically active and synthetic analogues of TRH were investigated and compared with the directly (without modification) obtained e.i. mass spectra (20). Molecular ions in the e.i. spectra were either completely absent or extremely weak, and thermal decomposition to diketopiperazine fragments was usually found as the predominant process. The f.d. spectra of TRH, Phe^2-TRH (Figure 1) and several related tripeptides showed abundant molecular ions and, in addition, a variety of fragment ions at emitter heating currents (e.h.c.) between 10 and approximately 20 mA[+)]. It is noteworthy that, at known amino acid composition of TRH (Glu, His, Pro) plus the finding that its C- terminus is blocked, the MH^+ ion at m/z 363 would give a significant lead to the correct structure assignment.

By contrast, conventional procedures including derivatization and e.i. mass spectrometry (23) long had failed because of the unusual N- and C- terminus, and the structure of TRH was not elucidated until synthesis of all six possible isomers (24). The appearance of a protonated, MH^+ ion instead of or in addition to the $M^{+\cdot}$ ion is caused by intermolecular and/or solvent interactions at the emitter surface (10) and is a typical feature in f.d. mass spectra of peptides like other polar compounds. Thus, m/z 363 = MH^+ is the predominant ion in the spectrum of TRH, possibly due to the presence of the His residue, while the spectrum of the analogue Pyr-Phe-Pro-NH_2 shows the $M^{+\cdot}$ ion (m/z 372) as base peak. This effect and the formation of further "cluster" ions, such as $2MH^+$ (10) may affect the unequivocal molecular weight determination of a completely unknown peptide, although some empirical rules to distinguish between $M^{+\cdot}$ and MH^+ ions have been worked out (25). In studies on the structure elucidation of alamethicin peptide antibiotics, an elegant solution to this problem was recently demonstrated by Rinehart et al. (26) who used the cationization by various alkali salts and corresponding shifts in the molecular cluster ions, e.g. MLi^+, MNa^+, MK^+ for the correct molecular weight identification. However, it should be noted that cluster ion formation by the presence of metal salts in the solvent may lead to problems in the case of small amounts of peptides isolated from biological material, e.g. for the sensitivity of the f.d. analysis and the assignment of fragment ions. More systematic work is necessary to better understand cationiza-

---

+) The temperature applied in f.d. analysis to a compound at the emitter surface can not be measured directly but is usually expressed by the heating current (mA) passing through the emitter (e.h.c.) (21). Melting point determinations of some reference substances suggest that 20 mA e.h.c. are roughly equivalent to 200 - 250 ^{0}C; however, e.h.c. and temperature are not related in a linear manner (22).

tion processes, and most model studies so far have been carried out under conditions which should avoid the formation of cationized cluster ions.

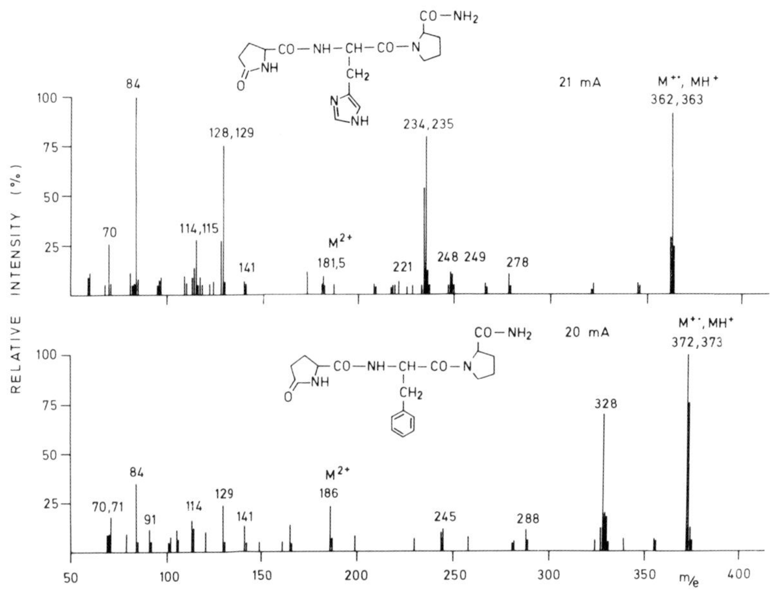

Figure 1. F.d. mass spectra of TRH (upper spectrum) at 21 mA and Pyr-Phe-Pro-NH_2 at 20 mA e.h.c. Solvent: H_2O; The BAT of both compounds was approximately 15 mA, at the conditions used (27).

Although abundant molecular ions have been obtained with nearly all synthetic and biological oligopeptides investigated so far, systematic studies yet need to be done to evaluate the certainty of molecular weight determination by f.d. mass spectrometry. This holds particularly for large, polyfunctional peptides that require relatively high temperatures for desorption. A critical example was found during the study of TRH derivatives, where tripeptides with α-Glu at the N-terminus (Glu-His-Pro, Glu-Pro-His) showed a high tendency for the thermal elimination of water and formation of Pyr (27). In their f.d. spectra most intensive $(M-H_2O)^+$ ions were found, while isomeric peptides with C- terminal Glu (e.g. Pro-His-Glu), which can not form Pyr, gave abundant MH^+ ions.

Nevertheless, the results available at present clearly indicate that the capability of f.d. mass spectrometry to directly identify the molecular weight of peptides is unique among other mass spectral methods. When the f.d. spectrum is recorded by electric registration at the BAT, many peptides show virtually a "one-peak" spectrum of the molecular ion even when high temperatures for desorption are required. By careful setting the desorption temperature, the analysis of oligopeptide mixtures should be possible and was demonstrated in first model studies by Schulten et al. (28). Figure 2 shows an example of an equimolar mixture of a tri-, tetra- and pentapeptide which, by automatic control of the BAT threshold (29),

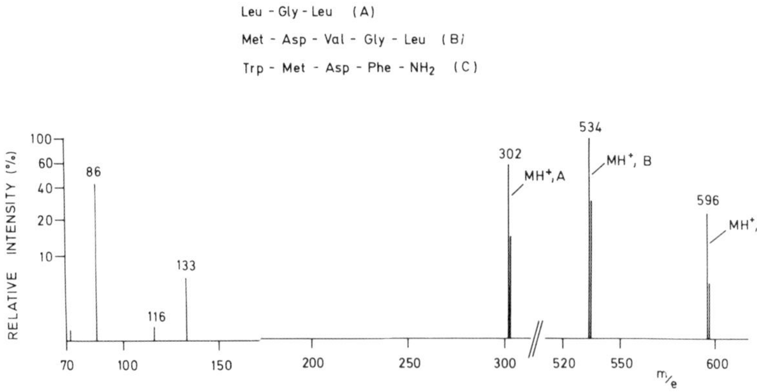

Figure 2. F.d. mass spectrum of an equimolar mixture of Leu-Gly-Leu, Trp-Met-Asp-Phe-NH_2 and Met-Asp-Val-Gly-Leu in aqueous solution. The spectrum was recorded photographically at the lowest possible threshold for molecular ion detection (28).

yielded almost exclusively the $M^{+\cdot}$ and MH^+ ions. Thus, mixture analysis should be promising both for biological peptides that are difficult to isolate in a pure form and as a complementary tool in combination with individual amino acid analyses; the latter has been applied already successfully during studies on the elucidation of peptide antibiotics (14).

The fact that f.d. mass spectrometry does not require any volatility of the sample to be analysed (15) has extended the molecular weight range usually amenable to mass spectrometry. Thus, intact molecular ions up to several thousand daltons were recently observed in f.d. mass spectra of synthetic polymers, such as polyethylene glykols and polystyrenes (30, 31). This exciting possibility has been shown to be applicable to the analysis of high molecular weight peptides; however, it makes severe demands on the performance of the mass spectrometer's analyzer and detection system and requires the development of suitable high mass reference compounds (19, 32). In the f.d. mass spectrum of somatostatin (Figure 3), a tetra-

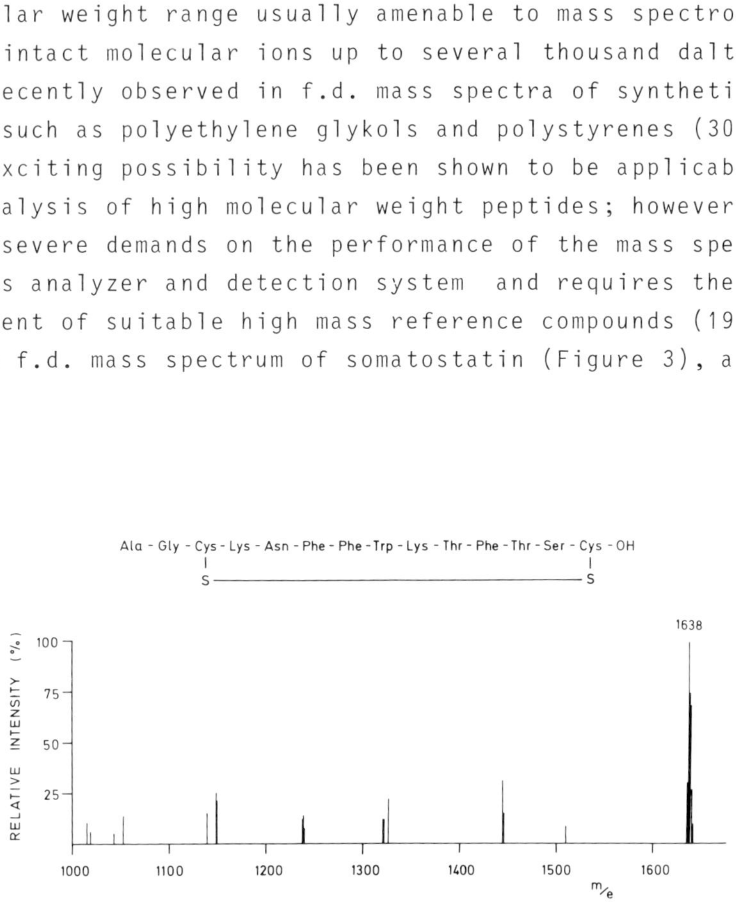

Figure 3. F.d. mass spectrum of somatostatin obtained at 22 mA e.h.c. in H_2O (33).

decapeptide hormone bridged between Cys^{3} and Cys^{14}, a most intensive MH^{+} ion at m/z 1,638 was obtained at 22 mA e.h.c. (33) using reduced ion accelerating voltage and a (perfluoroalkoxy) cyclotriphosphazene derivative (32) as reference compound. Molecular ions up to around 2,000 have been determined by Rinehart et al. for several peptide antibiotics (14, 34). Since the size of many biologically active peptides including peptide hormones is in the order of approximately 15 - 20 amino acid residues that seems practicable at present (35), the application of f.d. mass spectrometry to identify intact molecular species, as well as partial hydrolysates, will be highly useful as complementary tool for their structure determination. However, the use of very sensitive electrical ion detection (fast spectra accumulation) (15) or photographic registration devices will be a prerequisite of crucial importance for the successful f.d. analysis in this high mass range.

B. Structural Information by Fragment Ions in F. D. Mass Spectra of Oligopeptides

Fragmentation reactions in f.d. mass spectrometry can occur at or above the BAT (10). Until recently, f.d. fragment ion analysis has received relatively little attention because molecular weight determination was the major point of interest in most studies, and structural information was mainly sought from other ionization techniques (14). Since most oligopeptides yield f.d. spectra only at elevated temperature of the emitter, fragmentation can, in principle, occur, both, by thermal and field-induced pathways, so that their theoretical basis is even more difficult to understand than, for example, in e.i. fragmentation reactions. Nevertheless, careful control of the e.h.c. and fast electrical registration of f.d. spectra should make it possible to obtain purposely induced fragmentation, and first model studies indicate that at least some degree of useful structural information can be obtained in this way.

In the f.d. spectra of TRH and its derivatives obtained at various emitter temperatures, a series of fragments were observed that could be assigned to cleavages of the peptide backbone (27) (cf. Figure 1). In a first systematic attempt to evaluate a possible sequence-specific fragmentation, the tetrapeptide-amide Trp-Met-Asp-Phe-NH_2 which represents the biologically essential C- terminal sequence of human gastrin I (36) was investigated together with the corresponding di- and tripeptide partial sequences (37). Defined gradual heating of the emitter from ambient to above the BAT was employed in this study. It should be kept in mind that, due to the formation of protonated molecular ions, the assignment of f.d. fragment ions is possible only with a certainty of ± 1 amu. However, with this consideration, a series of fragmentation

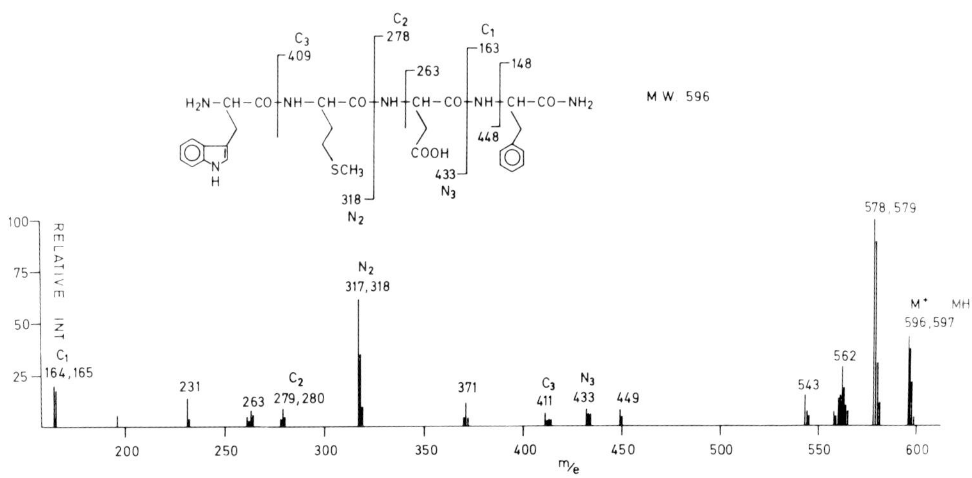

Figure 4. F.d. mass spectrum of Trp-Met-Asp-Phe-NH_2 at 14 mA e.h.c.; solvent: DMSO.

Table 1. Fragment ions formed by direct cleavage of peptide bonds in the f.d. mass spectra of Trp-Met-Asp-Phe-NH_2 (I), Met-Asp-Phe-NH_2 (II), Asp-Phe-NH_2 (III), and Met-Asp-OH (IV) (37).

m/z	partial sequence	N-, C-terminal sequence ion[a] (% relative intensity)[b] I	II	III	IV
433	Trp-Met-Asp	N_3 (8)			
411	Met-Asp-Phe-NH_2	C_3 (6)	MH^+(100)		
318	Trp-Met	N_2(35)			
280	Asp-Phe-NH_2	C_2 (6)	C_2 (13)	MH^+(100)	
247	Met-Asp		N_2 (11)		MH^+(100)
187	Trp	N_1 (-)			
164	Phe-NH_2	C_1(20)	C_1 (15)	C_1 (16)	
132	Met		N_1 (-)		N_1(8)
117	Asp			N_1 (20)	

[a] Partial sequences starting from the N- terminus are assigned N_1, N_2 etc., sequences starting from the C- terminus C_1, C_2 etc.

[b] Emitter heating currents, 14 mA (I), 10 mA (II), 10 mA (III), 9 mA (IV).

products leading to partly overlapping sequence information can be identified from the spectrum of Trp-Met-Asp-Phe-NH_2 (Figure 4). If ions arising by direct cleavage of peptide bonds from the molecular ion are assigned C_1, C_2 etc. (C- terminus) and N_1, N_2 etc. (N- terminus) all possible C- terminal "sequence" ions (38) are found (m/z 164,165, C_1; m/z 279,280, C_2; m/z 411, C_3). Of the three possible N- terminal ions, N_3 (m/z 432,433) and N_2 (m/z 317,318) are observed, while a fragment corresponding to N_1 (Trp) is missing and was not observed in repetitive scans from 12 to approximately 18 mA e.h.c. The C- terminal ions can be rationalized having amine end groups, whereas both acylium and keteneare possible end groups for the N- terminal ions but can not be determined exactly due to the uncertainty in f.d. fragment ion assignement mentioned above. However, a fairly good consistence with this pattern was found by comparison with the f.d. spectra of the model peptides, Met-Asp-Phe-NH_2, Asp-Phe-NH_2 and Met-Asp-OH (Table 1). Again, the C_1- and C_2- ions and N- terminal fragments were observed, except for the N_1- ion which is missing in the spectrum of Met-Asp-Phe-NH_2.

The fragmentation behaviour and possible sequence information of some higher oligopeptides was first studied by Asante-Poku et al. (39). The f.d. spectrum of bradykinin (Figure 5), a nonapeptide with free Arg residues at the N- and C- terminus, yielded extensive fragmentation reactions along the peptide chain at conditions that gave only a small molecular ion (m/z 1,060, MH^+). Two major points of interest are revealed from this and other peptide fragmentation analyses: a) degradation

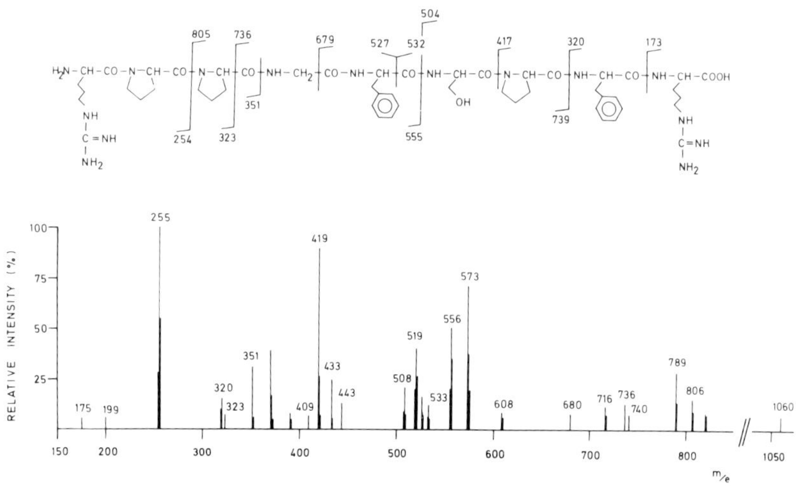

Figure 5. F.d. mass spectrum of bradykinin obtained by electric registration. Solvent, ethanol/water,3:1 (39).

pathways often yield two complementary fragment ions of the peptide; b) cleavages at both sides of the carbonyl groups are observed besides the fission of peptide bonds (see degradation scheme in Figure 5). N-Alkyl and C-C cleavage reactions might appear mainly by pyrolytic processes at increased temperatures. They may be of little value for obtaining sequence information as are the frequently observed eliminations of H_2O, CO_2 and NH_3. The formation of complementary fragment ions has been found as a characteristic pattern in f.d. mass spectrometry (10, 40) and may be quite useful especially for the analysis of biological peptides with unusual structures. Thus, although still very limited, the present data indicate that at least partly useful sequence information can be obtained from f.d. fragmentation of free peptides, and systematic studies of model compounds using more conclusive techniques such as stable isotope labelling and precise mass measurements will be interesting to further evaluate this potential of f.d. mass spectrometry.

C. Identification of Drug- Peptide Conjugates

In a number of recent studies, f.d. mass spectrometry has been successfully applied to the analysis of polar and thermally labile drugs and their metabolites from biological material (15, 41, 42). Thus, phase-I metabolites as well as intact biological conjugation products such as glucuronides (43) and sulfates (44), which are not amenable to e.i. mass spectrometry or gas chromatography- mass spectrometry, have been identified. The high intensity of molecular ions obtained in f.d. mass spectra of amino acids and peptides makes this method particularly suited for the analysis of drug metabolite-peptide conjugates. A detailed f.d. mass spectral investigation of glutathione (γ-Glu-Cys-Gly, GSH) and several GSH derivatives was carried out (37, 45) since GSH is one of the most important cellular "trapping" agents in metabolic detoxication pathways of drugs. The potential utility of f.d. mass spectrometry to directly identify GSH conjugates was of particular interest, because selective hydrolysis and separate analysis of the substrate metabolite is usually not possible with this type of conjugate. Figure 6 shows f.d. mass spectra of GSH and a conjugate which was identified as S-acetyl-glutathione during metabolism studies of monoacetylhydrazine, a toxic metabolite of the tuberculostatic agent, isoniazid (isonicotinic acid hydrazide, INH) (46). F.d. mass spectra mainly served to elucidate this and other GSH conjugates of highly reactive electrophilic intermediates formed upon metabolic oxidation of monoalkyl- and -acylhydrazines, which were found to be responsible for the hepatotoxic effects of INH and other related drugs (47). A clear structural identification is obtained by the MH^+ ions (m/z 308, GSH; m/z 350, S-acetylglutathione), and the typical fragmentation of the pep-

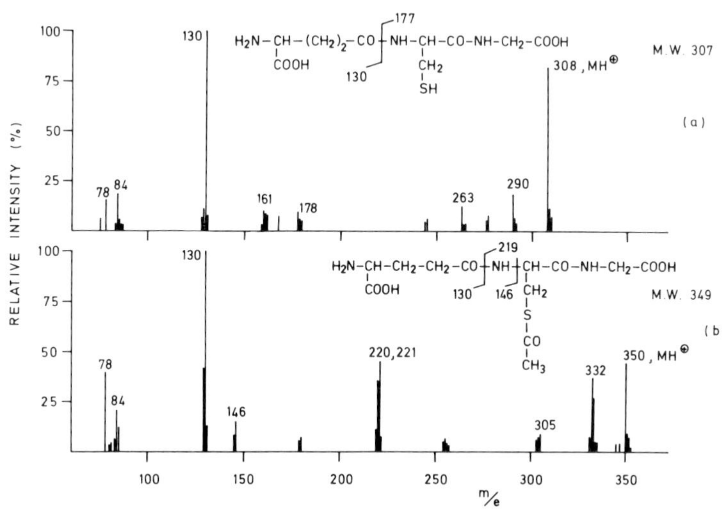

Figure 6. F.d. mass spectra of (a) glutathione at 13 mA and (b) S-acetylglutathione at 14 mA e.h.c. Solvent, DMSO (45).

Table 2. Fragment ions formed by cleavage of γ-Glu-Cys peptide bonds in the f.d. mass spectra of glutathione conjugates (37).

$$H_2N\text{-}\underset{\underset{\displaystyle SR}{|}}{\underset{\displaystyle CH_2}{\underset{|}{C}}}H\text{-}CO\text{-}NH\text{-}CH_2\text{-}COOH^{+\cdot}$$

R	e.h.c. (mA)	m/z	(% rel. intensity)
H	13	178	(11)
CH_3	14	192	(22)
C_2H_5	12	206	(100)
$CO\text{-}CH_3$	14	220	(42)
$CH(CH_3)_2$	15	220	(25)
$CH_2\text{-}C_6H_5$	14	268	(14)

tide bond between γ-Glu and Cys (see scheme in Figure 6). This pathway leads to characteristic ions at m/z 130 and the complementary fragment, Cys(SR)-Gly (m/z 220 for S-acetylglutathione). The latter fragments were consistently found in f.d. spectra of a series of GSH derivatives obtained slightly above the BAT and allow to specifically assign the linkage of the substrate metabolite (Table 2). Their high abundances might be explained by a tendency for the formation of pyroglutamyl (m/z 130) but, unlike for N- terminal α-Glu peptides (20), this fragmentation does not affect the molecular weight determination.

The successful application of f.d. mass spectrometry to the analysis of GSH conjugates was further demonstrated in recent metabolism studies of several chemotherapeutic agents and model substrates (48 - 50). In combination with high-pressure liquid chromatographic isolation, GSH and other thiol conjugates of the anticancer drugs m-AMSA (4'-(9-acridinylamino)-methanesulfon-m-anisidide) and cyclophosphamide have been identified in vitro and in vivo (50, 51). An example for the f.d. analysis of a conjugate of relatively complex structure is given in Figure 7. This compound was identified as a GSH conjugate of 2-hydroxyestradiol, a metabolite of estradiol (45). Its f.d. mass spectrum showed at 14 mA e.h.c. almost exclusively the MH^+ ion (m/z 594), whereas at 18 mA several fragment ions providing structural information were obtained (m/z 320,288,130). A Cys(SR)-Gly fragment ion at m/z 464 is not observed but an ion cluster centered at m/z 446 instead which is likely to be formed from additional loss of H_2O from the steroid skeleton.

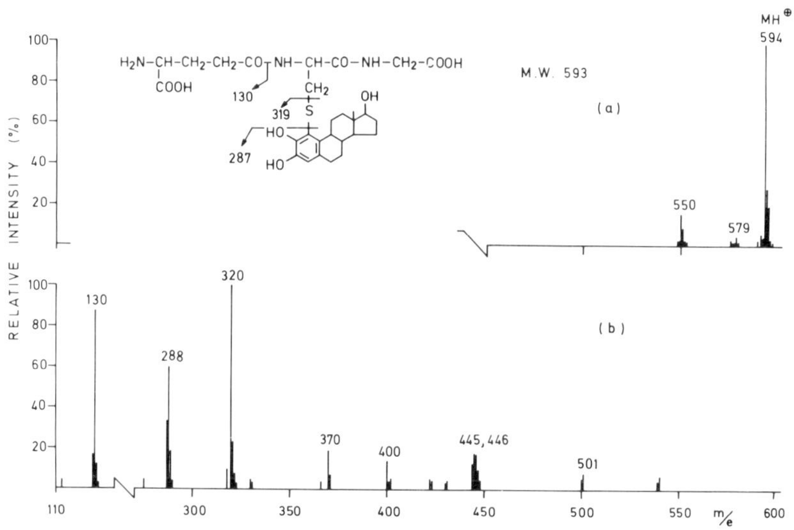

Figure 7. F.d. mass spectra of S-(2-hydroxyestradiol)- glutathione, a metabolite of estradiol, (a) at 14 mA e.h.c., and (b) at 18 mA e.h.c..Solvent: DMSO (45).

D. F.D. Mass Spectra of Oligopeptides Containing Protecting Groups

The qualitative and quantitative analysis of built-up peptide sequences is frequently a particular problem for the synthetic peptide chemist. The capability of f.d. mass spectrometry to exhibit intensive molecular ions should make this method a valuable tool for the rapid control of correct synthesis, and purity, e.g. for protecting groups that are introduced and removed during the manifold steps of classical peptide syntheses. In contrast to most special mass spectrometric and gas chromatographic derivatization techniques (52), protecting groups for functional sites of peptides used in synthetic procedures are often not very appropriate e.g. to improve volatilization properties for e.i. mass spectrometry, because

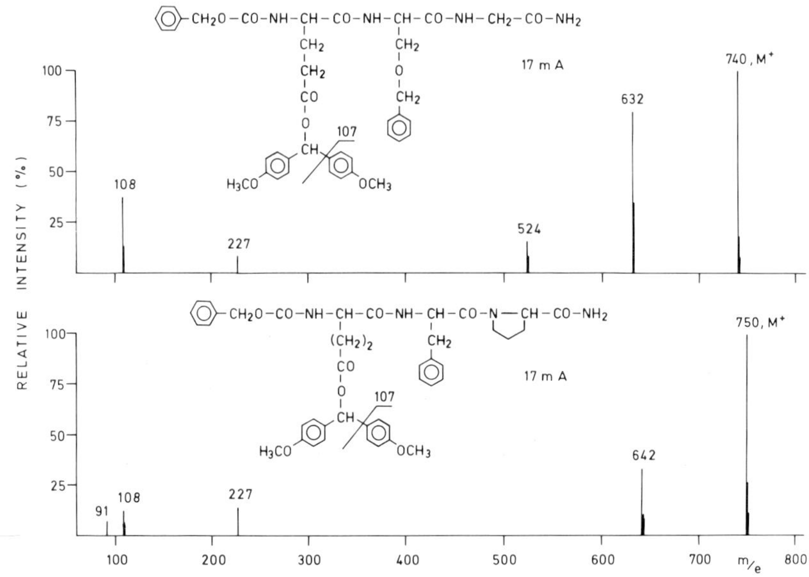

Figure 8. F.d. mass spectra of Z-Glu(Mbh)-Ser(Bzl)-Gly-NH_2 and Z-Glu(Mbh)-Phe-Pro-NH_2. Solvent: DMSO (27).

Table 3. F.d. fragment ions typical for protecting groups in partial sequences of somatostatin (33).

Protected peptide sequence[a]	Protecting group[b]	Fragment ions
Ala-Gly-Cys-OCH_3	Trt	243 ($(C_6H_5)_3C^+$)
	Boc	$(M-100)^+$
Lys-Asn-Phe-Phe-OCH_3	Boc	$(M-100)^+$
Trp-Lys-Thr-Phe-OCH_3	Bzl	91, $(M-108)^+$
	Boc	$(M-100)^+$
Ala-Gly-Cys-Lys-Asn-Phe-Phe-OCH_3	Trt	243 ($(C_6H_5)_3C^+$)
	Boc	$(M-100)^+$
Ala-Gly-Cys-Lys-Asn-Phe-Phe--Trp-Lys-Thr-Phe-OCH_3	Trt	243 ($(C_6H_5)_3C^+$)
	Bzl	91
	Boc	$(M-100)^+$

[a]Protected amino acids are underlined.

[b]Trt, triphenylmethyl; Boc, t-butyloxycarbonyl; Bzl, benzyl.

most of them are purposely chemically labile and lead to a drastic increase in molecular weight. In connection with the mass spectral investigations of derivatives of TRH and somatostatin (20, 33) some first comparative studies by f.d. mass spectrometry were carried out with protected oligopeptide sequences that were used in the stepwise partial and total synthesis of these peptides (53). The typical change in the f.d. mass spectral pattern caused by the presence of protecting groups is shown in Figure 8 by the example of two tripeptide derivatives (Glu-Ser-Gly-NH_2, Glu-Phe-Pro-NH_2) with dimethoxybenzhydryl (Mbh) and benzyloxycarbonyl (Z) groups. The introduction of protecting groups usually leads to a drastical enhancement of "volatility", i.e. desorption rate under f.d.

conditions, even when the derivatization is concomitant with a considerable increase in molecular weight. The molecular ion intensities are vastly enhanced while fragmentation of the peptide backbone is greatly reduced at or above the BAT (27). In the spectra of the two tripeptide derivatives (Figure 8) and other peptides with the Mbh group, only two typical fragment ions were observed that arise by splitting off one anisole moiety. The presence of this group is readily detected by the complementary ions at $(M-108)^+$ and m/z 108 (anisole).

The possibility of directly characterizing by f.d. mass spectrometry, some of the most common peptide protecting groups could be shown in a study of several protected partial sequences of somatostatin (33) (Table 3). In all f.d. spectra of these derivatives abundant molecular ions were obtained which usually appeared as base peaks. The presence of the triphenylmethyl (Trt) group for protecting Cys residues was always indicated by an ion at m/z 243 ($(C_6H_5)_3C^+$) of high intensity. Alkoxycarbonyl groups at the N-terminus or other free amino groups are readily split off from the molecule, yielding abundant $(M-100)^+$ ions in the case of the t-butyloxycarbonyl (Boc) group. The f.d. spectrum of a derivative of the heptapeptide, Ala-Gly-Cys-Lys-Asn-Phe-Phe carrying, both, the Trt and Boc groups is dominated by the Trt^+ and $(M-Boc)^+$ fragments (Figure 9). A similar behaviour was expected for Z-protected peptides, but the corresponding fragmentsformed by loss of the Z group were observed only in a few cases. In peptides carrying the benzyl group, an ion at m/z 91 ($C_6H_5-CH_2^+$) was found, but this fragment also arises from Phe- containing peptides and is therefore of little diagnostic value.

Generally, it appears from the f.d. spectra of the protected peptide derivatives studied that various common protecting groups can be readily identified. However, fragmentation of the peptide chain is minimized so that the presence of protecting groups does not seem to be advantageous for obtaining additional structural information.

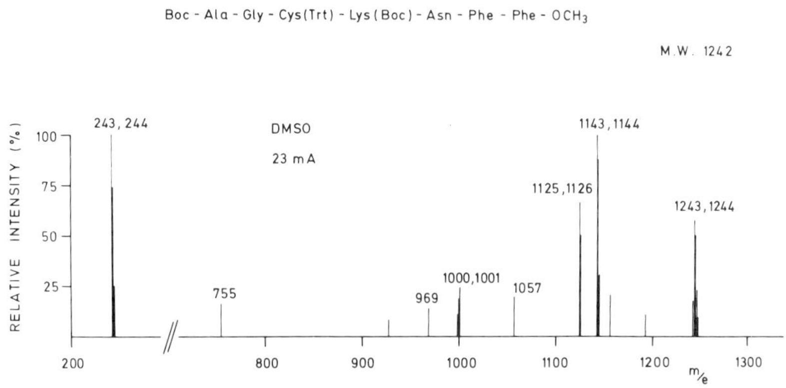

Figure 9. F.d. mass spectrum of a partial sequence of somatostatin, Ala-Gly-Cys-Lys-Asn-Phe-Phe-OCH_3, protected by Boc and Trt groups (33).

Conclusion

Although the major part of the investigations of peptides by f.d. mass spectrometry yet has been comprised of model studies with compounds of known structure, it appears from the rewarding results that this mass spectral method is developing into a highly useful tool among the analytical instrumentarium for oligopeptides. The advances that were achieved in recent years in the reproducible preparation of field ion emitters, sensitivity and reliability of f.d. mass spectrometric analyzer systems (10, 15), especially for electrical ion detection, have made this technique quite amenable to more than a few highly specialized mass spectrometry laboratories. In those studies in which f.d. mass spectrometry was applied to structure elucidation work of peptides from biological sources, its indispensable value as a complementary tool among the analytical methods employed has been demonstrated. This holds, in particular, for the combination of high-pressure liquid chromatography with f.d. mass spectrometry (15, 49, 50). However, at present further systematic model investigations are required to elucidate the many unsolved questions concerning f.d. molecular ion formation and fragmentation pathways, in order to fully evaluate the broad potential of this method in peptide analysis.

Acknowledgement

Financial support of a part of the work reported by a grant from the Bundesministerium für Forschung und Technologie, Bonn, West Germany and the St.-Denis-Stiftung, University of Mainz, is gratefully acknowledged. Mrs. B. Prommnitz is thanked for skilful technical assistance.

References

1. Das, B.C. and Lederer, E., in: New Techniques in Amino Acid, Peptide and Protein Analysis, A. Niederwieser and E. Pataki (eds.), Ann Arbor Sci. Publ., Ann Arbor, 175 (1971).

2. Morris, H.R., in: New Techniques in Biophysics and Cell Biology, R.H. Pain and B.J. Smith (eds.), Wiley-Interscience, New York, 149 (1973).

3. Biemann, K., in: Biochemical Applications of Mass Spectrometry, G.R. Waller (ed.), Wiley-Interscience, New York, 405 (1972).

4. Nau, H., Förster, H.J., Kelley, J.A. and Biemann, K.: Biomed. Mass Spectrometry 2, 326 (1975).

5. Dawkins, B.G., Arpino, P.J. and McLafferty, F.W.: Biomed. Mass Spectrometry 5, 1 (1978).

6. Milne, G.W.A. and Lacey, M.J.: Modern Ionization Techniques in Mass Spectrometry. Critical Reviews in Analytical Chemistry, CRC, Cleveland, Ohio 1974

7. Beckey, H.D.: Int. J. Mass Spectrometry Ion Phys. 2, 500 (1969).

8. Beckey, H.D.: Field Ionization Mass Spectrometry, Pergamon Press, Oxford 1971

9. Beckey, H.D.: Principles of Field Ionization and Field Desorption Mass Spectrometry, Pergamon Press, Oxford 1977

10. Beckey, H.D. and Schulten, H.R.: Angew. Chem. 87, 425 (1975).

11. Burlingame, A.L., Kimble, B.J. and Derrick, P.J.: Anal. Chem. 48, 368R (1976).

12. Burlingame, A.L., Shackleton, H.L., Howe, I. and Chizkov, O.S.: Anal. Chem. 50, 346R (1978).

13. Winkler, H.U. and Beckey, H.D.: Biochem. Biophys. Res. Commun. 46, 391 (1972).

14. Pandey, R.C., Cook Jr., C.J. and Rinehart, K.L.: J. Am. Chem. Soc. 99, 8469 (1977).

15. Schulten, H.R.: Int. J. Mass Spectrometry Ion Phys. 32, 97 (1979).

16. Beckey, H.D., Hoffmann, G., Maurer, K.H. and Winkler, H.U.: Adv. Mass Spectrometry 5, 626 (1971).

17. Schulten, H.R. and Beckey, H.D.: Org. Mass Spectrometry 6, 885 (1972).

18. Olson, K.L., Cook Jr. and Rinehart, K.L.: Biomed. Mass Spectrometry 1, 358 (1974).

19. McEwen, C.N. and Bolinski, A.C.: Biomed. Mass Spectrometry 2, 112 (1975).

20. Lüderwald, I., Przybylski, M., Ringsdorf, H., Silberhorn, D., Kalbacher, H. and Voelter, W.: Z. Naturforsch. 33b, 809 (1978).

21. Winkler, H.U. and Beckey, H.D.: Org. Mass Spectrometry 6, 655 (1972).

22. Kümmler, D. and Schulten, H.R.: Org. Mass Spectrometry 10, 813 (1975).

23. Desiderio, D.M., Burgus, R., Dunn, T.F., Vale, W., Guillemin, R. and Ward, D.N.: Org. Mass Spectrometry 5, 221 (1971).

24. Burgus, R., Dunn, T.F., Desiderio, D.M., Ward, D.N., Vale, W., Guillemin, R., Felix, A.M., Gillessen, D. and Studer, R.O.: Endocrinology 86, 573 (1970).

25. Schulten, H.R. and Beckey, H.D.: Org. Mass Spectrometry 9, 1154 (1974).

26. Rinehart, K.L., Cook Jr., J.C., Meng, H., Olson, K.L. and Pandey, R.C.: Nature (London) 269, 832 (1977).

27. Lüderwald, I., Przybylski, M., Ringsdorf, H., Silberhorn, D., Zech, K. and Voelter, W.: Adv. Mass Spectrometry 7B, 1513 (1978).

28. Beckey, H.D. and Schulten, H.R., in: Practical Spectroscopy Series, C.N. McEwen and C. Merritt (eds.), Marcel Dekker, New York, 145 (1979).

29. Schulten, H.R. and Nibbering, N.M.M.: Biomed. Mass Spectrometry 4, 55 (1977).

30. Matsuo, T., Matsuda, H. and Katakuse, I.: Anal. Chem. 51, 1329 (1979).

31. Lattimer, R.P., Harmon, D.J. and Welch, K.R.: Anal. Chem. 51, 1293 (1979).

32. Olson, K.L., Rinehart, K.L. and Cook Jr., J.C.: Biomed. Mass Spectrometry 4, 284 (1977).

33. Kalbacher, H., Przybylski, M. and Voelter, W.: Z. Naturforsch., in preparation.

34. Pandey, R.C., Cook Jr., J.C. and Rinehart, K.L.: J. Antibiotics 31, 241 (1978).

35. Winkler, H.U., Beuhler, R.J. and Friedman, L.: Biomed. Mass Spectrometry 3, 201 (1976).

36. Beacham, J., Bentley, P.H., Gregory, R.A., Kenner, G.W., MacLeod, J.K. and Sheppard, C.: Nature (London) 209, 586 (1969).

37. Przybylski, M., Lüderwald, I., Kraas, E., Voelter, W. and Nelson, S.D.: Z. Naturforsch. 34b, 736 (1979).

38. Senn, M., Venkataraghavan, R. and McLafferty, F.W.: J. Am. Chem. Soc. 88, 5593 (1966).

39. Asante-Poku, S., Wood, G.W. and Schmidt Jr., D.E.: Biomed. Mass Spectrometry 2, 121 (1975).

40. Schulten, H.R., in: Methods of Biochemical Analysis, Vol. 24, D. Glick (ed.), Wiley & Sons, New York, 368 (1977).

41. Adlercreutz, H., Soltmann, B. and Tikkanen, M.J.: J. Steroid Biochem. 5, 163 (1974).

42. Cortades, J., Macaya, G. and Bernardi, G.: Eur. J. Biochem. 76, 13 (1977).

43. Schulten, H.R. and Games, D.E.: Biomed. Mass Spectrometry 1, 120 (1974).

44. Games, D.E., Games, M.P., Jackson, A.H., Olavesen, A.H., Rossiter, M. and Winterburn, P.J.: Tetrahedron Lett., 2344 (1974).

45. Nelson, S.D., Vaishnav, Y. and Przybylski, M.: Proc. 25th Conf. Mass Spectrom. and All. Topics, Washington, D.C., 509 (1977).

46. Nelson, S.D., Hinson, J. and Mitchell, J.R.: Biochem. Biophys. Res. Commun. 69, 900 (1976).

47. Nelson, S.D., Mitchell, J.R., Timbrell, J.A., Snodgrass, W.R. and Corcoran, G.B.: Science 193, 901 (1976).

48. Brent, D.A., De Miranda, P. and Schulten, H.R.: J. Pharm. Sci. 63, 1370 (1974).

49. Przybylski, M., Cysyk, R.L., Shoemaker, D. and Nelson, S.D.: Adv. Mass Spectrometry, in press.

50. Przybylski, M., Adamson, R.H., Mead, J.A.R. and Cysyk, R.L.: Proc. Am. Ass. Cancer Res. 20, 287 (1979).

51. Przybylski, M., Ringsdorf, H., Lenssen, U., Peter, G., Voelcker, G., Wagner, T. and Hohorst, H.J.: Biomed. Mass Spectrometry 4, 209 (1977).

52. Blau, K. and King, G.S.: Handbook of Derivatives for Chromatography, Heyden & Son, London 1978

53. Gupta, D. and Voelter, W.: Hypothalamic Hormones-Structure, Synthesis and Biological Activity, Verl. Chemie, Weinheim 1975

CRYSTAL STRUCTURE ANALYSES OF POLYPEPTIDES

John J. Stezowski
Institut für Organische Chemie, Biochemie und Isotopenforschung der Universität Stuttgart, Pfaffenwaldring 55, D-7000 Stuttgart 80, Germany.

Introduction

Because of their obvious biological importance, scientists of many disciplines are investigating polypeptides in an effort to understand their various properties. One need only to examine a current issue of Chemical Abstracts to see that their study is a topic of broad interest to chemists.

An area of research that is basic to understanding many of the physical, chemical and biological properties of polypeptides at the molecular level is that of the elucidation of their structural properties. Characterization of polypeptide structure has interested crystallographers, spectroscopists and theoreticians for decades. The continuing discovery of new naturally occuring polypeptides and the synthesis of analogs with varying biological properties provides constant stimulation for a highly active level of study.

One often encounters controversy among investigators using different techniques over the question as to which conformation is the biologically "important" one. It is the opinion of this investigator that such discussion is of limited relevance to understanding polypeptide properties at the molecular level. Each of the three major disciplines of structural chemistry: crystallography, spectroscopy and theoretical chemistry, have made, and can be expected to continue to make, important contributions to our knowledge of polypeptide struc-

Structure and Activity of Natural Peptides

tural properties. It is the responsibility of experts in each field to pursue their research in a manner expected to yield the most comprehensive results possible.

The purpose of this report is to examine the contributions of crystallographic studies to understanding the structural properties of polypeptides. No attempt will be made to provide an extensive review of the literature. Rather, trends will be sought and suggestions made as to potential areas for further study.

Crystallographic studies of polypeptides can be appropriately devided into three areas: cyclic polypeptides, linear polypeptides and macromolecules (molecular weight greater than $\sim 10^4$). This discussion will be limited to the first two catagories.

Cyclic Polypeptides

Cyclic polypeptides have several properties which make them particularly attractive for study by X-ray diffraction methods. For example, they tend to be more easily crystallized than linear polypeptides with a comparable number of amino acid residues. The smaller cyclic peptides also have considerably less conformational freedom than their linear analogs. Consequently, the conformation observed in the crystal is more likely to be representative of the peptide in solution.

The crystal structure of the simplest cyclic peptide, diketopiperazine, anhydro-Gly-Gly, was reported by R. B. Corey (1) in 1938. I mention it here because it illustrates how long cyclic peptides have been of interest to crystallographers. Historically, the crystal structure of diketopiperazine has played a role in the development of models for both the *trans* (2,3) and the *cis* (4) peptide bonds.

A number of diketopiperazines have been examined by crystal-

Table I. The Distribution of *cis* and *trans* Peptide Bonds in Cyclo-(Sar_n) Homologs.

n	
2	*cis* - *cis*
3	*cis* - *cis* - *cis*
4	*cis* - *trans* - *cis* - *trans*
5	*cis* - *cis* - *cis* - *trans* - *trans*
6	*cis* - *cis* - *trans* - *cis* - *cis* - *trans*
7	*cis* - *cis* - *cis* - *cis* - *trans* - *trans* - *trans*
8	*cis* - *cis* - *trans* - *trans* - *cis* - *cis* - *trans* - *trans*
10	*cis* - *cis* - *cis* - *trans* - *trans* - *cis* - *cis* - *cis* - *trans* - *trans*

lographic techniques. Among them is N,N'-dimethyldiketopiperazine, the smallest of a series of cyclo-sarcosyl homologs investigated by P. Groth (5-12) over the last decade. The sarcosyl moiety, Sar, is an amino acid residue for which the *cis* and *trans* peptide bond conformational isomers are expected to be of similar internal energy. The six atom ring of the diketopiperazine allows only the presence of the *cis* peptide bond, but the higher homologs are clearly not so constrained. Nonetheless, the *cis* peptide bond is frequently observed, Table I.

Titlestad, *et al.* (13) have concluded that the conformation of cyclo-Sar_8 is the same in solution ($CHCl_3$) and in the crystal. They have also reported that the cyclo-depsipeptide cyclo-(Sar_3-OGl-Sar_3-OGl) adopts the same solution conformation. In the latter, a pair of *trans* ester bonds are assumed to occupy the positions of a pair of *trans* peptide bonds. Similar conformational integrity was found (14) upon comparison of the structures of recemic cyclo-(Ala-Sar_4) and cyclo-(Sar_5). An Ala residue of the former was found to occupy the position of a Sar residue in the latter that was involved in a *trans* peptide bond. The structural similarities in these analogs have been attributed to intrinsic conformational properties of the polypeptide chain.

A somewhat analogous form of conformational integrity has been reported by I. L. Karle (15) for two cyclic pentapeptides: cyclo-(Gly^1-L-Pro^2-L-Ser^3-D-Ala^4-L-Pro^5) and cyclo-(Gly^1-L-Pro^2-Gly^3-D-Ala^4-L-Pro^5). They differ in chemical structure only by substitution of a nonchiral Gly for an L-Ser. Their respective conformations are shown in Figure 1; their effective stereochemistry (effective, because the nonchiral amino

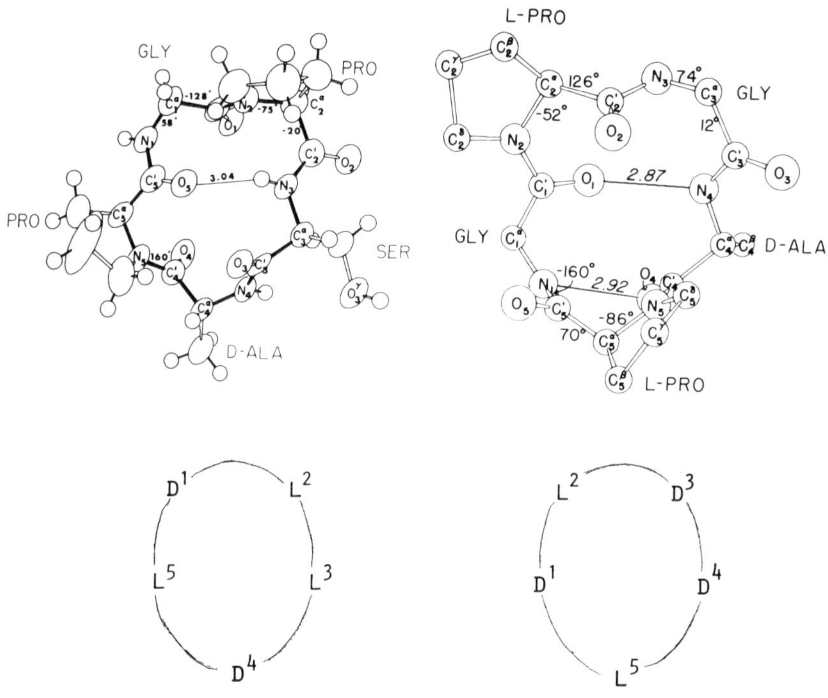

Figure 1. Interrelationships between backbone conformation and the position of chiral residues for cyclo-(Gly^1-L-Pro^2-L-Ser^3-D-Ala^4-L-Pro^5) and cyclo-(Gly^1-L-Pro^2-Gly^3-D-Ala^4-L-Pro^5) (15). Note the position of sequence superscripts.

acid Gly has been assigned D stereochemistry) is also presented schematically. Careful examination of the backbone conformations reveals that they are nearly mirror images, a property also displayed by their stereochemical schemes. It is important to note that the mirror related stereochemical sequence relationship and the approximate mirror symmetry in the peptide backbone conformation are achieved only by comparing amino acid residues after rotating the chemical sequence one residue in a counter clockwise direction. These interrelationships have led I. L. Karle to propose the possibility that the observed conformations are quite stable for each respective stereochemical sequence. This proposal is very intriguing and clearly deserves further investigation, particularly since cyclo-Gly_6, which is devoid of chiral centers, displays several conformations in the same crystal (16).

Over the past several years, considerable effort has been invested in an effort to ellucidate the conformational properties of peptide ionophores. Included among this group of biologically important molecules are two families of cyclic depsipeptides that have been investigated extensively by X-ray diffraction techniques. The first of these is the enniatin-beauvericin family of cyclic hexadepsipeptides. The antibiotic activity of this family appears to be coupled with membrane transport of alkali metal ions. The enniatins and beauvericin display the same depsipeptide backbone structure, Figure 2, but differ in the substituents of the α-amino acids.

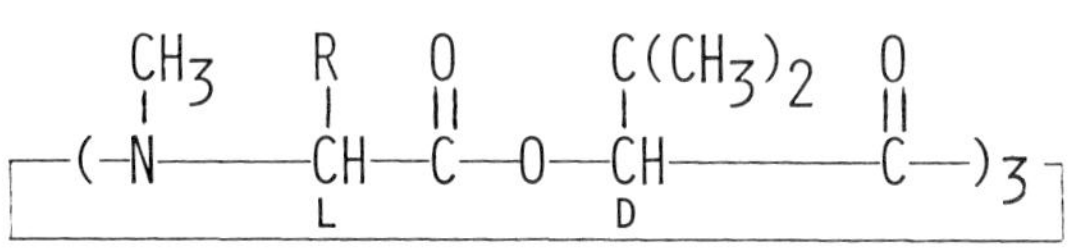

	R
E-A	$CH(CH_3)C_2H_5$
E-B	$CH(CH_3)_2$
E-C	$CH_2CH(CH_3)_2$
B	$CH_2C_6H_5$

Figure 2. Chemical Structures of the enniatin-beauvericin faimly. E-A = enniatin A, *etc.*, B = beauvericin.

Crystal structures for uncomplexed enniatin B (17) and hydrated beauvericin (18) have been reported. In each case, all peptide and ester bonds are *trans*. The carbonyl oxygen atoms of the amino acid residues are on one side of the mean plane of the molecule and those of ester groups are on the other. The molecular conformations of both display a fairly good approximation to C_3 symmetry, the maximum possible molecular symmetry. Deviations are minor in enniatin B, Figure 3, but are somewhat larger for beauvericin; the latter most likely results from the hydrogen bonding between the polypeptide and the waters of hydration.

Several crystal structure determinations have been reported for Group IIA metal ion complexes of beauvericin and Group IA metal ion complexes of enniatin B or derivatives thereof. The earliest of these is for a 1:1 K^+-enniatin B complex reported by Dobler, *et al.* (19). Unfortunately, the crystal was disordered to an extent that precluded a definitive determination. The authors proposed a model for the enniatin B-K^+ complex that was consistent with the general features of the X-ray diffraction data. The K^+ ion was placed in the midpoint of the hexadepsipeptide ring and was bonded to each of the six carbonyl oxygen atoms. The validity of this model has been questioned (20), however, it has been found to be

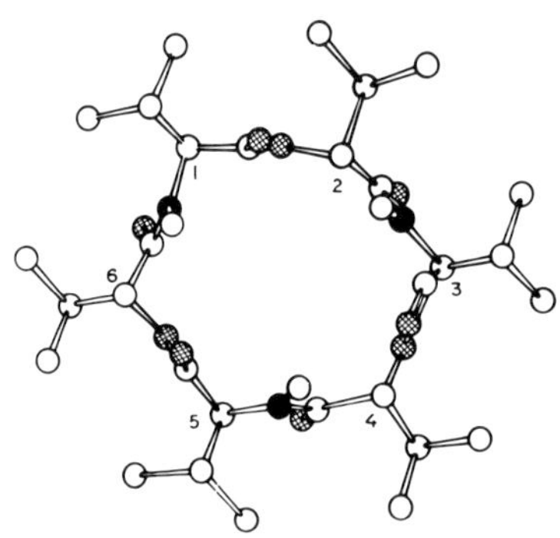

Figure 3. The conformation of enniatin B viewed down the pseudo-three-fold axis (17).

consistent with ^{13}C NMR data (21).

Evidence has been reported (22) that indicates that membrane transport of alkali metal ions is supported by a complex of stoichiometry other than 1:1. Complexes with 2:1 and 3:2 ionophore:metal ion stoichiometries have been proposed. To date, no crystal structure determination of such an enniatin complex has been reported. Dobler, *et al.* (19) have reported the space-group and unit cell volume for KNCS and NaNCS complexes with enniatin B. The unit cell volume of these compplexes is large enough to require two molecules of the ionophore per asymmetric unit. Consequently, the possibility exists that these crystals consist of a 2:1 enniatin B K^+ or Na^+ complex.

Tishchenko and colleagues have reported crystal structure determinations for stereoisomers of enniatin B. Among them is the noncomplexed DLLLLL stereoisomer (23) and the 1:1 Rb^+ complex of the LDLLDL analog (24). The latter, an active membrane transport ionophore, displays a "spiral sandwhich" structure of infinite order in the crystal. Each Rb^+ ion is octahedrally coordinated to 3 carbonyl oxygen atoms of one ionophore, to 2 carbonyl oxygen atoms of a crystallographically related ionophore and to the nitrogen atom of a thiocyanate anion. Whether or not the "head to tail" disposition of the two ionophores with respect to a single Rb^+ ion models the bonding geometry of either a 2:1 or 3:2 ionophore:metal ion complex remains to be determined.

Beauvericin differs from the enniatins in that it binds Group IIA as well as Group IA metal ions (20). Roeske, *et al.* (25) have reported five crystalline modifications for a barium picrate-beauvericin complex. Crystal structure determinations for two of them have been reported (20,26). In both cases, the stoichiometry of the complex is 2:2:3 (beauvericin:Ba^{++}: picrate); a fourth picrate ion, not directly bonded to a Ba^{++} ion, completes the charge balance. Each Ba^{++} ion is nine coordinate. Three of the coordination sites are filled by carbonyl oxygen atoms of the isovaleryl residues of the beauvericin ionophore; the remaining coordination sites are filled either by picrate hydroxyl or nitro oxygen atoms. The complex may be described as consisting of three picrate anions that bridge two Ba^{++} ions that are also each capped by a beauvericin ionophore. In the later crystal structure (26), two such complexes are present in the asymmetric unit. The beauvericin molecules are described as having similar backbone conformations but differing in the orientation of side groups.

There is also evidence (see ref (20)) that, in the transport of Ca^{++} by beauvericin, anions may be carried with the transporting complex. Therefore, the crystal structure determinations for the beauvericin-Ba^{++}-picrate complexes may have provided a model for the complexes responsible for membrane transport.

In view of the structural information that the crystal structure analyses of the enniatin-beauvericin family have thus far provided, it seems most appropriate to encourage investigators in this field to continue their efforts, particularly with respect to alkali metal complexes with stoichiometry other than 1:1. A further test of the beauvericin-alkaline earth ion transport model by examining the crystal structure of a Ba^{++} or Ca^{++} complex with anions other than picrate would also be of value.

The second cyclic depsipeptide ionophore system for which considerable crystallographic results are available is that of the valinomycin family, a series of related dodecadepsipeptides. The sequence of valinomycin is cyclo-$(-L-Val-D-Hyv-D-Val-L-Lac-)_3$. The presence of twelve residues in the chemical structure of these ionophores gives rise to a considerably more complicated problem for the crystallographer. Even so, as long ago as 1969, a preliminary crystal structure analysis was reported (27) for a K^+ ion complex of valinomycin. The structure determination was accomplished by introducing a heavy atom *via* the anion ($AuCl_4^-$).

In 1972, the first crystal structure determination for noncomplexed valinomycin was reported (28). For the structural chemist interested in biologically important molecules in the molecular weight range 10^2 - 10^4, this structure determination must be regarded as a breakthrough that demonstrated the potential of X-ray crystallography to deal with "large" equal atom problems. Further demonstration of that capability was provided by two additional crystal structure determinations for noncomplexed valinomycin (29,30). In both cases, two molecules of the ionophore were present in the crystallographic asymmetric unit, therefore the structure determinations were equivalent to that for a peptide with twenty four residues.

The three crystal structure determinations have provided five independent observations of the conformation of noncomplexed valinomycin. All five examples present the molecule in nearly the same conformation, one in which the ring has a somewhat elongated backbone with the carbonyl groups directed to the interior of the molecule and the hydrophobic side chains directed toward the surface. The conformation is clearly compatible with the fact that the crystals were obtained form largely hydrophobic solvents.

The crystal structure determination (31) of a meso analog of valinomycin, cyclo-$(D\text{-Val-}L\text{-Hyi-}L\text{-Val-}D\text{-Hyi})_3$ has revealed a second conformation that has been described as being more consistent with the results of solution studies. In view of the conformational integrity displayed by the crystalline natural product, it seem likely that the conformation observed therein is also present in solution at a significant population level. Figure 4 displays a schematic representation of a possible conformational equilibrium in solution. The elongated conformation is that found crystallographically for noncomplexed

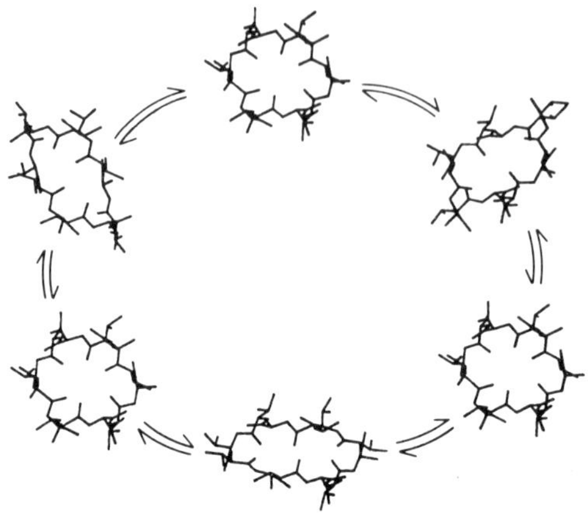

Figure 4. A plausible conformational equilibrium for noncomplexed valinomycin in solution.

valinomycin; the other is analogous to that found for the meso derivative.

A second crystal structure analysis has been reported (32) for a K^+ ion complex of valinomycin in which the complex has been more fully described. As in the earlier determination (27), the K^+ ion is "encapsulated" by the valinomycin molecule, Figure 5. The cation is coordinated to six carbonyl oxygen atoms with coordination geometry that closely approximates octahedral symmetry. A crystal structure determination of a

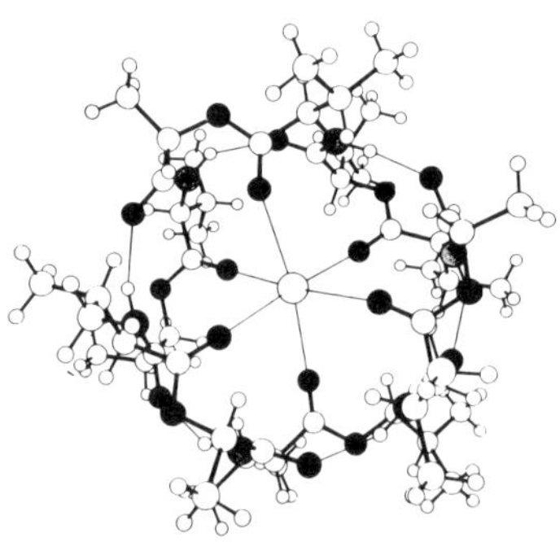

Figure 5. The valinomycin-K^+ complex (32).

Rb^+ complex with prolinomycin, cyclo-(D-Val-L-Pro-L-Val-D-Pro)$_3$, has recently appeared (33). The conformation of this all peptide analog of valinomycin is very similar to that of the complexed natural product. One interesting feature of prolinomycin is that it binds cations more strongly than valinomycin. It has been suggested (33) that this characteristic reflects the conformational properties of the noncomplexed ionophores.

A preliminary report (34) for a crystal structure determination of leucinomycin, a valinomycin analog in which each Val residue is replaced by an Ile residue of the same stereochemistry, reveals a conformation which may be appropriately described as being intermediate to those of complexed and noncomplexed valinomycin, Figure 6. The structure has been interpreted as providing insight into the mechanism of complex formation (34), Figure 7.

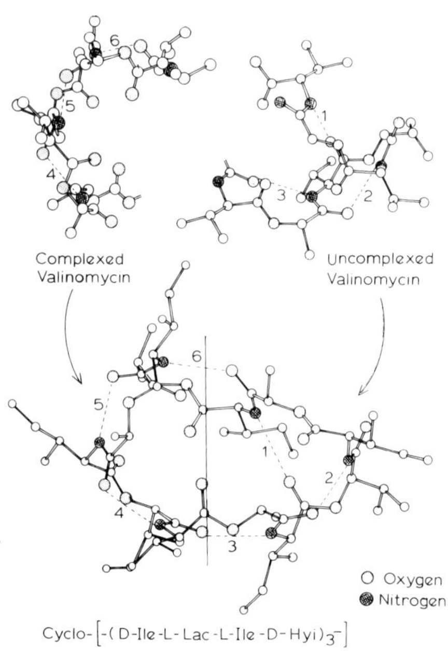

Figure 6. Conformational interrelationships between leucinomycin, noncomplexed and complexed valinomycin.

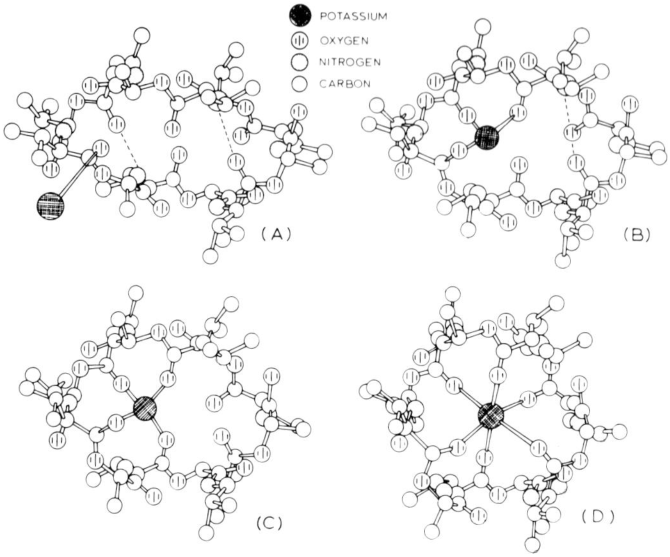

Figure 7. Potassium ion capture by valinomycin.

I'll conclude my discussion of cyclic polypeptides with a brief description of results from polypeptides from the poisonous mushroom *Amanita phalloides*. Antamanide, a cyclic decapeptide with sequence cyclo-(L-Val^1-L-Pro^2-L-Pro^3-L-Ala^4-L-Phe^5-L-Phe^6-L-Pro^7-L-Pro^8-L-Phe^9-L-Phe^{10}) and its Phe^4- Val^6 synthetic analog are both antagonists of the phallotoxins (cyclic hexapeptides). Crystal structure determinations have been reported for a Li^+ complex of antamanide (35), for a Na^+ complex of the Phe^4- Val^6 derivative (36) and for noncomplexed Phe^4-Val^6 antamanide((37). The metal ion complexes display the cyclic decapeptides in nearly the same conformation, Figure 8. In each case, the alkali metal ion is pentacoordinate with one coordination site occupied by a solvent O or N atom. These results alone provide a basis for understanding the preference of antamanide for Na^+ rather than K^+ ions. Like valinomycin, Phe^4-Val^6-antamanide, and almost certainly antamanide as well, displays a much more open conformation in the noncomplexed form, Figure 9.

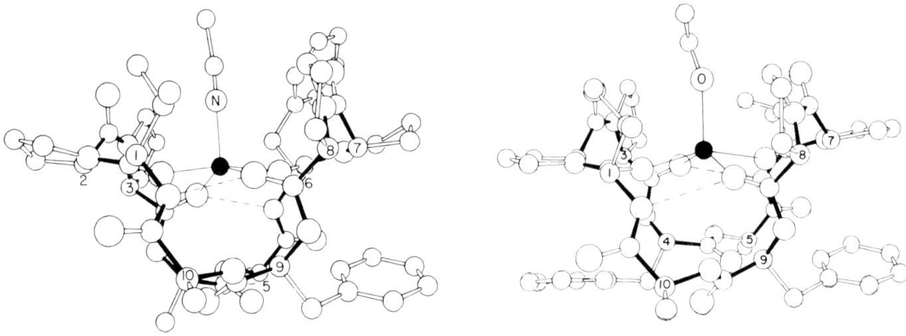

Figure 8. Conformations of the Li^+-antamanide complex and of the Na^+-Phe^4-Val^6-antamanide complex.

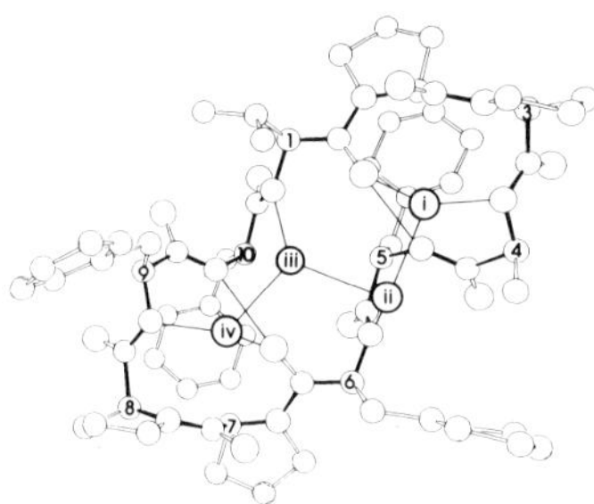

Figure 9. The conformation of hydrated Phe^4-Val^6-antamanide.

A crystal structure determination has also been reported (38) for β-amantin, an octapeptide toxin from *Amanita phalloides*. The toxin contains three unusual amino acids: hydroxylated isoleucine, 6-hydroxytryptophan and a cysteine sulfoxide transanular bridge which gives rise to a bicyclic structure. It is hoped that more structural data will become available for both the phallotoxins and amatoxins so that a better understanding of their unusual structural features can be gained.

The above discussion of the results of crystal structure determinations for cyclic polypeptides clearly demonstrates the depth of information that this technique can provide.

The studies I have described were selected because they were found to be illustrative of systems for which comprehensive crystallographic results are available. Taken as a whole, these studies nicely illustrate the importance of conformational characteristics of cyclic polypeptide to their physical and biological properties.

Linear Polypeptides

The body of crystallographic data available for linear polypeptides is substantially smaller than that for the cyclic polypeptides, particularly when one looks for structure determinations for peptides with more than two amino acid residues. Apparently, the greater conformational freedom of such molecules leads to difficulty in obtaining adaquate single crystals. There are, however, a few cases where interesting crystallographic results have been reported.

The *cis* peptide bond for normal amino acids has been estimated to be of about 2 kcal/mole/residue higher in internal energy than the *trans* isomer (4). However, in the presence of imino acids, the energy is expected to be similar for the two isomers. Standard dimensions for the *cis* peptide bond were proposed several years ago (4) and were compared with the bonding geometry of the Pauling-Corey (3) standard *trans* peptide unit after rotating the peptide bond by 180°. Until recently, the experimental values for the *cis* peptide bond were confined to cyclic peptides in which ring strain could be expected to influence the bonding geometry. A preliminary report (39) of the crystal structure determinations for Sar-Sar and N-methyl-Pro-Pro has presented accurate bond geometry for two examples of *cis* peptide bonds in linear dipeptides. The values are compared in Table II with the above mentioned standard models. The observed bonding geometry for the two peptides is generally similar and falls approximately midway between that of the

Table II. A Comparison of the Bonding Geometry in the *Cis* Peptide Bond in Two Linear Dipeptides with Standard Models.

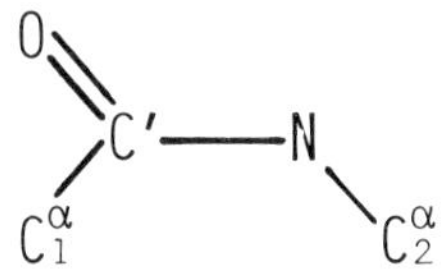

	Sar-Sar	N-Me-Pro-Pro	PC[a]	RV[b]
Bond Distances Å				
C_1^α-C'	1.526(2)	1.530(2)	1.53	1.53
C'-N	1.347(2)	1.340(2)	1.32	1.32
C'-O	1.226(2)	1.240(2)	1.24	1.24
N-C_2^α	1.451(2)	1.474(2)	1.47	1.47
Bond Angles Deg.				
C_1^αC'N	116.0(2)	116.6(1)	114	118
C_1^αC'O	119.8(2)	120.1(1)	121	119
NC'O	124.3(2)	123.3(1)	125	123
C'NC_2^α	124.2(2)	126.5(1)	123	126
Dihedral Angle				
C_1^αC'NC_2^α	-.6(2)	7.5(2)	0	0

[a]The Pauling-Corey (3) standard peptide model rotated 180° about the peptide bond

[b]The Ramachandran-Venkatachalam (4) standard *cis* peptide bond model.

two models.

Linear polypeptides with a small number of amino acid residues often display important biological properties. For example, thyroliberin (L-Pyr-L-His-L-Pro-NH_2), TRH, serves both as a thyrotropin releasing hormone and a stimulant of the central nervous system. The relative simplicity of the molecular structure of TRH has resulted in its receiving considerable attention from structural chemists. Numerous theoretical calculations and solution studies have been reported. Although several laboratories have expended considerable effort to crystallize TRH, no crystal structure has yet appeared.

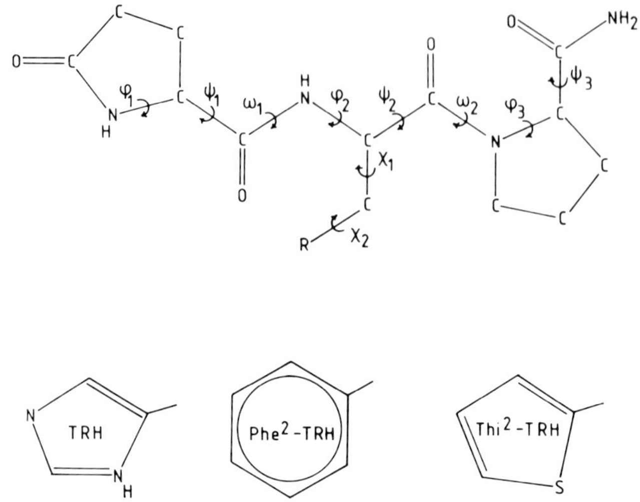

Dihedral Angle	TRH	Phe²-TRH	Thi²-TRH	
ϕ_1	-	101.7	104.3	107.1
ψ_1	-	143.8	8.2	10.0
ω_1	-	174.3	-179.3	-178.1
ϕ_2	-	-65.8	-91.7	-97.7
ψ_2	-	132.1	121.2	123.6
ω_2	-	-179.7	180.0	180.0
ϕ_3	-	-79.1	-77.7	-81.8
ψ_3	-	-24.3	144.4	141.7
χ_1	-	178.8	-63.9	-67.4
χ_2	-	-62.5	71.9	70.7

There are two symmetry independent molecules of Thi^2-TRH in the crystallographic asymmetric unit.

Figure 10. Conformational parameters for Phe^2-TRH (40) and Thi^2- TRH (41).

However, crystal structure determinations for two derivatives: Phe^2-TRH (40) and Thi^2-TRH (41) (Thi = β-(2-thienyl)-L-Ala) have been determined. Figure 10 displays the chemical structure of TRH, Phe^2-TRH and Thi^2-TRH and the conventional dihedral angles to characterize the conformations of the known structures. The dihedral angles indicate that there are conformational differences between the two derivatives, but that the two examples of Thi^2-TRH (two molecules per asymmetric unit) are very similar. Phe^2-TRH and Thi^2-TRH also differ in their biological activity in that the latter is a more active thyrotropin releasing hormone. It is, unfortunately, too early to attempt to correlate these differences with the observed conformations.

The enkephalins are another family of biologically important linear polypeptides for which an example of a successful crystal structure determination has been reported (42). The crystal structure of Leu^5-enkephalin (H-L-Tyr^1-Gly^2-Gly^3-L-Phe^4-L-Leu^5-OH) reveals the pentapeptide to be in a conformation that displays regions of spatial similarity with morphine, Figure 11. These similarities substantiate the possibility that the conformation found in the crystal is similar to that of the pentapeptide bound to the opiate receptor. Since the Leu^5 residue is included within the regions of spatial similarities between the enkephalin and morphine, a crystal structure determination for Met^5-enkephalin would be highly desirable.

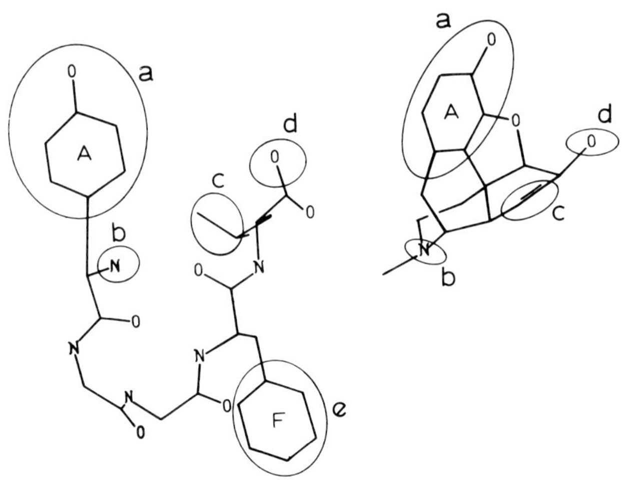

Figure 11. Regions of spatial similarities in the three dimensional structures of Leu^5-enkephalin and morphine (42).

To conclude, I want to put forth my opinion that X-ray crystallography has not yet fulfilled its potential with respect to structural investigations of polypeptides. It should be obvious from this presentation that very satisfying progress has been made and that many areas are receiving the level of attention they deserve. As an investigator with a growing interest in the structural properties of peptides, I can only request that colleagues working in the areas of peptide synthesis or isolation keep the crystallographer in mind when new peptides are encountered. From personal experience, I can be optimistic about the chances for a successful crystal structure determination for polypeptides with molecular weight well above 10^3 Daltons, even if heavy atoms are not present.

Acknowledgement

I particularly want to thank M. Dobler, W. L. Duax, I. L. Karle, G. D. Smith and G. N. Tishchenko for kindly providing photographic material for this article and B. Stensland for conformational data that has not yet been published.

References

1) Corey, R.B.: J.Amer.Chem.Soc. 60, 1598-1604 (1938).
2) Corey, R.B.: Chem.Rev. 26, 227-236 (1940).
3) Corey, R.B. and Pauling, L.: Proc.Roy.Soc.(London) Ser.B 141, 10-21 (1953).
4) Ramachandran, G.N. and Ventkatachalam, C.M.: Biopolymers 6, 1255-1262 (1968).
5) Groth, P.: Acta Chem.Scand. 23, 3155-3162 (1969).
6) Groth, P.: Acta Chem.Scand. A30, 838-840 (1976).
7) Groth, P.: Acta Chem.Scand. 24, 780-790 (1970).
8) Groth, P.: Acta Chem.Scand. 27, 3419-3426 (1973).

9) Groth, P.: Acta Chem.Scand. A31, 232-234 (1977).
10) Groth, P.: Acta Chem.Scand. 27, 3217-3226 (1973).
11) Groth, P.: Acta Chem.Scand. A29, 38-44 (1975).
12) Groth, P.: Acta Chem.Scand. A30, 840-842 (1976).
13) Titlestad, K., Groth, P., Dale, J. and Ali, M.Y.: J.C.S.Chem.Comm. 346-347 (1973).
14) Groth, P.: Acta Chem.Scand. A28, 449-454 (1974).
15) Karle, I.L.: J.Amer.Chem.Soc. 101, 181-184 (1979).
16) Karle, I.L. and Karle J.: Acta Cryst. 16, 969-975 (1963).
17) Tishchenko, G.N., Karimov, Z., Vainshtein, B.K., Evstratov, A.V., Ivanov, V.T. and Ovchinnikov, Yu.A.: FEBS Lett. 65, 315-318 (1976).
18) Geddes, A.J. and Akrigg, D.: Acta Cryst. B32, 3164-3171 (1976).
19) Dobler, M., Dunitz, J.D. and Krajewski, J.: J.Mol.Biol. 42, 603-606 (1969).
20) Hamilton, J.A., Steinrauf, L.K. and Braden, B.: Biochem. Biophys.Res.Comm. 64, 151-156 (1975).
21) Ovchinnikov, Yu.A., Ivanov, V.T., Evstratov, A.V., Mikhaleva, I.I., Bystrov, V.F., Portnova, S.L., Balashova, T.A., Meshcheryakova, E.N. and Tulchinksy, V.M.: Int. J. Peptide Protein Res. 6, 465-498 (1974).
22) Ivanov, V.T., Evstratov, A.V., Sumskaya, L.V., Melnik, E.I., Chumburidze, T.S., Portnova, S.L., Balashova, T.A. and Ovchinnikov, Yu.A.: FEBS Lett. 36, 65-71 (1973).
23) Shiskova, T.G. and Simonov, V.I.: Kristallografya 22, 515-525 (1977).
24) a) Tishchenko, G.N., Karimov, Z., Andrianov, V.I., Vainshtein, B.K., Evstratov, A.V., Ivanov, V.T. and Ovchinnikov, Yu.A.: Sov.J.Bioorg.Chem. 3, 352-355 (1977).
b) Tishchenko, G.N. and Karimov, Z.: Sov.Phys.Crystallogr. 23, 409-416 (1978).
25) Roeske, R.W., Isaac, S., King, T.E. and Steinrauf, L.K.: Biochem.Biophys.Res.Comm. 57, 554-561 (1974).
26) Geddes, A.J. and Akrigg, D.: 4th European Crystallog.

Meeting, Abstracts 302 (1977).

27) Pinkerton, M., Steinrauf, L.K. and Dawkins, P.: Biochem. Biophys.Res.Comm. 35, 512-518 (1969).

28) Duax, W.L., Hauptman, H., Weeks, C.M. and Norton, D.A.: Science 176, 911-913 (1972).

29) Karle, I.L.: J.Amer.Chem.Soc. 97, 4379-4386 (1975).

30) Smith, G.D., Duax, W.L., Langs, D.A., DeTitta, G.T., Edmonds, J.W., Rohrer, D.C. and Weeks, C.M.: J.Amer.Chem. Soc. 97, 7242-7247 (1975).

31) Vasil'ev, A.D., Shibanova, T.A., Andrianov, V.I., Simonov, V.I., Sanasaryan, A.A., Ivanov, V.T. and Ovchinnikov, Yu. A.: Sov.J.Bioorg.Chem. 4, 839-844 (1978).

32) Neupert-Laves, K. and Dobler, M.: Helv.Chim.Acta 58, 432-442 (1975).

33) Hamilton, J.A., Sabesan, M.N., Gisin, B.F. and Steinrauf, L.K.: Biochem.Biophys.Res.Comm. 80, 949-954 (1978).

34) Duax, W.L., Smith, G.D., Weeks, C.M., Pletnev, V.Z., Galitsky, N.M., Ivanov, V.T. and Ovchinnikov, Yu.A.: Acta Cryst. A34, S 74 (1978).

35) Karle, I.L.: J.Amer.Chem.Soc. 96, 4000-4006 (1974).

36) Karle, I.L.: Biochem. 13, 2155-2162 (1974).

37) Karle, I.L., Karle, J., Wieland, Th., Burgermeister, W. and Witkop, B.: Proc.Natl.Acad.Sci.USA 73, 1782-1785 (1976).

38) Kostansek, E.C., Lipscomb, W.N., Yocum, R.R. amd Thiessen, W.E.: Biochem. 17, 3790-3795 (1978).

39) Eckle, E., Stezowski, J.J. and Hughes, R.E.: Acta Cryst. B34, S 74 (1978).

40) Stezowski, J.J., Bürvenich, C. and Voelter, W.: Angew. Chem.Int.Ed.Engl. 18, 225-226 (1979).

41) Castensson, S. and Stensland, B.: 5th European Crystallog.Meeting, Abstracts, 16 (1979).

42) Smith, G.D. and Griffin, J.F.: Science 199, 1214-1216 (1978).

II. ISOLATION OF NEW PEPTIDES

NEW TOXIC CYCLOPEPTIDES FROM AMANITA VIROSA MUSHROOMS

Heinz Faulstich, Angeliki Buku, Heinz Bodenmüller,
Januz Dabrowski and Theodor Wieland

Max-Planck-Institut für medizinische Forschung
6900 Heidelberg

The green death cap Amanita phalloides produces a large variety of biologically active compounds, most of them being deadly poisonous [1]. Among those are the amatoxins, which to our best knowledge are solely responsible for lethal cases of human Amanita poisoning. The main component in this group of toxic cyclopeptides is α-amanitin (Fig. 1) which is capable of inhibiting RNA-polymerases of type II at concentrations as low as 10^{-9} M [2, 3, 4]. The subsequent decrease of m-RNA, followed by the lack of essential proteins, probably causes the necrosis of cells exposed to amatoxins. The high specificity of this inhibitory activity to RNA-polymerases of type II has made α-amanitin a valuable tool in molecular biology [5].

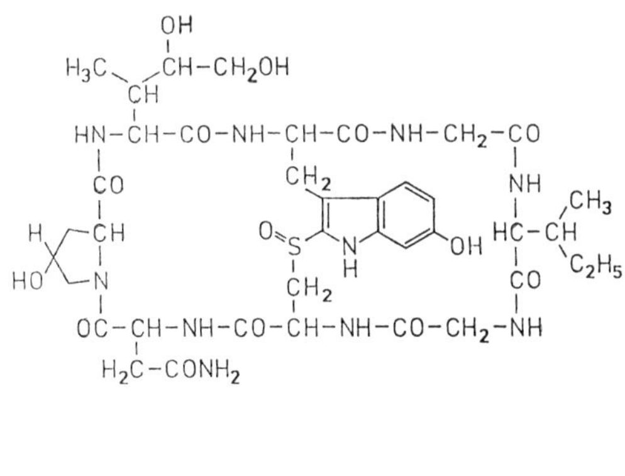

Fig. 1. Structure of α-amanitin.

A second group of toxic peptides present in Amanita phalloides are the phallotoxins, of which phalloidin (Fig. 2) is the best known component. These toxins have captured the interest of biochemists because of their affinity to the contractile protein actin [6]. Since phalloidin binds to the polymeric (or oligomeric) form of actin only, the toxin causes a stabilisation of the actin filaments [7]. As a consequence, the concentration of the monomeric actin decreases [8]. The hemorrhagic cell necrosis observed in livers of phalloidin-poisoned animals may be interpreted as a consequence of this decrease of monomeric actin in the liver cell [9]. Alternatively, the complexation to phalloidin, may change the affinity of filamentous actin toward components involved in cell architecture. As a consequence, rearrangements in the structure of plasma membranes may occur, similar to those described as the toxic effects of phalloidin in mammaliam liver cells.

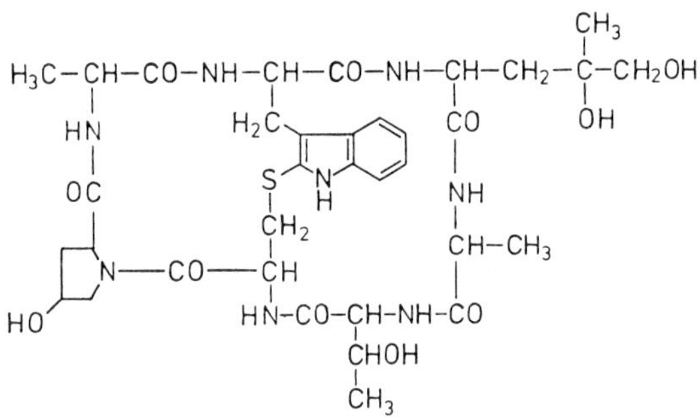

Fig. 2. Structure of phalloidin.

Most of the analytical work in the field of toxic Amanita peptides was performed with the green death cap A. phalloides. This is understandable since none of the other Amanitas produces such a range of biologically active components. Another reason is the ubiquitous availability of the green species in Western Europe. For these reasons studies with other Amanitas, like the less frequent white A.virosa, were initiated only recently.

A. virosa mushrooms though lacking a great many of the components present in the green species contain some toxins previously unknown: Among them amaninamide, another member of the amatoxin family [10], and a whole class of new toxins, the virotoxins [11]. These toxins were detected in methanolic extracts of A. virosa after development with water on long LH_{20}-Sephadex columns [12]. They attracted attention because they differed from the well characterized amatoxins and phallotoxins by several outstanding features:

1. The virotoxins occur in A.virosa mushrooms exclusively; they were detected neither in the green A. phalloides nor in the other white species A. verna.
2. The virotoxins have a monocyclic structure while exhibiting full biological activity. (All monocyclic derivatives of phallotoxins and amatoxins were found to be biologically inactive).
3. The virotoxins also represent the first natural peptides containing the chromophoric system of 2'-alkylsulfonyltryptophan (Fig. 3) (Sulfones were thus far obtained by semisynthesis only. Amatoxins are sulfoxides, phallotoxins are thioethers).
4. Virotoxins contain an isomer of 3,4-dihydroxyproline that until now had not been detected in nature. (Amatoxins and phallotoxins contain monohydroxylated prolines).

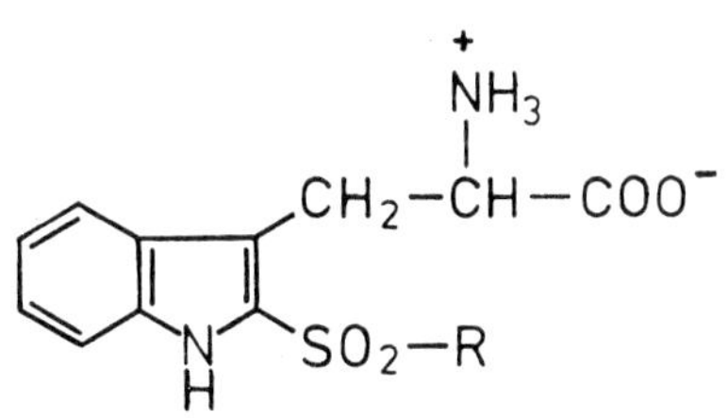

Fig. 3. Structure of 2'-alkylsulfonyltryptophan.

In the following, the isolation and characterisation of the main component of the virotoxins, viroisin, will be discussed in some detail.

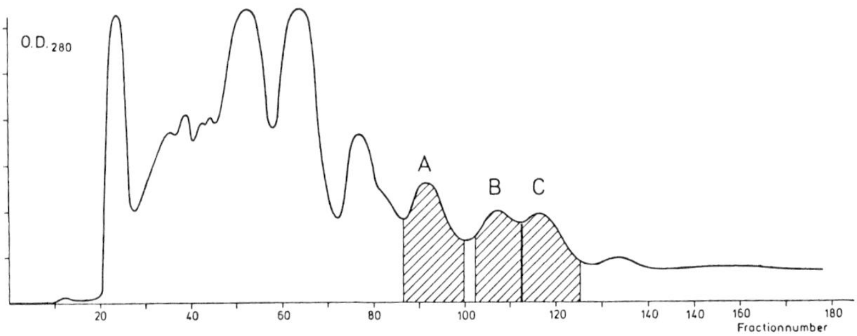

Fig. 4. Elution diagram from a Sephadex LH_{20} column of a methanolic extract of a North American sample of A.virosa as monitored by optical density at 280 nm. (Further data in the text).

The sample of A.virosa investigated in this study was collected in West Virginia, USA. The elution diagram of the chromatographic resolution (Fig. 4) showed that neither acidic toxins, such as β-amanitin or phallacidin, nor α- or γ-amanitin were present. Fraction A contained the virotoxins, fraction B amaninamide [10], while fraction C, which in similar experiments with A.phalloides consists of 3 neutral phallotoxins, in this case contained only phalloidin. Notably, two samples of A. virosa found in Europe, in addition to the fractions A - C, contained α-amanitin and phallacidin as well [12].

Further resolution of the crude virotoxin fraction revealed six components, all of which were recently elucidated in structure [11]. The structure of the main toxin, viroisin (accounting for about 50% of the toxin mixture) is illustrated in figure 5.

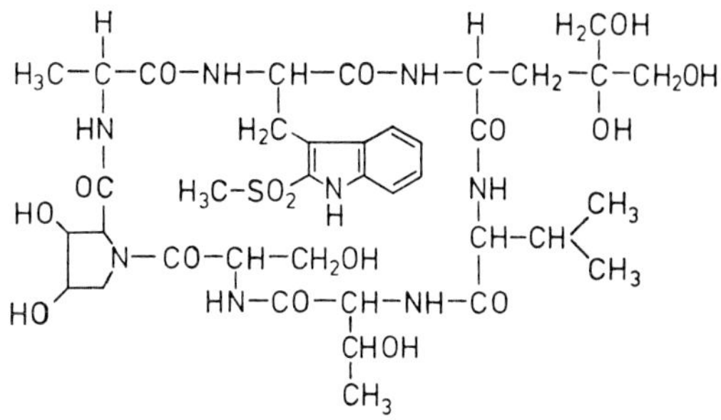

Fig. 5. Structure of viroisin .

Structural analysis of viroisin followed the methods established for the phallotoxins. Treatment with trifluoroacetic acid cleaved a single amide bond of the cyclic peptide adjacent to a γ-hydroxylated leucine moiety. The subsequent Edman-degradation showed in the first 2 steps amino acids known from the phallotoxins, but yielded serine in step 3. This result provided evidence that viroisin was a monocyclic peptide. In addition, the serine moiety proved to be D-configurated. Step4 of the degradation resulted in a unit, which was identified as a 3,4-dihydroxy-L-proline. Among the four possible stereoisomers, the one isolated from viroisin was determined by ^{1}H n.m.r. spectroscopy, to be 2,3-trans,3,4-trans,3,4-dihydroxy-L-proline [13] (Fig. 6).

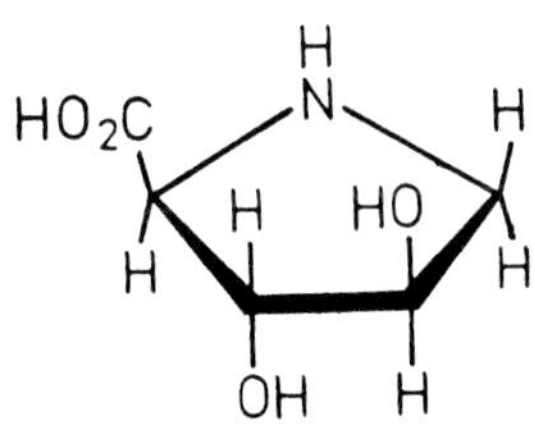

Fig. 6. Structure of the dihydroxylated proline isolated from viroisin.

Likewise by ^{1}H n.m.r., but also by independent synthesis, the moiety obtained in degradation step 6 was characterized as the 2'-methylsulfonyltryptophan, in agreement with the lack of any transannular sulfur bridge in the cyclic toxin. Finally, degradation step 7 yielded trihydroxyleucine, as was previously found in both phallisin [14] and phallisacin [15].

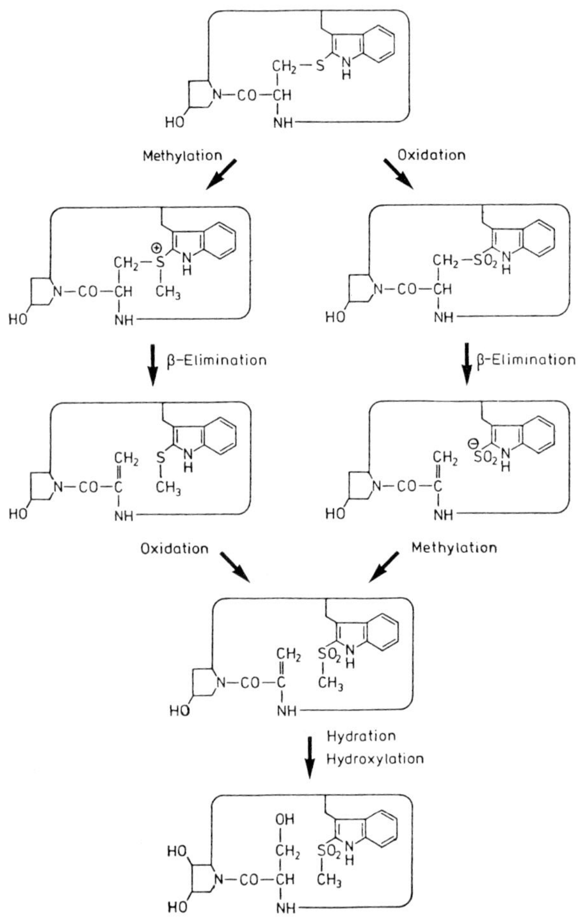

Fig. 7. Two sequences of physiological reactions which possibly convert phallotoxins to virotoxins.

Evidently, about 60% of the viroisin structure is composed of amino acids known from the phallotoxins. The remaining 40% amino acids contain new moieties, which, however, are closely related to those in phallotoxins. Accordingly, we assume that virotoxins may be generated from phallotoxins, or, at least, from common precursor molecules. In fact, reactions feasable under physiological conditions of the plant can be formulated, which would be capable of converting a phallotoxin into a virotoxin. Two possible reaction sequences are depicted in figure 7.

The biological activities of viroisin are comparable to those of the phallotoxins. Like phalloidin viroisin binds to rabbit muscle actin in the polymeric (or filamentous) form. Direct ecidence for such complexation was provided by the difference u.v. spectrum (Fig. 8) obtained in a similar way as for phalloidin [16] and for phalloidinsulfone [17].

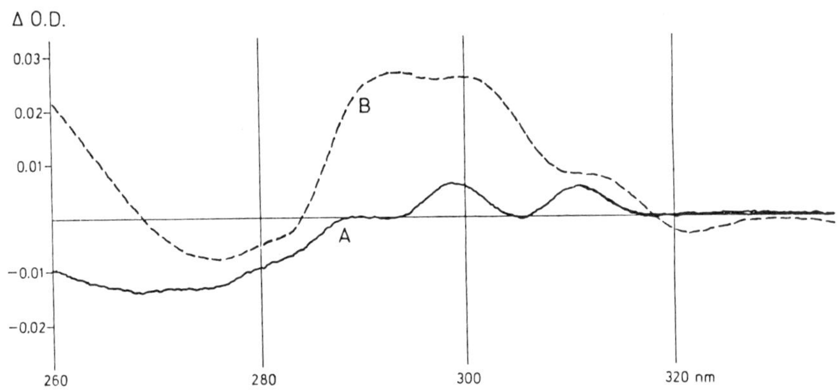

Fig. 8. Difference u.v. spectrum obtained by mixing $1x10^{-5}$M F-actin (in 0.1 M KCl 1 mM Tris pH 7.4) with an equimolar solution of viroisin (A) or phalloidin sulfone (B).

When the complexation reaction was quantitated by equilibrium dialysis experiments using [^{3}H]demethylphalloin, the apparent dissociation constants for both phalloidin and viroisin were found to be similar ($K_D \sim 2x10^{-8}M$). From the slope of the dose binding curves it appears, however, that the interaction of viroisin with actin is even somewhat tighter than that of phalloidin [11].

Likewise, the in vivo toxicity of viroisin, as tested by intraperitoneal administration of the toxin into the white mouse, was found indistinguishable from that of phalloidin. Both toxins had LD_{50} values of 2.5 mg/kg body weight and in both cases the animals died within 2 - 5 hours by severe hemorrhagia of the liver. Evidently, both toxins though different in structure exhibit very similar effects in vivo as well as in vitro, thus providing support to the concept that the interaction with actin indeed represents the mechanism of their in vivo toxicity.

In A. virosa, virotoxins and phallotoxins exist side by side. Evidently, nature has developed two groups of toxic peptides, which exhibit the same biological activity, but differ distinctly in structure. Since in A. virosa, no component or intermediate without biological activity has been detected thus far, it seems likely that nature produced and preserved these peptides for some specific physiological function. Reasonably, it may be suggested that actin-binding cyclopeptides could play a role in the physiology of Amanita mushrooms.

The structural prerequsites of phalloidin binding to actin have been established by conformational analysis and by synthesis of several phalloidin analogs. It was concluded that the bicyclic nature as well as the rigidity of the 15-membered ring represent criteria for the biological activity of the phallotoxins. Furthermore, 3 moieties could be identified which proved to be essential parts of the binding site: The methyl side chain of the 15-membered ring, the allo-hydroxy group of proline and the lipophilic indolylthioether moiety. In a model of phalloidin conformation in solution, as proposed by

Patel et al. [18], all three moieties mentioned are in close proximity and therefore, are likely to represent principal components of the binding site.

The structural moieties of phalloidin involved in binding to actin are bold faced in figure 9. This figure directly compares the structures of a phallotoxin, e.g. phalloidin, to that of one virotoxin, ala[1]-viroidin (representing about 10%

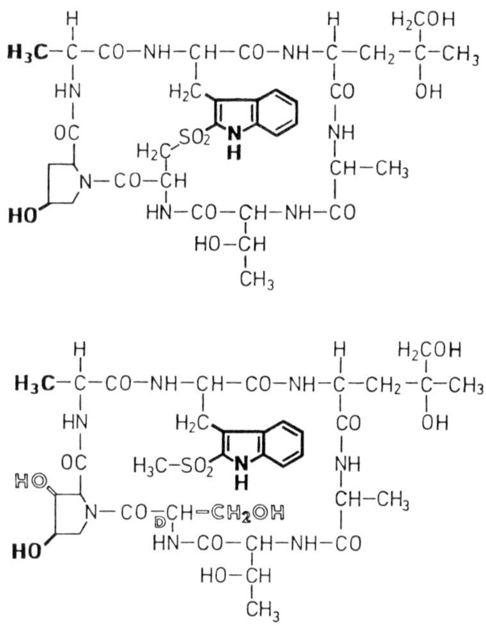

Fig. 9. Structure of phalloidin, as compared to ala[1]-viroidin. Moieties presumably involved in actin binding are in bold face type. Moieties assumed to contribute especially to viroidin binding are in open face type.

of the total virotoxin mixture). As was concluded from the difference u.v. spectrum observed under binding of viroidin to actin, also in the virotoxins, the indole moiety is part of the binding site. The two other moieties essential in phallotoxin binding, are also assumed to be involved, although there is no proof for this at the moment. However, the virotoxins are monocyclic in nature and therefore the topology of the binding site must be expected to possess more flexibility than in the phallotoxins. As a consequence, a lower affinity to actin should be expected.

In contrast to such predictions, viroisin possesses an equal or even higher affinity to rabbit muscle actin than phalloidin. This may be rationalized by the presence, in the virotoxins, of additional sites of interaction with actin which are absent in phallotoxins. Such sites may include the side chain of D-serine and the additional hydroxygroup in position 3 of proline (Fig. 9, open face typed). In fact, we could obtain some preliminary evidence for this hypothesis: A virotoxin analog, prepared from phalloidin by A. Buku, lacking the second hydroxygroup of proline and containing L-alanine instead of D-serine, was completely devoid of biological activity [19].

References

1. Wieland, Th., Faulstich, H.: Amatoxins, Phallotoxins, Phallolysin and Antamanide: The Biologically Active Components of Poisonous Amanita Mushrooms, in: Critical Reviews of Biochemistry, Fasman, G.D., ed., Vol. 5, Issue 3, 1978.
2. Fiume, L., Stirpe, F.: Biochim. Biophys.Acta 123, 643 (1966).
3. Lindell, Th.J., Weinberg, F., Morris, P.W., Roeder, R.G., Rutter, W.J.: Science 17o, 447 (197o).
4. Kedinger, C., Gniazdowski, M., Mandel, J.L., Gissinger, F., Chambon, P.: Biochem.Biophys.Res.Commun. 38, 165 (197o).
5. Faulstich, H.: The Amatoxins, in: Progress in Molecular and Subcellular Biology, Hahn, F.E. et al., eds., Vol. 7 (in press) Springer, 198o

6. Lengsfeld, A.M., Löw, I., Wieland, Th., Dancker, P., Hasselbach, W.: Proc.Natl.Acad.Sci. 71, 28o3 (1974).

7. Löw, I., Wieland, Th.: FEBS Lett. 44, 34o (1974).

8. Faulstich, H., Schäfer, A.J., Weckauf, M.: Hoppe Seyler's Z. Physiol.Chem. 358, 181, (1977).

9. Wieland, Th., Faulstich, H., The Phalloidin Story, in: Frontiers in Bioorganic Chemistry and Molecular Biology, Ovchinnikov, Y.A., Kolosov, M.N., eds., Elsevier, Amsterdam 1979

10. Buku, A., Wieland, Th., Bodenmüller, H., Faulstich, H.: Experientia, in press.

11. Faulstich, H., Buku, A., Bodenmüller, H., Wieland, Th.: submitted for publication.

12. Faulstich, H., Georgopoulos, D., Bloching, M., Wieland, Th.: Z. Naturforsch. 29c, 86, (1974).

13. Buku, A., Faulstich, H., Wieland, Th., Dabrovski, I.: In preparation.

14. Wieland, Th., Wieland, O.: The Toxic Peptides of Amanita Species, in: Microbial Toxins, Kadis, S., Ciegler, A., Ajl, S.J., eds., Vol. 8, Academic Press, New York, 1972

15. Faulstich, H., Brodner, O., Walch, St., Wieland, Th.: Justus Liebigs Ann.Chem. 2324 (1975).

16. Wieland, Th., de Vries, J.X., Schäfer, A., Faulstich, H.: FEBS Lett. 54., 73, (1975).

17. Faulstich, H., Wieland, Th., Jochum, C.: Justus Liebigs Ann.Chem., 713, 186 (1968).

18. Patel, D.J., Tonelli, A.E., Pfaender, P., Faulstich, H., Wieland, Th.: J.Mol.Biol., 79, 185, (1973).

19. Buku, A., unpublished results.

COMPETITIVE INHIBITION OF BENZODIAZEPINE BINDING BY FRACTIONS FROM PORCINE BRAIN

Stefan Fuchs, Christine Carlsohn, Fritz Seif
Medizinische Poliklinik der Universität Tübingen, D-7400 Tübingen, Liebermeisterstraße 14, Germany

Karl Folkers
Institute for Biomedical Research, University of Texas at Austin, Austin, Texas 78712, USA.

Wolfgang Voelter
Institut für Organische Chemie der Universität Tübingen, Auf der Morgenstelle 18, 7400 Tübingen, Germany

Introduction

The benzodiazepines are a group of compounds with wide therapeutic application such as relief of anxiety, muscle tension and anticonvulsant influences. Their main representative is diazepam.

Many centrally active drugs are assumed to act via specific membrane receptors, what can be demonstrated biochemically by binding studies using brain homogenates and a radiolabeled benzodiazepine as ligand. From at least six independent laboratories [^{3}H]benzodiazepine binding to brain homogenates (1,2,3,4,5,6) has been reported, and from all a single spesific binding site with non- specific binding of only 10 % or less is assumed. However, demonstration of binding is not a sufficent proof for a neurohormone receptor as binding site.

Other data as found in different laboratories, however, sup-

Structure and Activity of Natural Peptides

port the hypothesis that these binding sites are neurohormone receptors: High affinity (1), organ specificity (1), reversibility (6), stereospecificity (7), ligand specificity (4,7,8, 9,10), species specificity (11), synaptic localization (12) and increases in specific binding after preincubation indicating the possible dissociation of an endogenous ligand (2,13, 14).

In addition it was demonstrated that the binding sites in rat brain are highly specific, since drugs different from benzodiazepine and neurohormones have shown no competitive binding (7,10). Also, the distribution of benzodiazepine binding in the brain does not parallel the distribution of binding of any known drug or neurohormone (1).

If the benzodiazepines are interacting with a neurohormone receptor system, it should be possible to extract the neurohormone from the brain and demonstrate its competitive inhibition of [^{3}H]benzodiazepine binding. Marangos et al.(15) and Karobath et al.(16) have examined this possibility using acidified acetone extracts, containing mainly molecules below 2000 daltons. Both groups reported an inhibitory entity of around 500 daltons which was not affected by proteolytic enzymes. Marangos et al. demonstrated kinetically that their entity was a competitive inhibitor. Karobath et al. found that muscle extracts contained the highest activity of all tissues tested, while Marangos et al. reported lowest activity in muscle.

We have examined porcine brain fractions that could contain entities between 500 and 100,000 daltons for competitive inhibition of specific benzodiazepine binding in rat brain subfractions, and report inhibition by, both, high and low molecular weight extracts. The high M_r (40,000 - 70,000) fraction (termed benzodiazepine - competitive factor I (BCF-I)) was five times more potent than the low M_r (1000 - 2000) fraction (BCF-II). The high M_r fraction has been investigated for the nature of its binding inhibition, its purity and its structural class. Further characterization of the small M_r fraction is in progress.

Materials and methods

[Methyl-^{3}H]diazepam and [methyl-^{3}H]flunitrazepam were obtained from New England Nuclear and had specific activities of 40 Ci/mmol and 87.5 Ci/mmol, respectively. Labeled diazepam and flunitrazepam were gifts from Hoffmann-La Roche.

Since the highest densities of benzodiazepine - specific binding sites have been reported to be in the cortex cortex preparations were the most promising to give the highest yield of endogenous benzodiazepines (s). 40 g of lyophilized porcine cortices were extracted and fractionated as described earlier (17).

The hydrolysis of the active fractions and the trypsin digestion was carried out according to standardized peptide chemistry methods.

The binding assay was carried out with Sprague-Dawley rat homogenates (17). The second pellet (Fig.1) was resuspended in 0.1 M potassium phosphate buffer (pH 7.05) to a protein concentration of 1 mg/ml, determined according to Lowry et al.(18). This homogenate was preincubated at 37°C for 30 min to dissociate any potential endogenous ligands and frozen at -20°C for use the following day.

Brain extracts were screened for inhibition of benzodiazepine binding by incubating them with either 5 nM [^{3}H]diazepam or 2 nM [^{3}H]flunitrazepam. Bound [^{3}H]benzodiazepine was determined by filtration (17).

Specific binding was calculated by subtraction of nonspecific counts from the total cpm. Nonspecific binding was estimated

Sprague-Dawley rat brains in 0.1 M potassium phosphate buffer (pH 7.05)

homogenize

centrifuge 500 x g 10 min

pellet 1 — supernatant

supernatant: centrifuge 12, 000 x g

pellet 2 — supernatant

Fig.1. Preparation of rat brains for the binding assay.

by incubating a labeled benzodiazepine with a 500-fold excess of the corresponding unlabeled benzodiazepine. Inhibition of specific benzodiazepine binding was expressed in percent of control binding divided by mg of lyophilized brain extract introduced into the final assay mixture.

Electrophoresis of brain extracts was performed in slab gels (19) with the buffer system of Laemmli (20). The proteins were stained with 0.2 % Coomassie brillant blue R 250 in a solution of methanol/H_2O/acetic acid (7:7:1, vol/vol). The gels were destained in 7 % acetic acid. The slabs contained 10 % acrylamide as separating gel and 5 % acrylamide as stak-king gel.

Results and discussion

Our data indicate that maximal binding was obtained at pH 7,o and binding was reduced to 50 % at pH 6. The phosphate buffer tested had the highest buffering capacity at pH 7 and allowed maximal binding with the natural fractions.

Fig. 2 shows the inhibiting potency of a typical series of

Sephadex G-75 fractions. The 30,000 - 70,000 M_r fraction (BCF-I) and the 1000-2000 M_r fraction (BCF-II) consistently contained the most diazepam inhibitory activity. BCF-I was 5 times more potent than BCF-II on a weight basis.

According to gel electrophoresis BCF-I contains mostly polypeptides from 30,000 to 70,000 daltons and it may well be that BCF-I is composed of several molecular fractions. Also the BCF-II fraction may be a mixture of different compounds.

BCF-I was examined to determine the nature of its inhibition. Fig. 3 shows three double-reciprocal plots of the specific binding produced by various concentrations of labeled flunitrazepam in the presence or absence of 2 nM unlabeled flunitrazepam or 1 mg BCF-I. The inhibition produced by BCF-I was nearly identical to that produced by 2 nM unlabeled flunitrazepam and indicated competitive inhibition. However, this did not rule out substrate inhibition (inhibition by binding to the labeled flunitrazepam). Fig. 4 shows S curves of the fractional inhibition (i) of specific binding produced by various concentrations (plotted as logarithms) of unlabeled flunitrazepam or BCF-I in the presence of 2 nM labeled flunitrazepam. The inset of Fig. 4 shows the curves theoretically expected for this type of plot with a competitive inhibitor and substrate inhibitor according to Webb (21). A substrate inhibitor should produce a steeper rise in i than a competitive inhibitor (21). The main plot in Fig.4 shows that the BCF-I curve shows no hint of a steeper rise and is nearly identical to the unlabeled flunitrazepam curve. Noncompetitive inhibition, such as that which would be produced by an enzyme that degraded the receptor (e.g. a protease), was not indicated in a double-reciprocal plot which showed inhibition similar to that of unlabeled flunitrazepam (by definition a competitive inhibitor). The highest concentration of BCF-I, which was near saturation, only produced a 70 % (i = 0.7) inhibition of binding. Further purification should result in i=1.

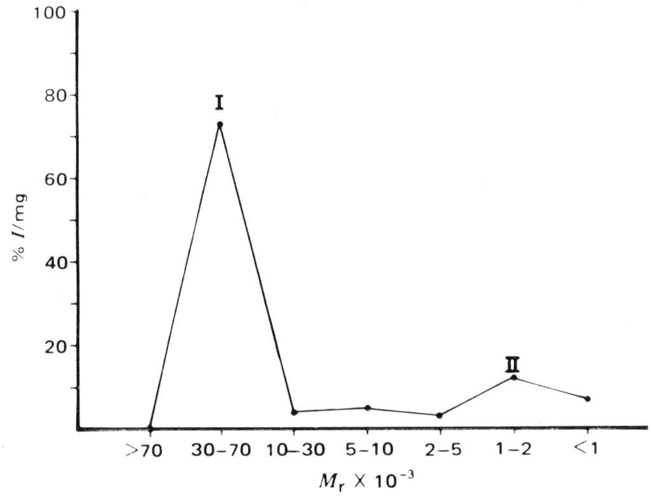

Fig. 2. The benzodiazepine binding inhibitory potency of a series of porcine brain fractions from a Sephadex G-75 column. Mr values were estimated by using prestandarized column flow rates. Potency was calculated as the percent inhibition of specific binding (% I) per mg of lyophilized extract fraction added to the final assay volume.

Table 1 shows that BCF-I is thermally stable, but it loses its inhibitory activity after trypsin digestion or acid hydrolysis. The loss of activity after trypsin digestion or acid hydrolysis indicates the presence of peptide bonds in the active portion of the BCF-I molecule(s).

We attented to extract factors from porcine brain that competitively inhibit benzodiazepine binding. Inhibitory activity was found in both, a high (BCF-I) and low (BCF-II) M_r fraction of a series of brain extract fractions. A molecule in the molecular range from 30,000 to 70,000 (BCF-I), however, would be a neurohormone of an unusual size. It is possible that such a large molecule may actually have a smaller active molecule as either part of its sequence or tightly bound to it in such a way that the activity of the smaller molecule

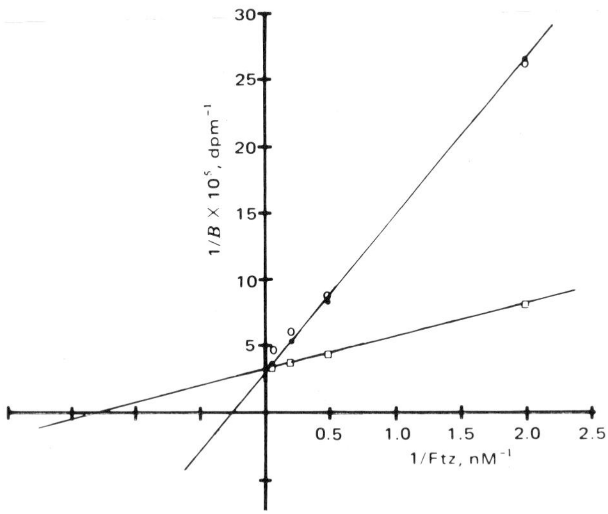

Fig. 3. Double-reciprocal plots of the specific binding (B) produced by various concentrations of [^{3}H]flunitrazepam (Ftz) in the presence or absence of inhibitors. □, Control; •, 2nM unlabeled flunitrazepam; o, 1.o mg BCF-I. Both, unlabeled flunitrazepam and BCF-I were dissolved in o.o5 M acetic acid. Control medium is o.o5 M acetic acid.

is retained or enhanced. The larger molecule could function as a carrier or prohormone, since molecules in the 30,000 - 100,000 molecular weight region with opiate activity have recently been discovered in the brain (22). Nevertheless this fraction needs extended further purifaction before meaningful results can be presented. The investigation of the smaller M_r fraction (BCF-II) is in progress.

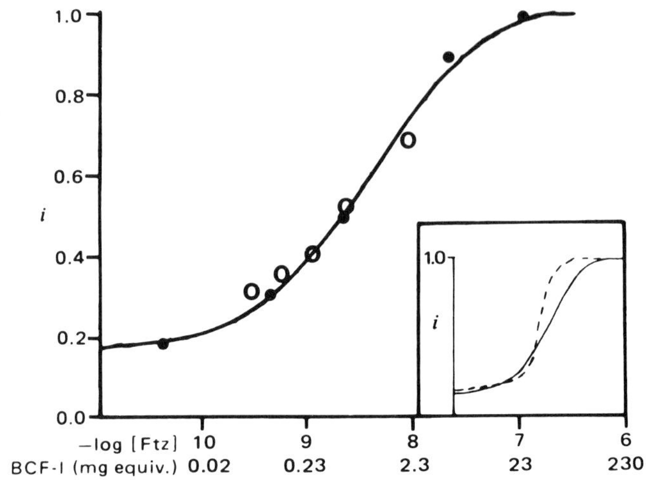

Fig. 4. Fractional inhibition (i) of the specific binding of 2nM [^{3}H]flunitrazepam produced by various concentrations of unlabeled flunitrazepam [Ftz] (•) or BCF-I (o). Inset, results that would theoretically be expected, if BCF-I were a substrate inhibitor (---) or a competitive inhibitor(——). BCF-I (mg equiv) = mg of BCF-I added to the final assay, plotted on a logarithmic scale. One miligram of BCF-I was set as being equivalent in inhibition to 2 nM unlabeled flunitrazepam on the basis of Fig. 3.

Table. Effects of heat, trypsin digestion or acid hydrolysis on BCF-I.

Sample	Specific binding, % control$^+$
Solvent control	100 ± 4
BCF-I ⊕ 0°C, 90 min	64 ± 1
25°C, 90 min	65 ± 3
100°C, 15 min	68 ± 4

Table (continued)

Sample	Specific binding, % control[+]
Trypsin-digested BCF-I[‡]	96 ± 1
Acid-hydrolyzed BCF-I[§]	98 ± 2

[+] Mean ± SD of triplates.

⊕ BCF-I (0.5 mg) was incubated at the indicated temperature for the indicated time before addition to the binding assay.

‡ BCF-I (0.5 mg) was incubated with 2 % trypsin at 25°C for 60 min followed by incubation at 100°C for 15 min to inactivate the trypsin before addition to the binding assay.

§ BCF-I (0.5 mg) was heated overnight at 100°C in 6 N HCl. Most of the HCl was then drawn off over NaOH crystals under reduced pressure. Any remaining HCl was carefully neutralized with dilute NaOH (pH 7.0) before addition to the binding assay.

References

1. Braestrup, C., Squires, R.F.: Proc.Natl.Acad.Sci.USA 74, 3805-3809 (1977).
2. Mohler, H., Okada, T.: Life Sci. 20, 2101-2110 (1977).
3. Bosmann, H.B., Case, K.R.,Di Stefano, P.: FEBS Lett. 82, 368-372 (1977).
4. Chang, R.S.L., Snyder, S.H.: Eur.J.Pharmacol. 48, 213-218 (1978).
5. Kochman, R.L., Mackerer, C.R.: Fed.Proc.Fed.Am.Soc.Exp. Biol. 37, 773 (1978).
6. Speth, R.C., Wastek, G.J., Hruska, R.E., Reisine, T.D., Kobayashi, R.M., Jamamura, H.I.: Fed.Proc.Fed.Am.Soc.Exp. Biol. 37, 773 (1978).
7. Mohler, H., Okada, T.: Science 198, 849-851 (1977).
8. Squires, R.F., Braestrup, C.: Nature 266, 732-734 (1977).
9. Braestrup, C., Albrechtsen, R., Squires, R.F.: Nature 269, 702-704 (1977).

10. Braestrup, C., Squires, R.F.: Eur.J.Pharmacol. 48, 263-270 (1978).

11. Nielsen, M., Braestrup, C., Squires, R.F.: Brain Res. 141, 342-346 (1978).

12. Bosmann, H.B., Penny, D.P., Case, K.R., Di Stefano, P., Averill, K.: FEBS Lett. 87, 199-202 (1978).

13. Colello, G.D., Case, K., Bosmann, H.B.: Pharmacologist 20, 273 (1978).

14. Simantov, R., Snowman, A.M., Snyder, S.H.: Mol. Pharmacol. 12, 977-986 (1976).

15. Marangos, P.J., Paul, S.M., Greenlaw, P., Goodwin, F.K., Skolnick, P.: Life Sci. 22, 1893-1900 (1978).

16. Karobath, M., Sperk, G., Schonbeck, G.: Eur.J.Pharmacol. 49, 323-326 (1978).

17. Colello, G.D., Hockenbery, D.M., Bosmann,H.B., Fuchs, S., Folkers, K.: Proc.Natl.Acad.Sci. 75, 6319-6323 (1978).

18. Lowry, O.H., Rosebrough, N.J., Farr, A.L., Randall, R.Y.: J.Biol.Chem. 193, 265-275 (1951).

19. Studier, F.W.: J.Mol.Biol. 79, 237-248 (1973).

20. Laemmli, U.K.: Nature 227, 680-685 (1970).

21. Webb, J.L.: Enzyme and Metabolic Inhibitors, Vol.1, Academic Press, New York 1963

22. Lewis, R.V., Stein, S., Gerber, L.D., Rubinstein, M., Udenfriend, S.: Proc.Natl.Acad.Sci. 75, 4021-4023 (1978).

ß-CASOMORPHINS —— EXORPHINS DERIVED FROM CASEIN PEPTONE

Agnes Henschen, Friedrich Lottspeich
Max-Planck-Institut für Biochemie, D-8033 Martinsried, München, Germany

Victor Brantl
Max-Planck-Institut für Psychiatrie, Abt. Neuropharmakologie, D-8000 München 40, Germany

Hansjörg Teschemacher
Pharmakologisches Institut der Justus Liebig-Universität, D-6300 Giessen 1, Germany

Introduction

In 1973 the presence of special receptors for opiates in the nervous tissues of vertebrates was demonstrated (1-3). During the following years several endogenous opioid ligands for these receptors, so called "endorphins", have been shown to be present in brain, pituitary and other tissues, and some of these ligands were identified as peptides (cf. ref. 4,5).

Recently, compounds which display opioid activity have also been found in milk and milk products, such as baby food and casein peptone (6-9). Some of these compounds proved to be resistant to the proteolytic enzyme pronase, which at first questioned their peptide nature (8,9). As these opioids obviously would develop their activity in another organism than the one where they were found, they may be described as exogenous opioids or "exorphins". The isolation (10) and identification (11) of opioids from casein peptone will be presented here.

Structure and Activity of Natural Peptides

Purification

Casein peptone is a proteolytic digest of bovine casein; it is commercially available. Originally, it was shown that chloroform-methanol extracts of this material displayed opioid activity in the guinea pig ileum bioassay (9). However, a multistep procedure, as outlined in Table 1, had to be used in order to isolate a homogeneous opioid compound (10). The fractionation of the casein peptone was started by a few batchwise extractions and adsorption-desorption steps. Thereafter, five consecutive high-pressure liquid chromatography steps were employed. In the first of these more than 50 ultraviolet ab-

Table 1. Purification procedure for opioid components in bovine casein peptone (from ref. 10).
Opioid activity, in nmol of normorphine equivalents, was determined in the guinea pig ileum bioassay.

Purification step	Opioid activity per g	Opioid activity total
Casein peptone	(9)	(4060)
Chloroform-methanol extraction	55	4060
Charcoal adsorption	400	2950
Amberlite XAD-2 adsorption	1000	1400
HPLC: µ-Bondapak C_{18}; isocratic, 0.1 M acetic acid-methanol	1750	700
HPLC: µ-Bondapak C_{18}; gradient, water to ethanol	3550	550
HPLC: µ-Porasil; isocratic, chloroform-methanol	4200	400
HPLC: µ-Porasil; gradient, water to methanol	5000	220
HPLC: µ-Porasil; isocratic, water-methanol	6500	180
Gel filtration: Bio-Gel P-2, 0.04% trifluoroacetic acid	6300	150

sorbing components could be discerned, in the last one a single component with highly similar ultraviolet absorption and opioid activity profiles (Fig. 1) was left. However, when this material was subjected to Bio-Gel P-2 chromatography it separated into several components, as shown in Fig. 2. Opioid activity was found in all components, but 90 % of the total activity in the chromatography was concentrated in the main peak. On re-chromatography of this peak no further heterogeneity was observed (Fig. 2, dotted line). The material in the main peak was therefore regarded as the purification end product and analysed for biochemical and pharmacological properties.

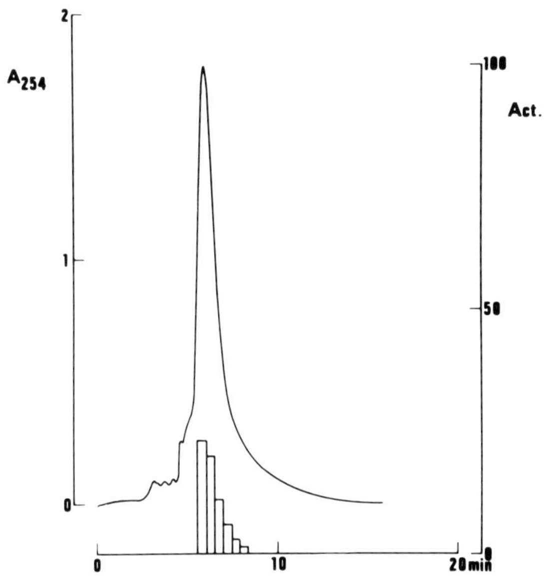

Fig. 1. High pressure liquid chromatography on μ-Porasil (from ref. 10).
Column dimensions, 0.79 x 30 cm. Eluant, methanol-water, 1:1 (v/v). —— Denotes absorption at 254 nm (left scale); vertical bars denote opioid activity in nmol of normorphine equivalents per fraction (right scale).

Biochemical Identification

The apparently homogeneous purification end product was first analysed for amino acid composition after acid hydrolysis. It contained only the five amino acids proline, glycine, isoleucine, tyrosine and phenylalanine, their molar ratio being 3:1:1:1:1. The yield of the amino acids accounted for most of the weight of the material analysed. The material was then subjected to manual Edman degradation (12,13). The unique N-terminal heptapeptide sequence, indicated in Table 2, upper part, was found. For C-terminal degradation carboxypeptidase Y was employed. The time course of the amino acid release as determined by amino acid analysis is plotted in Fig. 3, lower

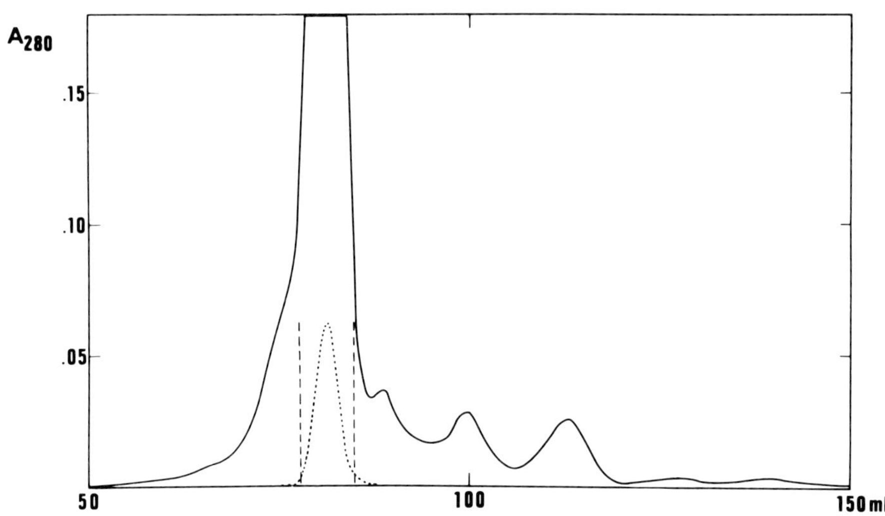

Fig. 2. Gel filtration chromatography on Bio-Gel P-2, <400 mesh (from ref. 10).
Column dimensions, 1 x 135 cm: Eluant, 0.04 % trifluoroacetic acid; —— Denotes first chromatography, 4.2 mg of material; ------denotes re-chromatography, 1/15 of the pool indicated by the hatched, vertical lines.

panel. It confirmed the order of the five most C-terminal residues. On C-terminal determination by means of hydrazinolysis isoleucin was found (Table 2), which proves that the carboxyl group is free. The ultraviolet absorption spectrum of the material in acid and alkaline solution showed absorption maxima at wavelengths characteristic of tyrosine with a free hydroxyl group and several waves characteristic of phenylalanine. Thus, all analytical data were in agreement with the heptapeptide sequence indicated (11).

In order to prove the identity between the heptapeptide structure and the principle of the opioid activity several tests were performed. Thus, it could be shown that on paper electro-

Table 2. Amino acid sequences of the peptide material and isolated carboxypeptidase Y (CPY) digestion products (from ref. 11).
⟵⟶ Denotes length of peptide, → amino acid residue located by direct Edman degradation, ← residue located by carboxypeptidase Y digestion, ⇤ residue located by hydrazinolysis.

Complete Sequence	Tyr-	Pro-	Phe-	Pro-	Gly-	Pro-	Ile
Isolated opioid material (ß-casomorphin-7)	⟵⟶	⟵⟶	⟵⟶	⟵⟶	⟵⟶	⟵⟶	⟵⟶
	→	→	→	→	→	→	→
			←	←	←	←	←
							⇤
CPY digestion Pool I (ß-casomorphin-7)	⟵⟶	⟵⟶	⟵⟶	⟵⟶	⟵⟶	⟵⟶	⟵⟶
	→						
CPY digestion Pool III (ß-casomorphin-5)	⟵⟶	⟵⟶	⟵⟶	⟵⟶	⟵⟶		
	→			←	←		
					⇤		
CPY digestion Pool V (inactive)	⟵⟶	⟵⟶	⟵⟶				
	→		←				
			⇤				

phoresis the only ninhydrin-positive component, which incidently had the heptapeptide amino acid composition, co-migrated with the opioid activity (11).

Several tests were aimed at modifying the structure and analysing for a concomitant change in activity. It was then found that, both, acid hydrolysis into free amino acids and, separately, acetylation with acetic anhydride of the α-amino group and/or phenolic hydroxyl group abolished all opioid activity (11).

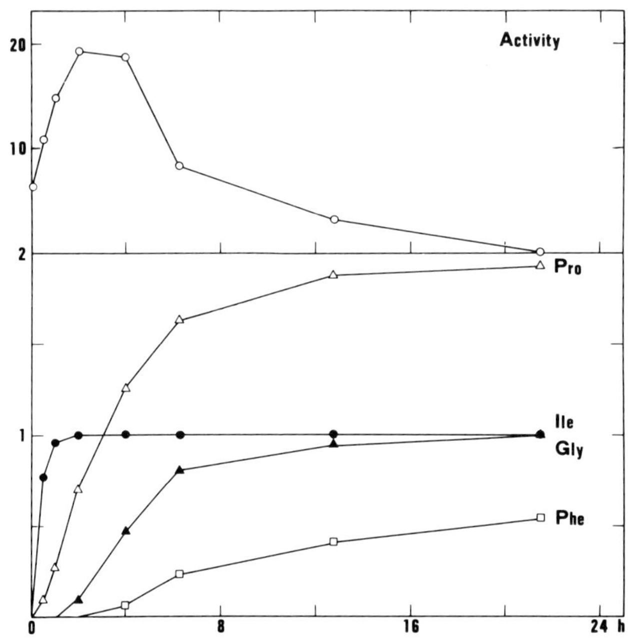

Fig. 3. Time course of digestion of the peptide material (ß-casomorphin-7) by carboxypeptidase Y (from ref. 11). o Denotes opioid activity in µmol normorphine equivalents as determined by the guinea pig ileum bioassay; ● denotes isoleucine, Δ proline, ▲ glycine and □ phenylalanine in mol amino acid per mol peptide, the released amino acids being determined by amino acid analysis.

Proteolytic Modification

Several proteolytic enzymes were tested for their ability to alter the peptide structure and/or the opioid activity. As it already was known that carboxypeptidase Y would modify the structure (see above), the effect of this enzyme on the activity was analysed first (11). The time course of the change in opioid activity may be compared with the release of amino acids in Fig. 3, upper and lower panel, respectively. At the same time as the C-terminal isoleucine residue and the ultimate proline residue are liberated there is a dramatic increase in opioid activity. The maximal activity in the digest is about 2.5-fold the initial activity. With the subsequent release of glycine and/or the next proline the activity drops gradually to zero. The release of phenylalanine, however, is of no importance for the activity, as half of the amount of phenylalanine still is attached to the peptide chain when all activity is gone.

In order to analyse the chain length-activity relationship more precisely an attempt was made to separate out the different peptide species in the carboxypeptidase Y digest at the time of the highest opioid activity, i.e. after 3 hours. The digest was fractionated by Bio-Gel P-2 chromatography, as shown in Fig. 4. Three main peaks and two smaller intermediate peaks were obtained. The first peak appeared in the same position as the starting material (cf. Fig. 2). As it has the same composition and N-terminal amino acid (Table 2), it represents the completely undigested heptapeptide. The second main peak appeared later in the chromatography (Fig. 4). The amino acid analysis together with N- and C-terminal determination (Table 2) prove that it corresponds to the pentapeptide. The third main peak appeared last in the chromatography (Fig. 4) and could be shown to correspond to the tripeptide (Table 2). The two small intermediate peaks contained, as judged by amino acid analysis, the hexa- and the tetrapeptide, respectively.

When the opioid activity of the hepta-, penta- and tripeptide were compared, it was found that the pentapeptide was considerably more active than the heptapeptide. The tripeptide, however, was inactive. These differences in activity are in complete agreement with the changes in activity observed during the carboxypeptidase Y digestion. The peptide chain length-opioid activity relationship will be described in more detail below.

The effect on structure and activity of several other proteases was also investigated. However, the heptapeptide showed a remarkable stability towards proteolytic attack. Thus, even at digestion for prolonged times with high concentrations of pronase, thermolysine, leucine aminopeptidase or carboxypeptidases A or B no change was observed either in structure or in activity.

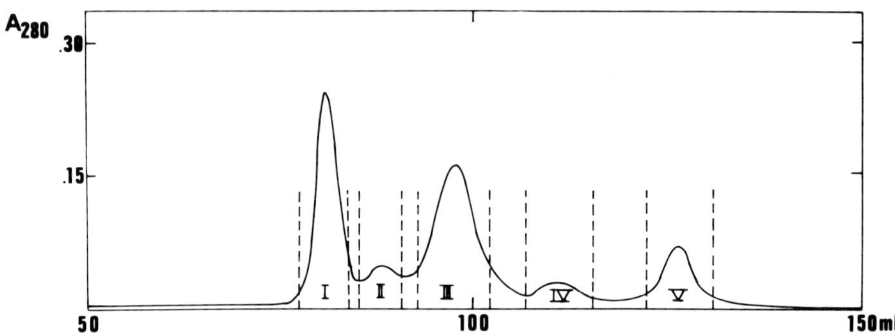

Fig. 4. Gel filtration chromatography on Bio-Gel P-2, <400 mesh, of 3h-digest of the heptapeptide (ß-casomorphin-7) by carboxypeptidase Y (from ref. 11).
Column dimensions, 1 x 135 cm. Eluant, 0.04 % trifluoroacetic acid. Sample, 2 mg of digest. Pooling is indicated by hatched lines.

It may be summarised that strong evidence for the identity between the heptapeptide structure and the opioid substance was obtained not only from the fact that they co-purified, but also from the fact that they could be destroyed or modified together, but not separately. As destructive or modifying reactions had been tried: Acid hydrolysis, acetylation and proteolysis. The final evidence, however, was obtained from the opioid activity elicited by synthetic peptides having the same amino acid sequence as those derived from casein peptone (14).

Pharmacological Characterisation

Throughout the purification and modification procedures the opioid activity was tested by means of the guinea pig ileum longitudinal muscle-myenteric plexus preparation by measuring the inhibitory effect on the electrically-induced contractions. The inhibition was regarded as an opioid effect when it

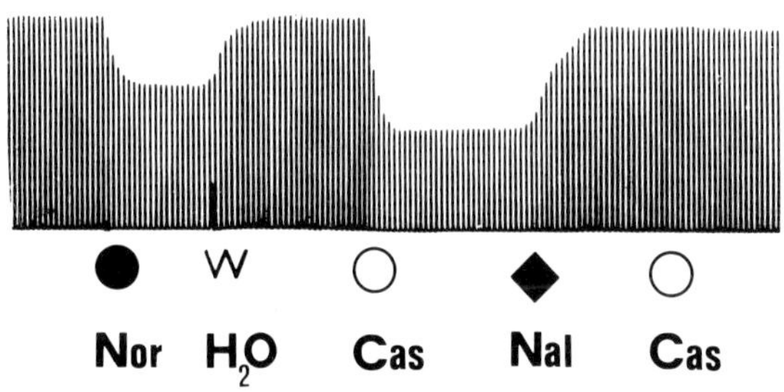

Fig. 5. Opioid activity of the heptapeptide (ß-casomorphin-7) in the guinea pig ileum bioassay (from ref. 10).
From left to right, • denotes the addition of 0.07 µM normorphine, W washing with buffer, o addition of 33 µM heptapeptide, ◆ addition of 0.3 µM (-)-naloxone, o new addition of 33 µM heptapeptide with naloxone still present.

could be specifically antagonised by (-)-naloxone, but not by the inactive isomer (+)-naloxone. The type of activity displayed by the purification end product is demonstrated in Fig. 5. It is obvious that the effect is opioid-specific. The dose-response relationship in the guinea pig ileum of the purification end product may be compared with those of normorphine and methionine-enkephalin in Fig. 6. The casein-derived opioid heptapeptide was found to be about 250 times less active than normorphine or methionine-enkephalin on a molar basis.

The influence of peptide chain length on opioid activity was determined not only in the guinea pig ileum bioassay but also in the mouse vas deferens bioassay, the results being compared with those obtained with normorphine and methionine-enkephalin in Table 3. As may be seen from the values listed, the potencies of all four casein-derived peptides are much higher in the guinea pig ileum than in the mouse vas deferens preparation. Thus, these peptides are morphine-like, but differ from

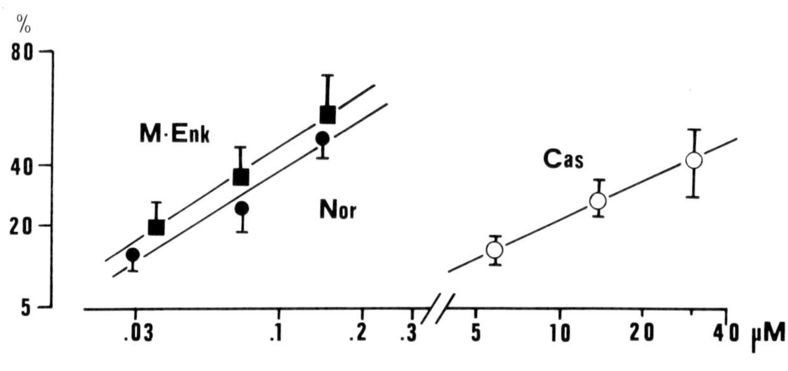

Fig. 6. Log-probit plot of dose-response relationships in the guinea pig ileum bioassay for opioid effects of normorphine, ●, methionine-enkephalin, ■, and the heptapeptide (β-casomorphin-7), o; mean values and standard deviations from three determinations (from ref. 10).

the peptide methionine-enkephalin, in their organ-specificity. The result is well compatible with the concept of different types of opiate receptors in these organs (15). The predominant type of opiate receptor in the guinea pig ileum has been designated the μ-receptor,the one in the mouse vas deferens the δ-receptor. Thus, the casein-derived peptides, like morphine, show higher effectson μ-receptors. The influence of peptide chain length within the group of casein-derived peptides is obvious (Table 3), the pentapeptide being the most active. The activity decreases then in the order tetrapeptide, hexapeptide and heptapeptide, the tripeptide being practically devoid of opioid activity (not shown).

Preliminary in vivo-tests have also been performed. At intracerebroventricular injection in the rat the heptapeptide elicited strong, opiate-specific analgetic and slight catatonic effects.

Table 3. Opioid activity in isolated and/or synthesised peptides and in methionine-enkephalin and normorphin. Concentration in μM effecting 50% inhibition of electrically-induced concentrations in the guinea pig ileum (GPI) and mouse vas deferens (MVD) bioassay; mean values and standard deviations from three determinations.

	GPI	MVD
Heptapeptide (ß-casomorphin-7)	57.0±7.5	>500
Hexapeptide (ß-casomorphin-6)	27.4±1.7	>500
Pentapeptide (ß-casomorphin-5)	6.5±0.6	50.8±7.1
Tetrapeptide (ß-casomorphin-4)	21.9±2.6	>500
Methionine-enkephalin	0.19±0.02	0.019±0.002
Normorphine	0.23±0.02	1.5±0.2

Relation to Casein

As the peptides with opioid activity had been isolated from an enzymatic digest of bovine casein, the amino acid sequences of α-, β-, and κ-casein were searched for the corresponding sequence. It was, indeed, discovered that the β-casein sequence (16) contained the heptapeptide sequence in the positions 60 to 66, as demonstrated in Fig. 7. It should be mentioned that the heptapeptide sequence so far never has been found in other connections than this one (17). Because of their origin and activity the peptides were now named β-casomorphins, with a number attached indicating their chain length, the N-terminal tyrosine being counted as residue number 1.

The sequences show that the β-casomorphins have the same relation to β-casein as the endorphins to β-lipotropin. The

1 Arg-Glu-Leu-Glu-Glu-Leu-Asn-Val-Pro-Gly-Glu-Ile-Val-Glu-
15 -Ser-Leu-Ser-Ser-Ser-Glu-Glu-Ser-Ile-Thr-Arg-Ile-Asn-Lys-
29 -Lys-Ile-Glu-Lys-Phe-Gln-Ser-Glu-Glu-Gln-Gln-Gln-Thr-Glu-
43 -Asp-Glu-Leu-Gln-Asp-Lys-Ile-His-Pro-Phe-Ala-Gln-Thr-Gln-
57 -Ser-Leu-Val-[Tyr-Pro-Phe-Pro-Gly-Pro-Ile]-Pro-Asn-Ser-Leu-
71 -Pro-Gln-Asn-Ile-Pro-Pro-Leu-Thr-Gln-Thr-Pro-Val-Val-Val-
85 -Pro-Pro-Phe-Leu-Gln-Pro-Glu-Val-Met-Gly-Val-Ser-Lys-Val-
99 -Lys-Glu-Ala-Met-Ala-Pro-Lys-His-Lys-Glu-Met-Pro-Phe-Pro-
113 -Lys-Tyr-Pro-Val-Gln-Pro-Phe-Thr-Glu-Ser-Gln-Ser-Leu-Thr-
127 -Leu-Thr-Asp-Val-Glu-Asn-Leu-His-Leu-Pro-Pro-Leu-Leu-Leu-
141 -Gln-Ser-Trp-Met-His-Gln-Pro-His-Gln-Pro-Leu-Pro-Pro-Thr-
155 -Val-Met-Phe-Pro-Pro-Gln-Ser-Val-Leu-Ser-Leu-Ser-Gln-Ser-
169 -Lys-Val-Leu-Pro-Val-Pro-Glu-Lys-Ala-Val-Pro-Tyr-Pro-Gln-
183 -Arg-Asp-Met-Pro-Ile-Gln-Ala-Phe-Leu-Leu-Tyr-Gln-Gln-Pro-
197 -Val-Leu-Gly-Pro-Val-Arg-Gly-Pro-Phe-Pro-Ile-Ile-Val

Fig. 7. Amino acid sequence of bovine β-casein A2 (from ref. 16). The sequence of the opioid heptapeptide (β-casomorphin-7) is boxed.

active peptides may in both cases be released from the inactive precursors by means of proteolytic enzymes. In the case of the ß-casomorphins the release might take place in the digestive tract, the casein of the ingested milk or milk product being cleaved by the various proteases present. Once formed, the ß-casomorphins would, because of their proline-rich sequence, be expected to be fairly stable and perhaps even reach concentrations of physiological significance in the intestine.

Relation to Other Opioid Peptides

The sequences of the ß-casomorphins may be aligned with those of the enkephalins (Fig. 8). It is obvious that the only features these two classes of opioid peptides have in common are the typrosine residue in the N-terminal position and the general hydrophobicity in the following part of the sequence. There is hardly any doubt that the N-terminal tyrosine is of decisive importance for the opioid character of the ß-casomorphins. However, the pronounced chain length-dependent differences in activity point to the additional importance of the following amino acids in the sequence.

In conclusion, the ß-casomorphins represent a new class of opioid peptides, since their amino acid sequence and also their pharmacological properties are quite different from those of the enkephalins.

Heptapeptide (ß-casomorphin-7)	Tyr-Pro-Phe-Pro-Gly-Pro-Ile
Pentapeptide (ß-casomorphin-5)	Tyr-Pro-Phe-Pro-Gly
Methionine-enkephalin	Tyr-Gly-Gly-Phe-Met
Leucine-enkephalin	Tyr-Gly-Gly-Phe-Leu

Fig. 8. Amino acid sequences of the opioid hepta- and pentapeptide and of the enkephalins.

References

1. Pert, C.B., Snyder, S.H.: Science 179, 1011-1014 (1973).

2. Simon, E.J., Hiller, J.M., Edelman, I.: Proc. Natl. Acad. Sci. U.S.A. 70, 1947-1949 (1973).

3. Terenius, L.: Acta Pharmacol. 33, 377-384 (1973).

4. Teschemacher, H.: in: Development in Opiate Research (Herz, A., ed.) pp. 67-151, Dekker, New York 1978

5. Teschemacher, H.: Arzneimittelforschung 28, 1268-1270 (1978).

6. Zioudrou, C., Klee, W.A.: in: Characteristics and Function of Opioids (van Ree, J.M., Terenius, L., eds.) pp. 243-244, Elsevier/North Holland, Amsterdam 1978

7. Wajda, I.J., Neidle, A., Ehrenpreis, S., Manigault, I.: in: Opiates and Endogenous Opioid Peptides (Kosterlitz, H.W., ed.) pp. 129-136, Elsevier/North Holland, Amsterdam 1976

8. Teschemacher, H., Csontos, E., Westenthanner, A., Brantl, V., Kromer, W.: in: Endorphins in Mental Health Research (Usdin, E., Bunney, W.E., Kline, N.S., eds.), pp. 203-208, MacMillan, London 1979

9. Brantl, V., Teschemacher, H.: Naunyn-Schmiedeberg's Arch. Pharmacol. 306, 301-304 (1979).

10. Brantl , V., Teschemacher, H., Henschen, A., Lottspeich, F.: Hoppe-Seyler's Z. Physiol. Chem., 360, 1211-1216 (1979).

11. Henschen, A., Lottspeich, F., Brantl, V., Teschemacher, H.: Hoppe-Seyler's Z. Physiol. Chem. 360, 1217-1224 (1979).

12. Edman, P., Henschen, A.: in: Protein Sequence Determination (Needleman, S.B., ed.) 2nd edn , pp. 232-279, Springer, Berlin 1975

13. Peterson, J.D., Nehrlich, S., Oyer, P.E., Steiner, D.F.: J.Biol.Chem. 242, 4866-4871 (1972).

14. Lottspeich, F., Henschen, A., Brantl , V., Teschemacher, H.: in preparation (1980).

15. Lord, J.A.H., Waterfield, A.A., Hughes, J., Kosterlitz, H.W.: Nature (London) 267, 495-499 (1977).

16. Ribadeau Dumas, B., Brignon, G., Grosclaude, F., Mercier, J.-C.: Eur. J. Biochem. 25, 505-514 (1972).

17. Dayhoff, M.O., Hunt, L.T., Barker, W.C., Schwartz, R.M., Orcutt, B.C.: Protein Segment Dictionary 78, National Biomedical Research Foundation, Washington 1978

STRUCTURE OF SAUVAGINE, A VASOACTIVE PEPTIDE FROM THE SKIN OF A FROG

Pier Carlo Montecucchi

Farmitalia Carlo Erba SpA Ricerca & Sviluppo Chimico
Via dei Gracchi, 35 - 20146 Milano, Italia

Agnes Henschen

Max-Planck-Institut für Biochemie
8033 Martinsried, Germany

Vittorio Erspamer

Istituto di Farmacologia Medica I, Università di Roma
00100 Roma, Italia

Introduction

During our systematic screening for active polypeptides in the amphibian skin, it was found that methanol extracts of the skin of *Phyllomedusa sauvagi*, a frog of South America, contain in addition to peptides belonging to the bradykinin (1), caerulein (2), physalaemin (3) and bombesin families, a new peptide which possesses a number of pharmacological actions on diuresis, the cardiovascular system and endocrine glands.
The substances, called sauvagine, has now been isolated in pure form (4) and identified as a polypeptide constituted by forty amino acid residues.

Structure and Activity of Natural Peptides

Pharmacological Actions

Sauvagine displays a long-lasting hypotensive action in all animal species so far examined: rat, dog, monkey and man. Hypotension was particularly striking in the rat, both in intensity and duration (5). Threshold dose was 0.5 μg/Kg, subcutaneously, and 5-10 ng/Kg, intravenously. In conscious rats hypotension was accompanied by marked tachykardia, in part attributable to adrenergic stimulation.

The pronounced lowering of systemic blood pressure is probably the main cause of the intense antidiuresis seen in hydrated rats, following sauvagine administration. Reduction in urine flow, accompanied by conspicuous tubular reabsorption of sodium, was closely correlated with reduction in glomerular filtration rate.

In the dog, the moderate pressure fall elicited by sauvagine appeared to be mainly due to vasodilatation in the area of the mesenteric artery. Only the vessels of the intestinal mucosa were affected (6).

In these effects on blood pressure and diuresis, sauvagine showed a close biological resemblance to urotensin I, a peptide isolated from teleost urophysis, which is considered a caudal neurosecretory system (7,8).

In addition to its properties on the cardiovascular system, sauvagine displayed conspicuous effects on the rat anterior pituitary, provoking both, release and inhibition of release of pituitary hormones (9).

The following effects could be observed, both <u>in vivo</u> and <u>in vitro</u> on isolated pituitary tissue:

1) Intense stimulation of ACTH release, with consequent adrenocortical activation and corticosterone liberation (threshold subcutaneous dose was 0.5 μg/Kg).
2) Concomitant potent stimulation of β-endorphin release.
3) Inhibition of PRL, TSH and GH release, the inhibition of TSH release

being apparently the strongest. The fall in plasma levels of TSH caused not only a sharp reduction of ^{125}I uptake by the thyroid tissue *in vivo*, but also a reduction in the weight of the thyroid gland.

4) No apparent effect on gonadotropin release.

Release of adrenocortical hormones was confirmed in the dog and monkey, inhibition of ^{125}I uptake by the thyroid gland in the dog and increase in plasma immunoreactive ACTH also in man.

It is difficult at present to interpret the pleiotropic effects of sauvagine. Pharmacological research on this polypeptide is only in the initial stages and it is easy to predict that the spectrum of activity of sauvagine will expand in the near future.

There are good reasons to believe that sauvagine is the prototype of a new peptide family, having representatives not only in the amphibian skin and in bony fish urophysis, but also in the mammalian gut and brain (figure 1). This assumption is based on preliminary results obtained in the rat gastrointestinal tract and brain with an anti-sauvagine serum raised in rabbits with a semipurified *Phyllomedusa sauvagei* extract. Should the existence of a sauvagine-like peptide in the CNS be confirmed, its candidature as a physiological releaser of ACTH and β-endorphin would be conceivable.

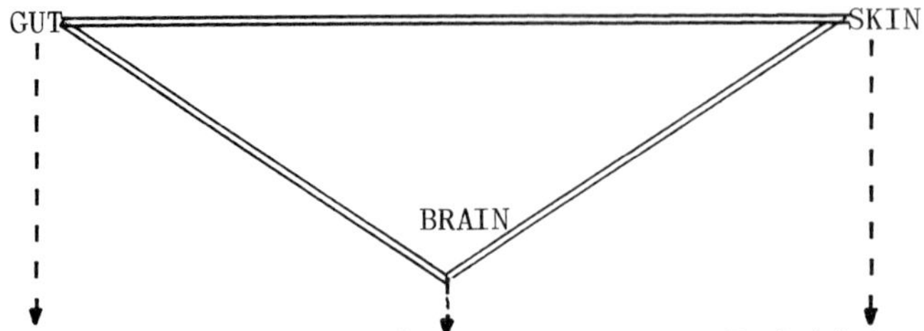

Substance P	Substance P	Tachykinins
CCK	CCK	Caeruleins
Gastrin	Gastrin (?)	
Neurotensin	Neurotensin	Xenopsin
Enkephalins	Enkephalins	Dermorphins
TRH	TRH	TRH
Somatostatin	Somatostatin	?
Bombesin-like pept.	Bombesin-like pept.	Bombesin
VIP	VIP	VIP
?	Angiotensin-like pept.	Angiotensin-like pept.
?	(Urotensin I)	Sauvagine
Indolealkylamines	Indolealkylamines	Indolealkylamines
Imidazolealkyl-amines	Imidazolealkyl-amines	Imidazolealkyl-amines
Catecholamines	Catecholamines	Catecholamines

Figure 1. The gut-brain-skin triangle.

Purification Scheme

The purification sequence which was established after pilot studies consisted of several steps, as follows : (i) extraction with methanol from the fresh skins of several hundred specimens of Phyllomedusa sauvagei, collected in the region of Tucuman (Argentina); (ii) countercurrent distribution between petroleum ether and 80 % ethanol, in order to remove fats; (iii) dialysis; (iv) precipitation with 90 % ethanol; (v) precipitation of the ethanolic supernatant with acetone; (vi) purification of the acetonic residue on a column of Sephadex LH-20 in 90 % ethanol; (vii) chromatography on a DEAE Sephadex A-50 column; (viii) gel filtration on a column of Sephadex G-100; (ix) chromatography on a DEAE cellulose column in the presence of 6M urea; (x) gel filtration on Bio--Gel P-6 or Sephadex G-50 sf. The rat antidiuresis test was routinely used to follow the purification procedure.

Isoelectric focusing of the purest sample of sauvagine obtained as described showed that the product was still heterogeneous. In fact it split again into two bands: a small band with pI 5.05, a large band with pI 5.32 (4). Densitometer scanning of the gel showed that the major component accounted for 90 % of the total. After elution only the spot focalized at pH 5.32 appeared to be potently active. The material obtained from Sephadex G-50 column (step x) was considered as a standard peptide in all our pharmacological studies and for sequence analysis.

The molecular weight of sauvagine appeared lower than that of bovine insulin by gel filtration technique on Bio-Gel P-6 and very close to that of porcine ACTH by SDS gel electrophoresis, where the extrapolated value was about 4500 (4).

Amino Acid Sequence

After total acid hydrolysis the molecule of sauvagine appeared to be constitued by 40 amino acid residues (table). The amino acid composition as determined by amino acid analysis is in good agreement with that determined by sequence analysis.

When this polypeptide was analysed for primary structure, the N-terminus was found to be blocked. However, after treatment with calf liver pyroglutamate aminopeptidase (EC 3.4.11.8) the larger part of the sequence

Amino acid	mol residues/mol of peptide	
	Amino acid analysis	Sequence analysis
Asp	3.7	2
Asn	-	2
Thr	0.9	1
Ser	2.1	2
Glu	8.8	5
Gln	-	3
⌐Glu*	-	1
Pro	1.8	2
Gly	1.3	1
Ala	2.0	2
Met	0.6	1
Ile	4.7	5
Leu	7.0	7
Lys	3.7	4
Arg	2.0	2
	-	40

* ⌐Glu, pyrrolidonecarboxylic acid.

Table - Amino acid composition of sauvagine

could be directly determined in the Edman sequenator (10). Phenylthiohydantoins were identified by thin-layer (10) and high-performance liquid chromatography (11). The released amino terminal pyrrolidonecarboxylic acid could be identified after separation on Sephadex G-10 and hydrolysis by amino acid analysis as glutamic acid. The automatic sequence of the deblocked N-terminal fragment was unambiguously extended to lysine 27 (figure 2).

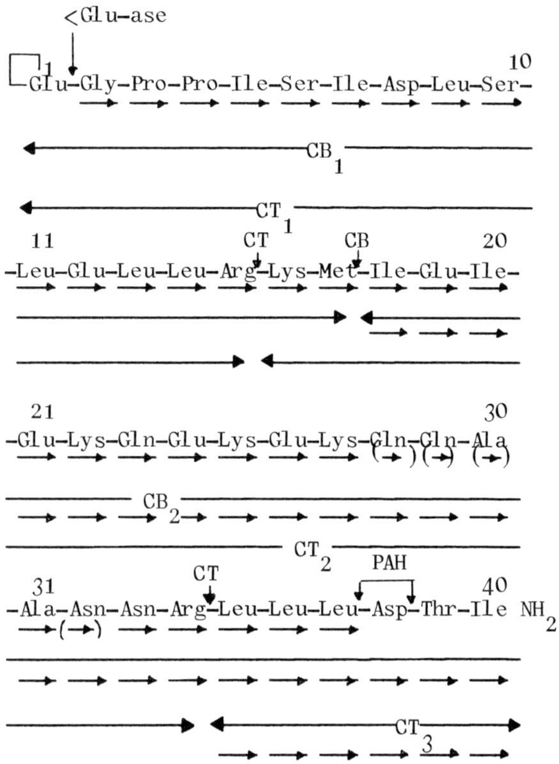

Figure 2. Complete amino acid sequence of sauvagine.
<Glu-ase ↓, action of pyroglutamate aminopeptidase; CT ↓, trypsin cleavage point after citraconylation; CB ↓, cyanogen bromide cleavage point; →, amino acid residue located by direct thiazolinone degradation; (→) residue located with lower degree of confidence ←→, denotes lenght of peptide; CB denotes a peptide isolated after cyanogen bromide cleavage, CT, after arginine--directed trypsin cleavage; the vertical arrows in CT_3 between Leu (37), Asp (38) and Thr (39) indicate the sites of partial acid hydrolysis (PAH).

The C-terminal part of the sequence was obtained by analysing in the sequenator the fragment CB_2 from cyanogen bromide cleavage at the single methionine residue. The cleavage products were first fractionated by gel filtration on a column of Sephadex G-50 sf. (figure 3).

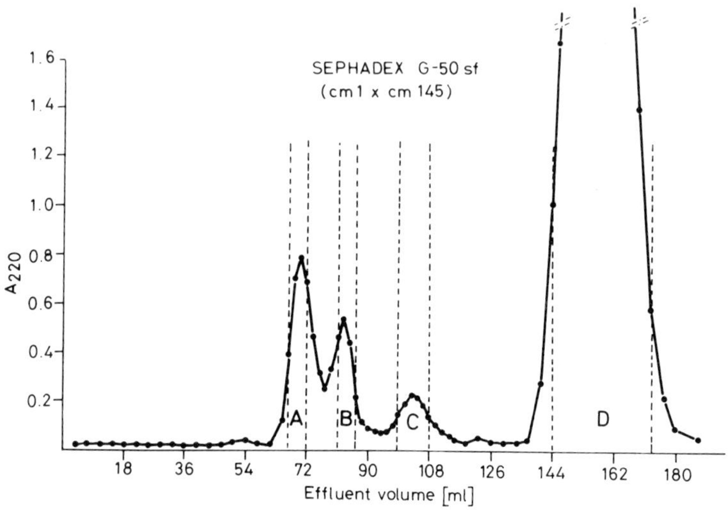

Figure 3. Gel filtration of sauvagine on Sephadex G-50 sf. column (eluent 0.05 % trifluoroacetic acid; flow rate 1.8 ml/h).

Four peaks were isolated. After total acid hydrolysis and amino acid analysis it was possible to establish that pool A contained the intact molecule, probably because the methionine residue was in the sulfoxide form; pool B was recognised as the C-terminal part of the molecule; pool C could be identified as the N-terminal fragment by the presence of methionine (or really homoserine residue) in its amino acid composition; pool D did not contain peptidic material. Upon rechromatography on the same column, pool B (fragment CB-2) appeared as a single peak with the same effluent volume. This fragment was analysed in the sequenator and it was possible to establish the complete sequence (figure 2).

Under the action of carboxypeptidase A (EC 3.4.12.2) on the sauvagine molecule no amino acid was released. However, with carboxypeptidase Y (EC 3.4.12.-) isoleucine was found. This indicated a blocked C-terminus.

For this purpose the sauvagine molecule was selectively cleaved at the arginyl bonds by trypsin (EC 3.4.21.4) after blocking of the ε-amino groups of lysine residues by citraconylation. The cleavage mixture was fractionated on Sephadex G-10 and G-25. The C-terminal hexapeptide (fragment CT_3) was isolated and submitted to sequence analysis by automatic Edman degradation and partial acid hydrolysis with 0.25 M acetic acid. The fragment CT_3, submitted to partial acid cleavage, was examined on high voltage paper electrophoresis at pH 6.5 (figure 4). The cathodic fragment AC_3, yellowish with ninhydrin, was eluted. After total acid hydrolysis and amino acid analysis it was possible to observe the liberation of threonine and isoleucine in the same molar ratio. The fragment AC_3 behaved identical with the synthetic dipeptide H-Thr-Ile-NH_2, both, in electrophoresis and thin-layer chromatography on silica gel.

As a result of amino acid analysis and sequence studies on the intact molecule and on the peptide treated with proteolytic enzymes the amino acid sequence proposed for sauvagine is indicated in figure 2.

Should this structure be confirmed by synthesis, sauvagine would represent one of the few examples of natural active polypeptides lacking in aromatic amino acid residues in their molecule. No active peptide of amphibian skin shows this peculiarity.

Inspection of the primary structure reveals that this molecule exhibits a lot of repetitive sequences. In addition,two basic adjacent amino acids (Arg-Lys) are present at position 15-16; similar sequences are known to be the cleavage site for generation of active peptides (12-13). Two distinct regions with a strongly hydrophobic character can be recognised from residues 1 to 20 and from 30 to 40 (figure 5). The remaining sequence is rich in Glx and lysine residues and probably exists as an extended

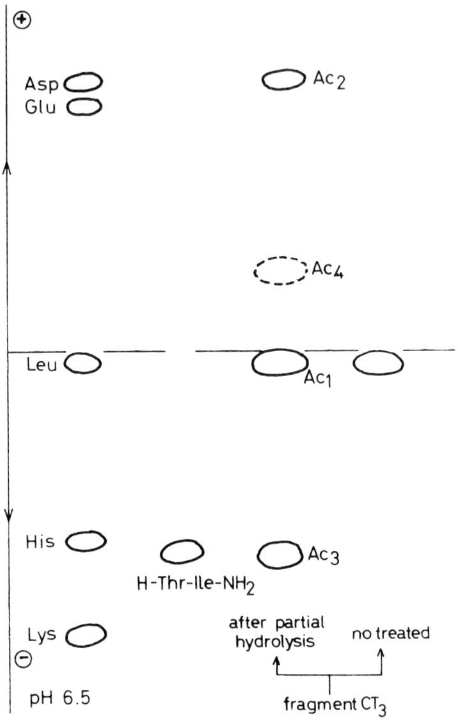

Figure 4. Electrophoretic pattern of fragment CT_3 after partial acid hydrolysis. Ac denotes a peptide isolated after PAH; Ac_1 H-Leu-Leu-Leu-OH; Ac_2 Asp; Ac_3 H-Thr-Ile-X; Ac_4 H-Leu-Leu-Leu--Asp-OH (this spot was very faint).

chain. The significance of these findings is however difficult to evaluate.

There is no doubt that the hydrophobic interactions play an important role in biological systems (14,15,16). We can speculate that the biological activity of sauvagine is linked to an interaction between hydrophobic groups of the molecule and hydrophobic pockets of the membrane receptors by a mutual biospecific affinity.

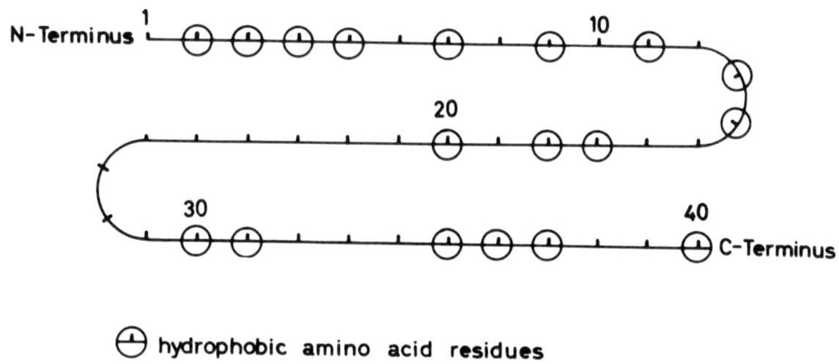

Figure 5. Schematic drawing of the sauvagine molecule.

Acknowledgment

The authors thank Dr. de Castiglione R., Farmitalia-Carlo Erba, S.p.A., Chemical Research and Development Laboratories, Milan (Italy) for his interest in this work and for the supply of synthetic dipeptide H-Thr--Ile-NH_2.

References

1. Anastasi, A., Erspamer, V., Bertaccini, G., Cei, J. M. : in "Hypotensive Peptides" (Erdös, E. G., Back, N., Sicuteri, F., Wilde, A. F., Eds.) Springer-Verlag, New York Inc., pp. 76-85(1966).

2. Anastasi, A., Bertaccini, G., Cei, J. M., De Caro, G., Erspamer, V., Impicciatore, M. : Brit. J. Pharmac. 37, 198-206 (1969).

3. Anastasi, A., Falconeri-Erspamer, G. : Experientia 26, 866 (1970).

4. Montecucchi, P. C., Anastasi, A., de Castiglione, R., Erspamer, V. : Int. J. Peptide Protein Res., submitted for publication.

5. de Castiglione, R., Erspamer, V., Falconeri-Erspamer, G., Improta, G., Negri, L. : Brit. J. Pharmac., submitted for publication.

6. Melchiorri, P. : Personal Communication.

7. Lederis, K., Medakovic, M. : Brit. J. Pharmac. 51, 315-324 (1974).

8. Medakovic, M., Chan, D. K. O., Lederis, K. : Pharmacology 13, 419-426, (1975).

9. Erspamer, V., Melchiorri, P., Montecucchi, P. C., Sopranzi, M. : Brit. J. Pharmac., submitted for publication.

10. Edman, P., Henschen, A. : in "Protein Sequence Determination" (Needleman, S. B., Ed.) 2nd edn., Springer-Verlag, Berlin, Heidelberg, New York, pp. 232-279 (1975).

11. Zimmerman , C. L., Appella, E., Pisano, J. : Anal. Biochem. 77, 569--573 (1977).

12. Kimura, S., Lewis, R. V., Gerber, L. D., Brink, L., Rubinstein,M., Stein, S., Udenfriend, S. : Proc. Natl. Acad. Sci. (USA) 76, 1756-1759 (1979).

13. Folkers, K., Sakura, N., Leban, J., Dahmen, J., Lundanes, E., Ohta, M., Chang, D., Wan, Y.-P. : This book.

14. Ochoa, J. L. : Biochimie 60, 1-15 (1978).

15. Mufson, R. A., Laskin, J. D., Fisher, P. B., Weinstein, I. B. : Nature 280, 72-74 (1979).

16. Knöppel, E., Eisenberg, D., Wickner, W. : Biochemistry 18, 4177-4181, (1979).

HERBICOLINS - NEW PEPTIDE ANTIBIOTICS FROM ENTEROBACTERIA

Günther Winkelmann, Christel Glück
Institut für Biologie I, Mikrobiologie I der Universität Tübingen
D-7400 Tübingen, Germany

Rudolf Lupp, Günther Jung
Institut für Organische Chemie der Universität Tübingen
D-7400 Tübingen, Germany

Introduction

Most of the 3000 antibiotics described so far have been isolated from Streptomyces strains. Eubacteria are a minor source for antibiotics, among which the genus Bacillus is the most important reservoir of peptide antibiotics [1]. As defined recently [2], peptide antibiotics have molecular weights of 270 - 4500 and are generally produced as a family of closely related peptides differing from each other by one or more amino acids. Peptide antibiotics may contain other constituents and frequently contain amino acids which are unique and are not found in proteins. Furthermore, most peptide antibiotics are cyclic and are generally resistant to hydrolysis by peptidases or proteases of animal and plant origin.

Enterobacteriaceae are assumed to be insignificant in antibiotic production, but are known to produce proteins such as bacteriocins [3] and cytolytic toxins [4]. A common feature of some high molecular weight toxins and some low molecular weight peptide antibiotics is their mode of action on membrane integrity, permeability and function. Their lipid membrane modifying properties can be easily assayed by measuring the hemolytic activities [5,6].

Structure and Activity of Natural Peptides

In the present paper we report the isolation of two new bacterial peptide antibiotics - the herbicolins A and B - produced by a strain of Erwinia herbicola, a member of the Enterobacteriaceae, and describe some preliminary chemical and biological properties.

Results

1. Fermentation and isolation

The herbicolin producing strain was isolated from lake water during 1977 and was later identified as Erwinia herbicola. This strain can be grown on agar medium containing per liter 10 g malt extract, 4 g yeast extract, 4 g glucose (pH 6). For antibiotic production a chemically defined medium was used containing per litre 7 g K_2HPO_4, 3 g KH_2PO_4, 0,5 g tri-Na citrate · 2 H_2O, 0,1 g $MgSO_4$ · 7 H_2O, 2 g $(NH_4)_2SO_4$ and 10 g glucose autoclaved separately. Submerged cultures grow optimally at 110 rev/min in a rotary shaker at 25° C. 5-liter-culture flasks, containing 2 liter chemically defined medium were inoculated with a 24 h preculture (100 ml) and incubated on a rotary shaker at 22° C. Antibiotic production was accompanied by an intensive foam secretion. The initially white suspension becomes yellowish after a three day incubation period by a simultaneous production of yellow pigments. The herbicolins were generally isolated from the culture centrifugate after 3 days. The culture fluid was separated from bacterial cells by centrifugation in a Sorvall centrifuge (Rotor GSA) at 6000 g for 20 min or by continuous separation (Rotor SZ-14 GK). The culture fluid was then passed through a Servachrome XAD-2 column allowing adsorption of all lipophilic components including the herbicolins. Two volumes of water, followed by one volume of acetone/water (1:1) removed most of the yellow pigments and other slightly lipophilic components.

The herbicolins were desorbed by one volume of purified dry methanol. The obtained methanolic solution contained more than 90 % of the total herbicolins. A further purification was achieved by gel filtration on LH-20. Attempts to isolate the herbicolins from silica gel columns were unsuccessful as decomposition occured. Ion-exchange gels like CM-Sephadex or DEAE-Sephadex in methanol/water (1:1) revealed no further purification. Generally, a 10 liter culture contained 100 - 200 mg herbicolin A, and finally yielded 30 mg of a chromatographically pure product. Crude preparations could also be obtained by TLC, after spraying with water and eluting with methanol.

2. Physico-chemical characterisation

Herbicolin A and B could be detected and isolated in small amounts by TLC on silica gel-60 plates using the following solvent systems: chloroform/methanol/water (65:25:4), Rf (A) = 0.05, Rf (B) = 0.22 and 1-butanol/acetic acid/water (4:1:1), Rf (A) = 0.20, Rf (B) = 0.35. The herbicolins were localized using chlorine/4,4'-tetramethyldiamino-diphenylmethane or by spraying with water. As herbicolin spots were ninhydrin negative and esterification with diazomethane revealed no change of the chromatographic behaviour on TLC, free amino groups and free carboxyl groups are supposed to be absent.

The amino acid content of herbicolin A was analyzed as follows. 0.25 mg of purified herbicolin A was heated with 1 ml 6 n HCl at 110° C for 24 h in a hydrolysis vial. The hydrolysate was evaporated to dryness and dissolved in 0.8 ml citrate buffer (pH 2.2). 100 µl samples were analysed with a Biotronic amino acid analyser (System LC 6000 E) and gave the following amino acid ratio: 2 glycines, 2 threonines, 1 glutamic acid, 1 leucine and 1 arginine. No further ninhydrin positive component was detected in this routine analyzer program.

In order to determine the absolute configuration of the amino acids, the dried hydrolysate was esterified with 200 µl 2 n HCl in isopropanol at 110° C for 1 h. After evaporating with nitrogen, the residual amino acid isopropyl esters were acylated with 50 µl pentafluoropropionic acid anhydride in 250 µl dichloromethane for 1 h at room temperature. Finally the reaction vial was heated to 60° C for 5 min. After evaporating with nitrogen the N-pentafluoropropionylamino acid isopropyl esters were dissolved in 50 µl dichloromethane. The amino acid derivatives were analyzed on a capillary column coated with a chiral phase (N-propionyl-L-valin-tert.-butylamide polysiloxan) [7] using hydrogen as a carrier gas (Carlo-Erba gaschromatograph, model 2101, FID detector, temperature program 87-200° C, 4° C/min). The peaks (Fig. 1) obtained by GLC on the chiral stationary phase were identified by mass spectrometry (Varian, CH-7) and by comparison with authentic samples according to their retention times and mass spectra.

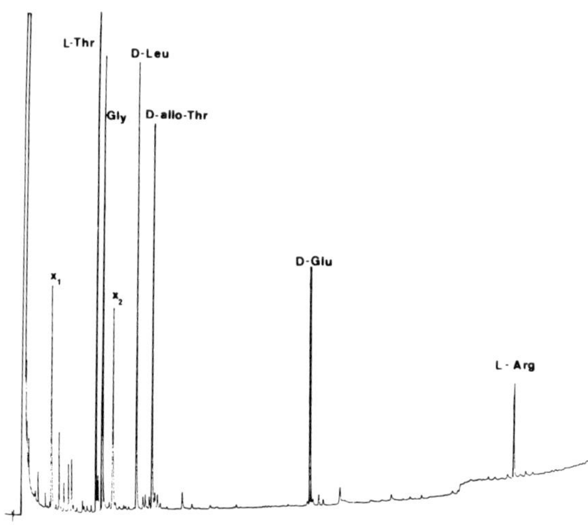

Fig. 1. Gas chromatogram of an esterified and acylated amino acid hydrolysate of herbicolin A on a chiral stationary phase.

Assuming the lowest molar ratio herbicolin A contains the following amino acids: 1 L-threonine, 1 D-allo-threonine, 1 D-glutamic acid, 1 glycine, 1 D-leucine and 1 L-arginine. Thus the peptide moiety of herbicolin A could be a heptapeptide which may be cyclic or linear. As the common sugars could not be detected, the non-peptidic components contain probably no carbohydrates. An aromatic or heteroaromatic residue of herbicolin A could also be excluded, as the UV spectrum (Fig. 2) revealed no absorption band in the region 220-320 nm.

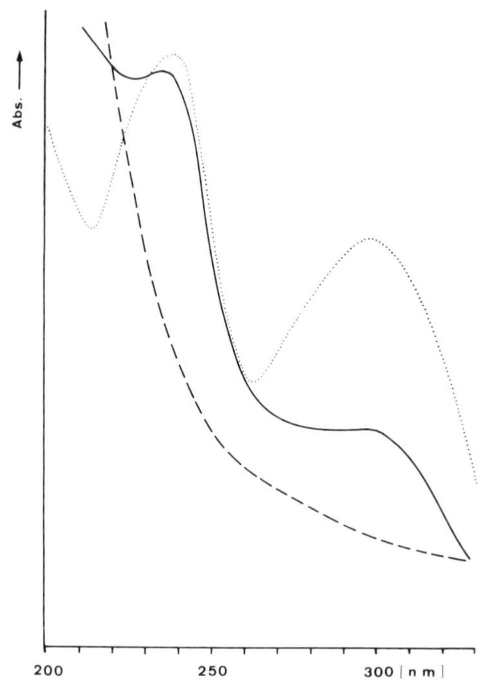

Fig. 2. Ultraviolet absorption spectrum of herbicolin A (---), N^{δ}-(4,6-dimethylpyrimidyl-2)ornithine (.....) and N^{δ}-(4,6-dimethylpyrimidyl-2)ornithyl-herbicolin A in methanol.

Treatment with 1,3-diketones confirmed the presence of the free guanidino group of the arginine residue, as the condensation reaction led to a N^{δ}-(4,6-dimethylpyrimidyl-2)-L-ornithine derivative, which could be visualized in the UV absorption spectrum (λmax = 237 nm and 300 nm).

The circular dichroism spectrum of herbicolin A is shown in Fig. 3. As the exact molecular weight is not yet known, the reduced circular dichroism $\Delta D_{1\ cm}^{1\ \%}$ is given. A solution of herbicolin A in methanol (c = 0.323 mg/ml) was placed in a cell with an optical path of 0.1 mm, measured with a JASCO J-20A CD spectrometer. From the CD spectrum a periodically ordered conformation (α-helix or β-structure) of the peptide portion of herbicolin A could be excluded.

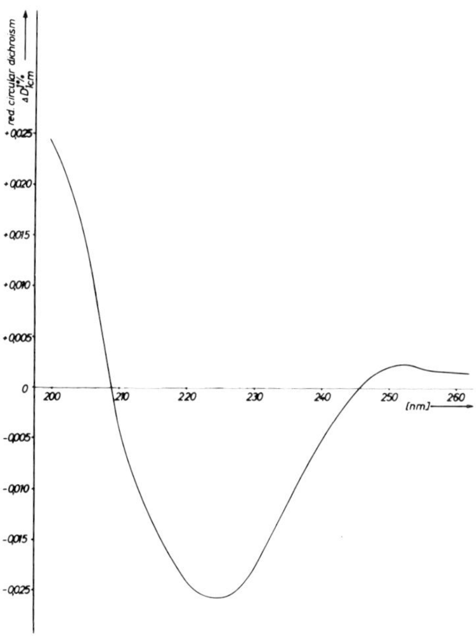

Fig. 3. Circular dichroism spectrum of herbicolin A in methanol solution.

After boiling a sample of herbicolin A with anhydrous methanolic HCl for 60 min and the addition of one volume of water, fatty acid methyl esters could be extracted with hexane. The extract was analyzed by GLC on a packed column coated with OV-1 (temperature program: 100-200° C, 6° C/min).

The main peak (Fig. 4) revealed a retention volume which was identical with a sample of pure D-β-hydroxy myristic acid methyl ester (Applied Science, Inc.). The fatty acid methyl-

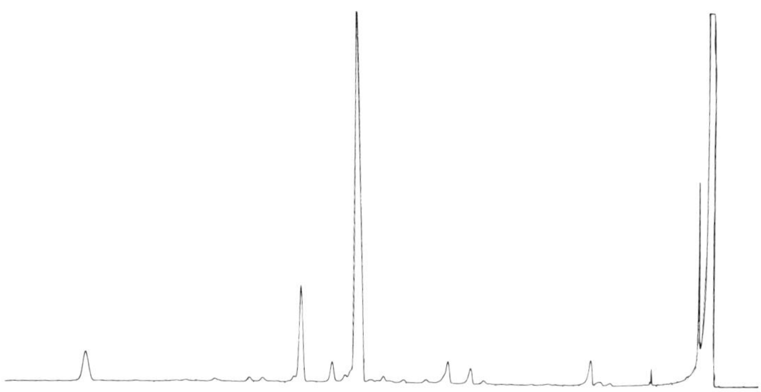

Fig. 4. Gas chromatogram of fatty acid methyl esters from herbicolin A (OV-1, temperature program: 100-200° C, 6° C/min).

esters obtained from herbicolin A were further analyzed by GLC/MS. The mass spectra of an isolated and a reference sample (Fig. 5) showed a common base peak, m/e = 103 attributable to the fragment ion caused by β,γ-fragmentation of β-hydroxy alkanoic acids and a molecular ion peak at m/e = 258. The fragment m/e = 103 was found to be characteristic for β-hydroxy-alkanoic acid methyl esters in a systematic study by Ryhage and Stenhagen [10]. It is assumed, that part of the minor fatty acid residues obtained during transesterification in methanol represent a microheterogeneity. The absolute configuration of the hydroxy alkanoic acid is still unresolved.

Although some important constituents of herbicolin A could be established, others remain uncertain at present. In addition to the 7 amino acids and a β-hydroxy alkanoic acid the GLC/MS measurements indicate further constituents which are to be elucidated.

Fig. 6 shows the ^{13}C NMR spectrum of herbicolin A recorded in $^{12}CD_3OD$. Of the 13 carbonyl signals between 150-180 ppm 10 signals can be assigned to the seven amino acids (including the γ-carboxyl group of glutamic acid and the guanidino carbon of arginine) and to the carboxyl group of the hydroxy alkanoic

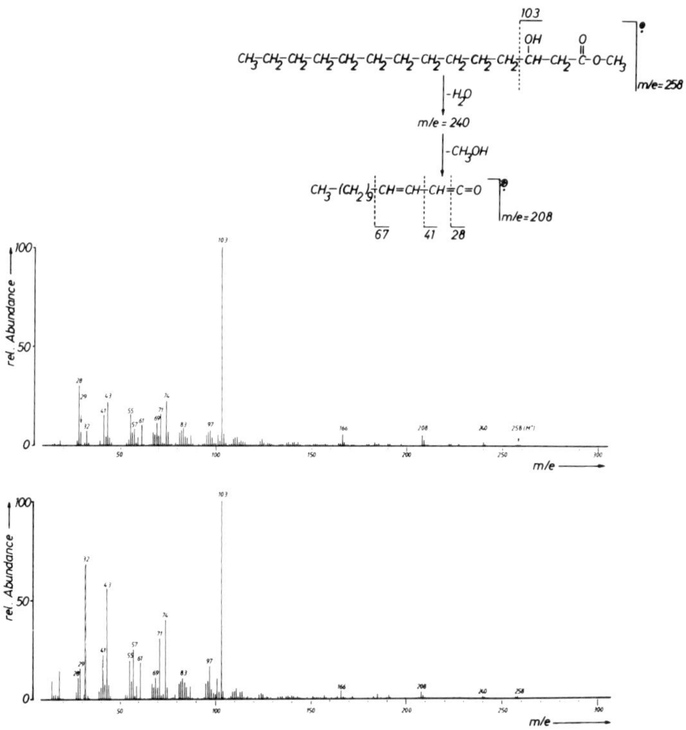

Fig. 5. Mass spectra of β-hydroxy myristic acid methyl esters from herbicolin A (below) and from a reference sample (above).

acid residue. Thus 3 signals of the carbonyl region remain unresolved. In the aliphatic region (10-80 ppm) the ^{13}C NMR signals of the seven amino acids could be assigned by comparison with known data. Integration of the main signal at 30.7 ppm indicates the presence of only one long chain fatty acid. The signals at 97.9, 122.4 and 132.5 suggest the presence of double bonds. The exact assignments of these and the remaining signals will be reported when further structural data are available.

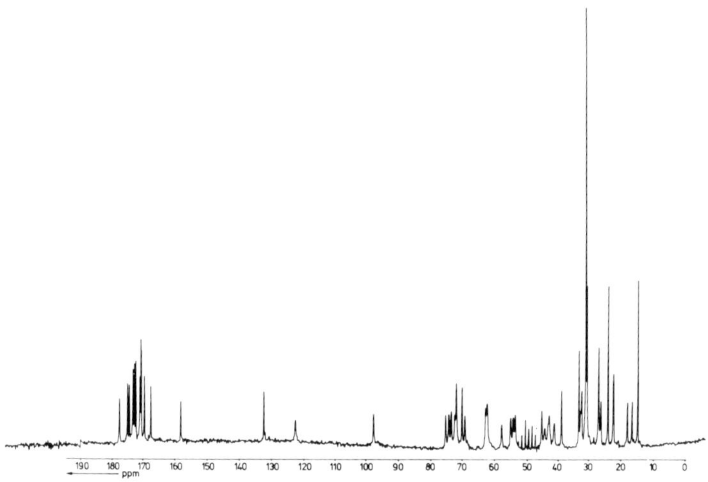

Fig. 6 . ^{13}C NMR spectrum of herbicolin A (c = 50 mg/ml in $^{12}CD_3OD$, 20.1 MHz).

3. Biological properties

The most prominent property of herbicolin A is the pronounced antifungal activity against yeasts and filamentous fungi. The minimal inhibitory concentration (MIC) using the plate diffusion test on YM-agar (4 g yeast extract, 4 g glucose, 10 g malt extract per liter) was 1 - 5 µg per filter disc for

Candia albicans and for a variety of filamentous fungi, such as *Neurospora crassa*, *Aspergillus fumigatus*, *Fusarium dimerum*, *Botrytis cinerea*. An increased sensitvity was found for dermatophytes, such as *Trichophyton rubrum*, *Trichophyton mentagrophytes* or *Epidermophyton floccosum* on Sabouroud medium.

In a serial dilution test using Vogels minimal medium (salts, asparagine and glucose) the MIC of herbicolin A for *Neurospora crassa* was 0.25 µg/ml.
Bacteria seem to be generally insensitive to herbicolins. All the bacteria such as *Bacillus subtilis* (ATCC 6633), *Enterobacter cloacae* (ATCC 13044), *Escherichia coli* (ATCC 25922), *Proteus vulgaris* (ATCC 6380), *Staphylococcus* aureus (ATCC 25923) and even *Mycoplasma laidlawii* were insensitive to herbicolins.

There are several indications that herbicolins act solely on eucaryotic organisms. Microalgae, such as *Chlorella* and *Scenedesmus* were affected at concentrations of 10 - 20 µg/ml. Protozoa, for example, *Paramecium caudatum* and *Nägleria* were even more sensitive to herbicolins and were killed at 1 - 2 µg/ml.
A comparison of herbicolin A with two well-known antifungal agents, amphotericin B and griseofulvin in the plate diffusion test on YM-agar with conidiospores of Neurospora crassa revealed the following inhibition zones: Griseofulvin (none) amphotericin (10 mm) and herbicolin (18 mm).

The inhibition of mycelial growth of Neurospora crassa was antagonized by sterols, such as cholesterol, ergosterol and stigmasterol. Among the phospholipids tested, with phosphatidyl cholin (PC), phosphatidylethanolamine (PE), phosphatidylserine (PS) and diphosphatidylglycerol (DPG), only diphosphatidylglycerol was able to antagonize the antifungal activity of herbicolin A.

Table 1 . Reversal of herbicolin A inhibition by lipids .

Lipid	Ratio herbicolin A/lipid (w/w)								
	9:1	8:2	7:3	6:4	5:5	4:6	3:7	2:8	1:9
Deoxycholate	+	+	+	+	+	+	+	+	+
Cholesterol	+	+	+	+	+	-	-	-	-
Ergosterol	+	+	+	+	+	+	-	-	-
Stigmasterol	+	+	+	+	+	+	+	-	-
PC	+	+	+	+	+	+	+	+	+
PE	+	+	+	+	+	+	+	+	+
PS	+	+	+	+	+	+	+	+	+
DPG	+	+	+	+	+	+	+	+	-

+ Inhibition, - no inhibition (reversal)

It is suggested that herbicolins act on fungal membranes by interference with internal lipid/lipid or lipid/protein interactions. The cytolytic effect of herbicolin A was assayed by measuring the 50 % hemolysis value. Human blood, diluted 1:10 with 3.8 % Na-citrate solution was centrifuged for 1 min at 120 g. The sedimented cells were washed twice with isotonic buffer (10 mM phosphate, 154 mM NaCl, pH 7.2) and finally resuspended in the same buffer to give a 1:10 diluted erythrocyte suspension. 0.5 ml of this suspension was added to 4.5 ml buffer containing 2.5 mM phosphate, 154 mM NaCl, 0.6 mM $K_3Fe(CN)_6$, 1 mM KCN and increasing amounts of herbicolin in methanolic solution. Absorption was read after 15 min (37° C) at 546 nm against a control. Complete hemolysis was determined using the hemolysis assay from Merck, Darmstadt. A typical hemolysis curve is shown in Fig. 7. 50 % hemolysis was observed at 4.3 µg/ml.

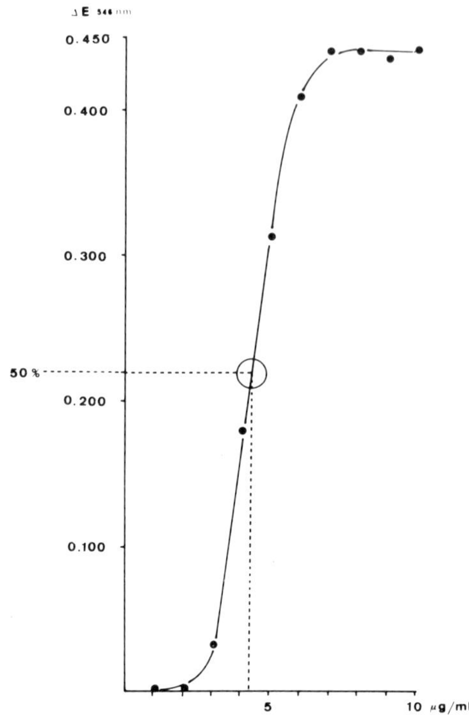

Fig. 7 . Hemolysis of human erythrocytes by herbicolin A .

4. Discussion

Although the complete structure of herbicolin A is unknown at present, there are some characteristic features which should be discussed in connection with properties of other peptide antibiotics.
The occurence of peptide antibiotics in a member of the enterobacteriaceae is surprising, as enterobacteriaceae are assumed to be insignificant in antibiotic production. The only exception reported so far are the serratomolides [10] which are depsipeptides composed of serine and 3-hydroxy-alkanoic acid and which exert antibacterial and antifungal activity.

Depsipeptides containing 3-hydroxy-alkonoic acid and an oligopeptide chain are wide-spread among the genus bacillus, e.g., surfactin [11], esperin [12], EM 49 [13] or cerexin [14]. The fatty acid chain may be different as in polymyxin [15], brevistin [16] or amphomycin [17]. However, viscosin [18], produced by Pseudomonas viscosa, is a further example that such lipopeptide antibiotics are not restricted to the genus bacillus.

Besides these lipopeptide antibiotics, which act primarily on bacteria, there are other lipopeptide antibiotics possessing predominantly antifungal activity, such as mycosubtilin [19, 20], bacillomycin [21] and iturin A [22], all isolated from Bacillus subtilis. As shown for iturin A, the heptapeptide chain is linked to a 3-aminoalkanoic acid of varying chain length [23]. The reason why these antibiotics are inactive against most bacteria but highly active against fungi is unknown at present.
Herbicolin A is a powerful antifungal agent. The antifungal acitivty seems to be correlated with its hemolytic and cytolytic property which probably depends on the ability to penetrate or integrate into the eucaryotic membrane. Despite much information on intrinsic lipid/protein interactions in eucaryotic and procaryotic membranes [24] the behaviour of lipopeptide antibiotics within the membrane is unresolved.

Membrane modifying peptides, which are composed exclusively of amino acids, like alamethicin, trichotoxin or suzukacillin possess antibacterial, antifungal and hemolytic activity [6]. The presence or the formation of a lipophilic helical structure is regarded to promote their integration into the membrane. Thus an additional lipid residue is not required.
On the other hand there are a variety of cytolytic toxins [5], which can disrupt eucaryotic membranes without penetrating the membrane. Because of their high molecular weights (5 000 - 400 000), the mode of action may be completely different from that observed with low molecular membrane active peptides.

The involvement of cholesterol as a target for membrane disruption has been discussed [5].
The antifungal activity of herbicolin A is specifically decreased in the presence of exogenous lipids, indicating a possible involvement of lipids during herbicolin action. Herbicolin differs from most other lipopeptide antibiotics by its arginine content. Glutamic acid, threonine, allo-threonine, leucine and even glycine residues are found in many other lipopeptide antibiotics. However, the presence of the positively charged guanidinium function of arginine in herbicolin is a novel finding, which may have consequences in regard to its membrane modifying properties. The elucidation of the structure using mainly selective cleavage conditions, isolation of fragments and GLC/MS analysis is in progress. Further biological work has to be done to answer the question why herbicolins act solely on eucaryotic membranes.

Acknowledgments

We wish to thank Dr. W. Jung and Dr. M. Busse (Südd. Versuchs- und Forschungsanstalt für Milchwirtschaft, Weihenstephan) and Prof. U. Ullmann (Hygiene Institut, Tübingen) for their help in determining and testing some bacterial strains and Dr. P. Wookey for reading the manuscript.
The assistence of Grame Nicholson and Gerlinde Reichle is gratefully acknowledged. This work was supported by the Deutsche Forschungsgemeinschaft (SFB 76).

References

1. Hopwood, D.A., Merrick, M.J.: Bacteriol. Rev. 41, 595-635 (1977).
2. Katz, E., Demain, A.L.: Bacteriol. Rev. 41, 449-474 (1977).
3. Reeves, P.: The Bacteriocins. Molecular Biology, Biochemistry and Biophysics, Vol. 11, Springer-Verlag, New York 1972
4. Bernheimer, A.W.: Biochim. Biophys. Acta 344, 27-50 (1974).
5. Alouf, J.E.: The Specificity and Action of Animal, Bacterial and Plant Toxins, in: Receptors and Recognition, Series B, Vol. 1 (P. Cuatrecasas, ed.) pp. 219-270, Chapman and Hall, London 1977
6. Irmscher, G., Jung, G.: Eur. J. Biochem. 80, 165-174 (1977).
7. Frank, H., Nicholson, G.J., Bayer, E.: J. Chromatogr. Sci. 15, 174-176 (1977).
8. Tjoeng, F.-S., Kraas, E., Stark, E., Breitmaier, E., Jung, G.: Chem. Ber. 108, 862-874 (1975).
9. Ryhage, R., Stenhagen, E.: Arkiv för Kemi 15, 545-574 (1960).
10. Wasserman, H.H., Keggi, J.J., McKeon, J.E.: J. Amer. Chem. Soc. 84, 2978-2982 (1962).
11. Arima, K., Kakinuma, A., Tamura, G.: Biochem. Biophys. Res. Comm. 31, 488-494 (1968).
12. Ovchinnikow, Yu.A., Ivanov, V.T., Kostetsky, P.V., Shemyakin, M.M.: Tetrahedron Lett. 1966, 5285-5290 (1966).
13. Parker, W.L., Rathnum, M.L.: J. Antibiot. 26, 449-466 (1973).
14. Shoji, J., Kato, T., Sakazaki, R.: J. Antibiot. 29, 1268-1274 (1976).
15. Hayashi, K., Seketa, Y., Tsukamoto, K., Suzuki, T.: Experientia 22, 354-355 (1966).
16. Shoji, J., Kato, T.: J. Antibiot. 29, 380-389 (1976).
17. Bodanszky, M., Sigler, G.F., Bodanszky, A.: J. Amer. Chem. Soc. 95, 2352-2357 (1973).
18. Hiramoto, M., Okada, K., Nagai, S.: Tetrahedron Lett. 1970, 1087-1090 (1970).
19. Walton, R.B., Woodruff, H.B.: J. Clin. Invest. 28, 924-426 (1949).

20. Peypoux, F., Michel, G., Delcambe, L.: Eur. J. Biochem. 63, 391-398 (1976).

21. Landy, M., Warren, G.H., Roseman, S.B., Colio, L.G.: Proc. Soc. Exp. Biol. and Med. 67, 539-541 (1948).

22. Besson, F., Peypoux, F., Michel, G., Delcambe, L.: J. Antibiot. 29, 1043-1049 (1976).

23. Peypoux, F., Guinand, M., Michel, G., Delcambe, L., Das, B.C., Lederer, E.: Biochemistry 17, 3992-3996 (1978).

24. Chapman, D., Gómez-Fernández, J.C., Goni, F.M.: FEBS Lett. 98, 211-223 (1979).

III. METHODS OF PURIFICATION, ISOLATION AND CHARACTERIZATION OF PEPTIDES

SIMULTANEOUS DETERMINATION OF THE AMINO ACID, AMINO SUGAR AND NEUTRAL SUGAR CONTENT IN GLYCOPROTEINS

Hermann Bauer, Wolfgang Voelter

Institut für Organische Chemie der Universität Tübingen, Auf der Morgenstelle 18, D-7400 Tübingen, FRG

Mikhael M. Tikhomirov

U.S.S.R Academy of Sciences, Shemyakin Institute of Bioorganic Chemistry, Vavilova 32, 117312 Moscow, USSR

Introduction

Since years liquid chromatographic methods with sufficient sensitivity and resolution for the qualitative and quantitative analysis of amino acids (1,2) and sugars (3) are known. The main disadvantage of these chromatographic procedures is the necessity of two different separations (one for the determination of amino acids and amino sugars and another for neutral sugars) and a double waste of material.

First, ion-exchange resins for the separation of these two classes of substances are used. The further separation into components is performed on two different columns. Thus, neutral sugars and basic compounds are separated immediately after injection with the aid of a small pre-column which is filled with a strongly acidic, protonated ion-exchange resin. Basic compounds are retarded and neutral sugars pass without interaction and thus can be transferred to the carbohydrate column. Amino acids are thereafter eluted with sodium citrate buffers and brought on to the amino acid column. The successive analysis of the different compounds is controlled with a high-pressure stable motor valve.

Experimental

A Biotronik liquid chromatograph, LC 6000 E, was modified for these investigations. The chromatographic system is shown schematically in Fig. 1.

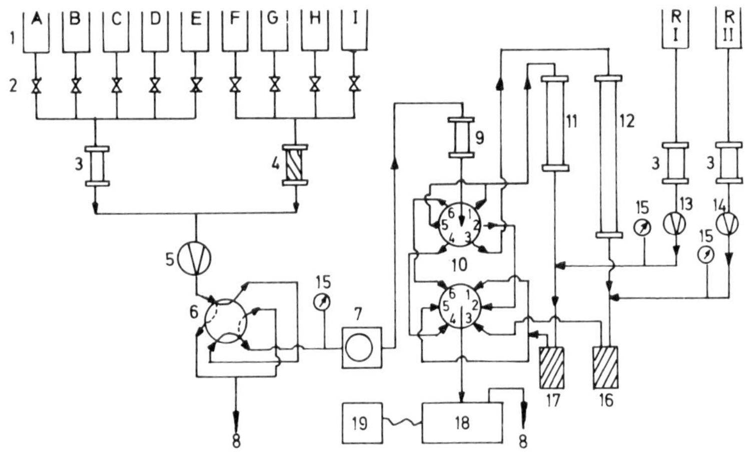

Fig.1. Schematic flow sheme of the combined amino acid-sugar analyser. 1: Eluent reservoirs, 2: Electric valves, 3: Bubble trap, 4: Ammonia filter, 5: Milton-Roy Dosapro mini-pump, 6: Manual valve, 7: Sample injection, 8: Out-flow, 9: Pre-column, 10: Motor valve, 11: DA-X-8-8 column for sugars, 12: DC-8-9 column for amino acids and amino sugars, 13: Modified acid-resistant Milton-Roy mini-pump (orcinol-H_2SO_4), 14: Ninhydrin pump, 15: Pressure gauge, 16: 30 m x 0.3 mm PTFE reaction coil (100°), 17: 20 m x 0.5 mm PTFE reaction coil (100°), 18: Two-channel photometer (420 and 570 nm); 19: Recorder.

All connections are made of 0.3 x 2 mm PTFE tubes and are resistant up to 120 bar. Neutral sugars and amino acids are separated in a small pre-column (5 mm and I.D. 4 mm). Successive chromatographic analyses are performed with a 250 x 4 mm column for amino acids and amino sugars and a 160 x

4 mm column for neutral sugars. All glass columns used are high-pressure stable tensile discharged (maximum pressure 120 bar), glass jacketed and connected with a program-controlled constant-temperature water circulator. A commercially available automated sampler was modified and is used for control of flow for the successive analysis. For development of the colour reactions adjustable heating baths, equipped with 30 m x 0.3 mm (for amino acids) and 20 m x 0.5 mm (for sugars) PTFE coils are used. Detection was carried out with a two-channel filter photometer at 420 and 570 nm and a cell path length of 16 mm. Control of time-dependant processes is carried out with a micro processor (15 time stores, 20 functions per store).

Resins: As ammonia filter a Dowex 50 W-X 8 resin (50-100 mesh) is used. The precolumn and the amino acid column are filled with DC-8-9 resin (8 % cross-linked sulphonated polystyrene, particle size 8 ± 2 µm, Durrum, Palo Alto, Calif., USA.). The sugars are separated on DA-X-8-8 resin (8 % cross-linked basic polystyrene, particle size 8 ± 2 µm, Durrum).

Chemicals and reagents: Sugars and amino acids of the highest available purity are obtained from Merck (Darmstadt, G.F.R.). Boric acid, sulphuric acid, orcinol, sodium citrate, sodium chloride, hydrochloric acid, cellosolve and ninhydrin are of analytical reagent grade(Merck). The buffers are prepared by dissolving boric acid, sodium citrate and sodium chloride in doubly distilled and deionized water and adjusting the pH to the desired value with concentrated sodium hydroxide solution or hydrochloric acid. For the regeneration of the sugar column 10 % potassium tetraborate solution, for the amino acid column 0.5 N sodium hydroxide solution and for the pre-column 0.5 N hydrochloric acid are used. As reagents for the colour reactions orcinol (0.1 %) in concentrated sulphuric acid and the Stein and Moore reagent are applied.

Procedure: All important data for the separation procedure are given in the table. The injected sample is brought onto the precolumn and neutral sugars are eluted with water and transferred to the DA-X-8-8 column in borate form, whereas basic components are retarded on the protonated DC-8-9 resin. Within 5 min all neutral carbohydrates are eluted and thereafter all connections from the buffer valves to the precolumn are rinsed with starting buffer for the amino acid separation. A period of about 10 min is sufficient for rinsing and converting the pre-column into the sodium form without releasing the amino acids. This step is necessary in order to avoid disturbing the equilibrium conditions of the DC-8-9-column. After changing the flow to the DC-8-9 column amino acids and amino sugars are separable with high resolution and sensitivity. A further rinsing step with the starting buffer for elution of neutral sugars is applied before changing the flow to the DA-X-8-8 column. After elution of the neutral sugars the precolumn is converted into the protonated form with 0.5 N hydrochloric acid and equilibrated with water. Thereafter the whole system is ready for the next sample injection.

Results and discussion

The chromatograms which are performed with the combined amino acid-carbohydrate analyzer demonstrate clearly the advantage of the simultaneous determination of carbohydrates and amino acids. Sensitivity and resolution of this combined system are equal to those of separate analyses procedures for sugars and amino acids. Thus, only half amount of sample is necessary for the analysis which especially is of importance if only small amounts of isolated substances are available. As all procedures are performed automatically with high reproducibility the direct quantitative comparison of sugars and amino acids becomes possible without additional internal standard.

Table. Data of separation procedure.

Step	Flow way	Eluent	Position of motor valve	Temperature of columns	Reagent	Time
1	To DA-X-8-8 column	Water	1	45°C	-	5
2	To waste	0.2 N, pH 3.25 +)	2	45°C	-	4
3	To waste	0.2 N, pH 3.25 +)	2	45°C	-	6
4	To DC-8-9 column	0.2 N, pH 3.25 +)	3	45°C	Ninhydrin	30
5	To DC-8-9 column	0.2 N, pH 4.5 +)	3	59°C	Ninhydrin	45
6	To DC-8-9 column	1.6 N, pH 4.5 +)	3	68°C	Ninhydrin	45
7	To DC-8-9 column	0.5 N sodium hydroxide	3	68°C	-	30
8	To DC-8-9 column	0.2 N, pH 3.25	3	45°C	-	30
9	To waste	0.2 m borate buffer, pH 8.3	4	45°C	-	5
10	To waste	0.2 m borate buffer, pH 8.3	4	45°C	-	10
11	To DA-X-8-8 column	0.2 m borate buffer, pH 8.3	5	45°C	Orcinol/ H_2SO_4	60
12	To DA-X-8-8 column	0.4 m borate buffer, pH 9.4	5	68°C	Orcinol/ H_2SO_4	135
13	To DA-X-8-8 column	0.2 m borate buffer, pH 8.3	5	45°C	-	30
14	To waste	0.5 N hydrochloric acid	6	45°C	-	15
15	To waste	water	6	45°C	-	30

+) sodium citrate buffer

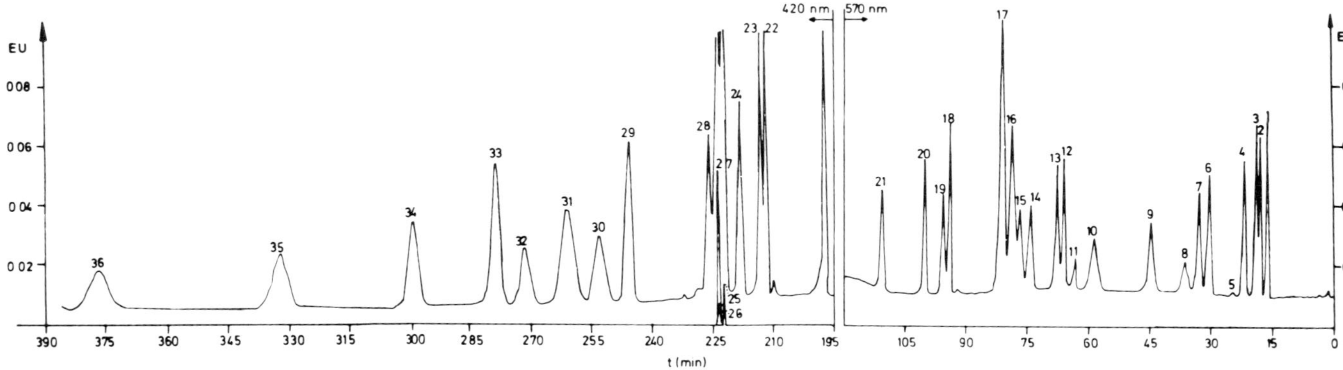

Fig.2. Chromatogram of a mixture of amino acids, amino sugars and neutral sugars. Pre-separation column, 5 x 4 mm, DC-8-9 resin; amino acid and amino sugar column, 250 x 4mm, DC-8-9 resin; neutral sugar column, 160 x 4 mm, DA-X-8-8 resin; flow-rate,0.24 ml/min; pressure 80-110 bar; recorder range, 0-0.2 A (ninhydrin, 570 nm) or 0-0.1 A (orcinol-sulphuric acid, 420 nm); for further separation conditions, see table. Peaks: 1=aspartic acid; 2=threonine; 3=serine; 4=glutamic acid; 5=proline; 6=glycine; 7=alanine, 8=cystine; 9=valine; 10=methionine; 11=unknown; 12=isoleucine; 13=leucine; 14=tyrosine; 15= phenylalanine; 16=galactosamine; 17=glucosamine; 18=lysine; 19=ammonia; 20=histidine; 21=arginine; 22=sucrose; 23=trehalose; 24=cellobiose; 25=maltose; 26=rhamnose; 27=lactose; 28=ribose; 29=mannose; 30=fructose; 31=arabinose; 32=galactose; 33=xylose; 34= glucose; 35=gentiobiose; 36=melibiose; 37=fucose.
Each peak represents 2.05 nmole per amino acid and 4.40 nmole per amino sugar and neutral sugar.

In Fig. 2 a chromatogram of a mixture of amino acids, amino sugars and neutral sugars (each peak represents 2.05 nmole per amino acid and 4.40 nmole per sugar)and in Fig. 3 a chromatogram of 7.8 µg of a hydrolysate of ovalbumin which was optimized especially for the elucidation of the sugar part in glycoproteins are shown. The calculated amino acid, amino sugar and neutral sugar contents are in good accordance with the literature (1). For sample preparation of glycoproteins for the simultaneous determination of the amino acid and sugar content a new hydrolysis method had to be introduced: A hydrolysis of 24 hours in 6 N trifluoroacetic acid at 95°C proved to be suitable for our purpose. Contrary to the usually applied method with 6 N hydrochloric acid, which destroys all sugars completely, the degradation of the amino sugars and neutral sugars is less than 20 %. The yield of amino acids is nearly the same (± 5 %).

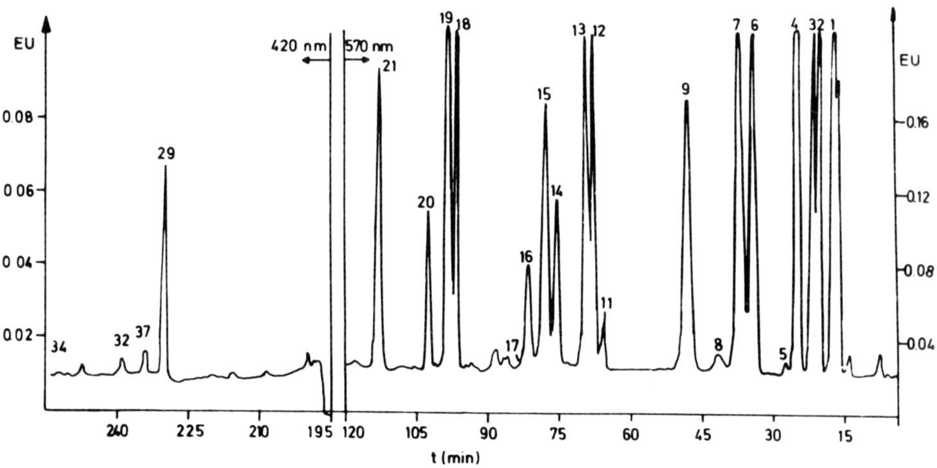

Fig. 3. Chromatogram of a hydrolysate of ovalbumin. Sugar column: 65 x 4 mm; neutral sugars are eluted isocratically (0.5 M boric acid, pH 8.5); for further separation conditions see fig. 2 and the table; amount injected: 7.8 µg.

References

1. Gottschalk, A.: Glycoproteins, Second revised and expanded edition, Elsevier Publishing Company, Amsterdam-London-New York 1972

2. Spackman, D.H., Stein, W.H., Moore, S.: Anal.Chem.30, 1190 (1958).

3. Bauer, H., Voelter, W.: Chromatographia 9, 433 (1976).

DIFFERENTIAL EVOLUTION OF THE α- AND β-CHAINS OF BIRD HEMOGLOBINS. INVESTIGATIONS ON THE GREYLAG GOOSE (*Anser anser*)+

Gerhard Braunitzer , Walter Oberthür

Max-Planck-Institut für Biochemie, Abteilung Proteinchemie
D-8033 Martinsried bei München, Germany

Introduction

For the determination of the primary structure of hemoglobin of the greylag goose (*Anser anser*) the chains were cleaved and the cleavage products isolated. The chains and peptides were degraded automatically in the sequenator: the complete primary structure of the α- and β-chains was obtained and compared with the hemoglobin of the chicken.

Experiments

After hemolysis of erythrocytes and removal of nuclei and membranes the hemoglobin solution was obtained. The globin was adsorbed on Servacel CM 52 (column: 2.5 X 15 cm up to 400 mg, 5.0 X 15 cm up to 4 g) and eluted with a pH and salt gradient between pH 5.0 to 5.8 in 8 M urea (1). The chains were isolated in yields of 60 - 80 %.

Abbreviations:

Quadrol = N,N,N',N'-Tetrakis(2-hydroxypropyl)ethylenediamine, I = sodium 4-(isothiocyanato)benzenesulfonate, IV = trisodium 7-(isothiocyanato)naphthalene-1,3,5-trisulfonate.

+ Source: Dissertation W. Oberthür, Fachbereich Chemie, Universität Tübingen, in preparation.

Enzymatic digestion, chemical cleavages and peptide separation

1) Cyanogen bromide cleavage (2) in 70 % formic acid.
2a) Tryptic digestion: The native or oxidized (with peroxi-formic acid) globin was used. The peptides, first, were fractionated by gel filtration on Sephadex G-25 in 0.1 M acetic acid. The material from the single peaks was subsequently chromatographed on Dowex 50X2 and sometimes further purified with paper electrophoresis.
2b) The chains, modified with iodoacetamide, were blocked with maleic acid anhydride (3), and after digestion (40 min, 37° C) the peptides were isolated by chromatography on Sephadex G-50 (6 M guanidine, 0.1 N Tris/HCl, pH 8.5).
3) Asp-Pro cleavage: Native chains were treated for 60 h (42° C) with 6 M guanidine/70 % formic acid (4) and the two peptides were separated by gel filtration with Sephadex G-50 (10 % formic acid, 8 M urea). The smaller peptide (α95Pro→ α141Arg) was oxidized and treated 30 h at 37° C with protease from Staphylococcus aureus, V8 (5).

Amino acid analysis
The analyses were performed with a Beckman Model 121 C amino acid analyzer (Palo Alto, Calif., USA).

Sequence determination
Degradation was done automatically in the sequenator. Two programmes were applied:
a) A quadrol programme was used for sequencing the intact polypeptide chain, large peptides and some lysyl peptides which had been reacted with reagent I or IV (6,7).
b) An N,N-diethylaminopropyne programme was used for arginyl peptides. Conversion was done in 3 M trifluoroacetic acid 14 min, 80° C.

The phenylthiohydantoin derivatives of the amino acids were identified by thinlayer chromatography and high-performance liquid chromatography (8) (with gradients).

Results and discussion

The sequences of the α- and β-chains, the main components of greylag goose hemoglobin, are given in illustration 1 and 2.

When the sequences from chicken (9, 10) and greylag goose (11) are compared, 30 positions in the α-chain are different but surprisingly only 6 positions in the β-chain. In the α-chain mutations are clustered at the end of the E-helix (from E13-E20 6 exchanges) and from the end of the G-helix to the beginning of the H-helix (α109-α120 8 exchanges). The last 20 residues of the C-terminus are identical and the remaining 16 exchanges are distributed randomly throughout the α-chains. Neither of the α-chains contain tryptophan. The exchanges in the β-chains Thr→Ser in β4, Glu→Asp in β22, Ala→Ser in β43, Ser→Ala in β86 and β119 as well as Asp→Glu in β121 are neutral. The C-terminal is identical from β122 onwards.

Sequences and palaeontology

An estimate of the rate of biological evolution can be made by the use of sequence comparisons (12) in conjunction with palaeontological data to provide a suitable time scale (13). In hemoglobins a mutation rate of 6 - 10 million years per substitution has been estimated. Whether this rate of evolution remains constant has often been discussed and particularly which parameters could affect this rate. From the data of primate hemoglobins it has been established that in the last 60 million years the rate of mutation has decreased by approximately 30 % i.e. evolution has become slower (14). Furthermore it is already known that in hemoglobins the mutation rate for the β-chains is usually higher than that for the α-chains.

However, by a comparison between the primary structures of chicken and goose hemoglobin we have established that the mutation rate in the β-chains is only 1/5 of that of the

```
        A______________________________________________________________ AB  B______________________
            Asn             Asn                 Ile     Thr         Ala
Val-Leu-Ser-Ala-Ala-Asp-Lys-Thr-Asn-Val-Lys-Gly-Val-Phe-Ser-Lys-Ile-Gly-Gly-His-Ala-Glu-Glu-Tyr-Gly-
                                     10                                      20

_________________________________________C __________________________CD
                                 Ile-Gly-Phe     Thr                                         Ser
-Ala-Glu-Thr-Leu-Glu-Arg-Met-Phe-Thr-Ala-Tyr-Pro-Gln-Thr-Lys-Thr-Tyr-Phe-Pro-His-Phe-Asp-Leu-Gln-His-
                  30                                      40                                      50

     E______________________________________________________________________________ EF
                         Gly E7                      Leu     Ile-Thr-Asn     Ile-Glu
-Gly-Ser-Ala-Gln-Ile-Lys-Ala-His-Gly-Lys-Lys-Val-Ala-Ala-Ala-Leu-Val-Glu-Ala-Val-Asn-His-Ile-Asp-Asp-
                                      60                                      70

                 F__________________________________ FG                  G__________________________
     Ser                                     F8      His                                         Leu
-Ile-Ala-Gly-Ala-Leu-Ser-Lys-Leu-Ser-Asp-Leu-His-Ala-Gln-Lys-Leu-Arg-Val-Asp-Pro-Val-Asn-Phe-Lys-Phe-
                  80                                      90                                     100

________________________________________________ GH                  H______________________________
         Gln                     Leu-Val-Ala     Leu     Ala-Glu     Ala     Lys
-Leu-Gly-His-Cys-Phe-Leu-Val-Val-Val-Ala-Ile-His-His-Pro-Ser-Ala-Leu-Thr-Pro-Glu-Val-His-Ala-Ser-Leu-
                                     110                                     120

____________________________________________________

-Asp-Lys-Phe-Leu-Cys-Ala-Val-Gly-Thr-Val-Leu-Thr-Ala-Lys-Tyr-Arg
                 130                                     140
```

Fig. 1. Sequence of greylag goose hemoglobin α-chains. Where the chicken and greylag goose sequences differ the chicken amino acids appear in the top line (18). A, B, C, ... helical regions etc.

```
    A --------------------------------------------------- B ------------------
    Thr                                                           Glu
Val-His-Trp-Ser-Ala-Glu-Glu-Lys-Gln-Leu-Ile-Thr-Gly-Leu-Trp-Gly-Lys-Val-Asn-Val-Ala-Asp-Cys-Gly-Ala-
                                     10                                      20

---------------------------------------------------- CD                       D --
                                                         Ala
-Glu-Ala-Leu-Ala-Arg-Leu-Leu-Ile-Val-Tyr-Pro-Trp-Thr-Gln-Arg-Phe-Phe-Ser-Ser-Phe-Gly-Asn-Leu-Ser-Ser-
                 30                                      40                                      50

------------------------- E --------------------------------------------------------------------
                                      E7
-Pro-Thr-Ala-Ile-Leu-Gly-Asn-Pro-Met-Val-Arg-Ala-His-Gly-Lys-Lys-Val-Leu-Thr-Ser-Phe-Gly-Asp-Ala-Val-
                                     60                                      70

---- EF                         F ------------------------------ FG                   G ------
                                     Ser                  F8
-Lys-Asn-Leu-Asp-Asn-Ile-Lys-Asn-Thr-Phe-Ala-Gln-Leu-Ser-Glu-Leu-His-Cys-Asp-Lys-Leu-His-Val-Asp-Pro-
                 80                                      90                                      100

------------------------------------------------------------- GH               H -----------
                                                                     Ser     Asp
-Glu-Asn-Phe-Arg-Leu-Leu-Gly-Asp-Ile-Leu-Ile-Ile-Val-Leu-Ala-Ala-His-Phe-Ala-Lys-Glu-Phe-Thr-Pro-Glu-
                                     110                                     120

------------------------------------------------------------------ HC

-Cys-Gln-Ala-Ala-Trp-Gln-Lys-Leu-Val-Arg-Val-Val-Ala-His-Ala-Leu-Ala-Arg-Lys-Tyr-His
                 130                                     140
```

Fig. 2. Sequence of greylag goose hemoglobin β-chains. Where the chicken and greylag goose sequences differ the chicken amino acids appear in the top line (18). A, B, C, ... helical regions etc.

α-chains. How can this surprising result be interpreted?

Because fossil finds from birds (Archeopteryx) are extremely rare, palaeontology can tell us little about the bird ancestral tree. Today, the common ancestor for goose and chicken is assumed to have arisen between 80 and 120 million years ago. A mutation rate of 6 million years per substitution would indicate a normal mutation rate for the α-chains (2 X 15 substitutions ≈ 90 million years) but a considerably reduced and thus abnormal mutation rate for the β-chains.

Differences in mutation rate by comparison of different genes? By studies of abnormal human hemoglobins it has been found that two identical α genes code for the α-chains whereas only one gene exists for the β-chains (15, 16, 17).

From the two chicken hemoglobin types, AI and AII it is known that the tetrameric molecule contains two different α-chains but that the β-chains are identical. By comparing the main components of goose and chicken (AII) a comparison can be made between chains that are products of different genes.

The results from two other bird hemoglobins can only be explained by postulating different mutation rates for the α- and β-chains. It appears clear that in the birds in spite of sequence homologies, almost identical tertiary structure and similar functions of the peptide chains that selection is operating singificantly different on the α- and β-chains. This phenomenon can now be explained by recent experimental results.

Phosphate binding sites

The affinity of hemoglobin for oxygen is regulated by phosphates. In the birds myoinositol-1,3,4,5,6-pentaphosphate (18, 19) is found to be the heterotropic, allosteric effector. The binding sites in the chicken can be established using X-ray data from human hemoglobin and comparison of the β-chain sequences (20). As can be seen from the sequence and the table, the binding sites in chicken and goose are formed from the same amino acid side chains.

In all birds so far investigated (except for one case of a slight functional modification) no substitutions have been found in these positions. The larger number of 12 phosphate contacts in the birds, compared to 6 contacts in mammals, leads to a reduction in the degrees of freedom and therefore a more conserved structure of the β-chains, whereas the α-chains, which contain no binding sites, are free to evolve at the normal rate.

Position in		chicken	greylag goose
helix	β-chains		
NA1	1	Val	Val
NA2	2	His	His
EF6	82	Lys	Lys
H13	135	Arg	Arg
H17	139	His	His
H21	143	Arg	Arg

Table. Phosphate contacts of the non-α-chains.

Summary

We have sequenced the hemoglobins from the greylag goose (Anser anser). Comparison with chicken sequences shows 30 substitutions in the α-chain, but only 6 in the β-chain. The question arises that in birds in general the β-chain has a lower mutation rate. We have confirmed this surprising result by the sequence analysis of another two birds. The possibility that two different genes have been compared has been ruled out. We are of the opinion that the reduced substitution rate in the β-chain indicates a much higher selection pressure. This could be a consequence of the interaction of the β-chains with the heterotropic allosteric effector myoinositol-1,3,4,5,6-pentaphosphate, which has 12 contacts in all birds so far investigated.

We thank Mrs. Dr. Sibylle Kallas, Institut für Verhaltensforschung und Tiersoziologie der Österreichischen Akademie der Wissenschaften, Grünau (Österreich) for first samples of blood of the greylag goose. Dr. J. Lamprecht, Max-Planck-Institut für Verhaltensphysiologie, Seewiesen (Deutschland) was so kind as to make us a gift of two greylag geese.

References

1. Clegg, J.B., Naughton, M.A. and Weatherall, D.J.: J. Mol. Biol. 19, 91-108 (1966).
2. Gross, E. and Witkop, B.: J. Am. Chem. Soc. 83, 1510-1511 (1961).
3. Butler, P.J.G., Harris, J.I., Hartley, B.S. and Leberman, R.: Biochem. J. 103, 78P-79P (1967).
4. Jauregui-Adell, J. and Marti, J.: Anal. Biochem. 69, 468-473 (1975).

5. Houmard, J. and Drapeau, G.R.: Proc. Natl. Acad. Sci. 69, 3506-3509 (1972).

6. Braunitzer, G., Schrank, B. and Ruhfus, A.: Hoppe-Seyler's Z. Physiol. Chem. 351, 1589-1590 (1970).

7. Braunitzer, G., Schrank, B., Ruhfus, A., Petersen, S. and Petersen, U.: Hoppe-Seyler's Z. Physiol. Chem. 352, 1730-1732 (1971).

8. Zimmermann, C.L. and Pisano, J.: Methods Enzymol. 47, 45-51 (1977).

9. Matsuda, G., Takei, H., Wu, K.C. and Shiozawa, T.: J. Protein Res. 3, 173-174 (1971).

10. Matsuda, G., Maita, T., Mizuno, K. and Ota, H.: Nature New Biol. 244, 244 (1973).

11. Braunitzer, G. and Oberthür, W.: Hoppe-Seyler's Z. Physiol. Chem. 360, 679-683 (1979).

12. Matsuda, G., Gehring-Müller, R., Braunitzer, G.: Biochem. Z. 338, 669 (1963).

13. Zuckerkandl, E. and Pauling, L.: Horizons in Biochemistry, Acad. Press, New York 1962

14. Goodman, M., Barnabas, J., Matsuda, G. and Moore, G.W.: Nature 233, 604-613 (1962).

15. Bunn, H.F., Forget, B.G. and Ranney, H.M.: Human Hemoglobins, W.B. Saunders Comp., Philadelphia, London, Toronto 1977

16. Christomanos, A.A.: Molekulare Biologie der Hämoglobine, Bd. I., Scientific Press, Greg. Parisianos, Athens 1972

17. Lehmann, H., Huntsman, R.G.: Man's Hemoglobins, North-Holland Publ. Comp., Amsterdam, Oxford 1974

18. Braunitzer, G., Gehring-Müller, R., Hilschmann, N., Hilse, K., Hobom, G., Rudloff, V. and Wittmann-Liebold, B.: Hoppe-Seyler's Z. Physiol. Chem. 325, 283-286 (1961).

PURIFICATION OF FLUORESCEIN-LABELED CHYMOTRYPSIN BY ISOELECTRIC FOCUSING AFTER BIOSPECIFIC ADSORPTION TO SOYBEAN TRYPSIN INHIBITOR

Ingeborg Büchtmann, Volker Kasche
Fachbereich Biologie, Universität Bremen, D-2800 Bremen-33, FRG

Introduction

Proteins labeled with radioactive nuclides, fluorochromes or enzymes are used extensively for analytical purposes (radio-, fluoro-, or enzyme-immunoassays) (1-3). They are also used to localize proteins in cells and receptors and their movements in the cell membrane (4,5). In analytical and quantitative studies nonhomogeneous preparations of these labeled molecules are normally used. They contain molecules with a different number of labels/molecule (1-4).

From studies on protein modification we may infer that the biological activity of the labeled molecule depends on the number of labels/molecule (6). Therefore, the data obtained in studies where these labeled molecules are used depend on the distribution of the number of labels/molecule that normally is not determined. In quantitative studies with labeled molecules it would be valuable to use homogeneous preparations of labeled molecules. Affinity chromatography cannot be used here as it does not separate labeled and non-labeled protein with the same biological activity (7). As most labels change the isoelectric point of the protein electrophoresis or isoelectric focusing could be used to prepare fractions with different label/protein ratios. This procedure, however, gives no direct information on the biological activity of the different labeled protein fractions.

Structure and Activity of Natural Peptides

We have recently published a method how this information can be directly obtained by using a biospecific sample application in the isoelectric focusing procedure (8). Then only the labeled protein molecules that retain their biological activity are separated. This procedure is used here to obtain information on optimal labeling conditions, distributions of labels/molecule and the biological activity of homogeneously labeled protein fractions. The labeling of the protease α-chymotrypsin with the fluorochrome fluorescein was used as a model system.

Materials

Bovine α-Chymotrypsin* E.C. 3.4.4.5 (Worthington Biochem. Corp. CDI), soybean trypsin inhibitor (Worthington Biochem. Corp. SI), fluoresceinisothiocyanate, isomer I (Sigma Chem. Corp.), fluorescein (Riedel - De Haen AG), N-acetyl-L-tyrosine ethylester (Serva, Feinbiochemica), Ampholine PAG plate (pH 3,5 - 9,5) (LKB-Products), Coomassie brilliant blue R 250 (Serva, Feinbiochemica), Ampholine (pH 5-8, 5-7, 7-9) (LKB-Products), Sephadex G-75 superfine (Pharmacia Fine Chemicals), Biogel P-6, 100-200 mesh, wet (BIO-RAD Laboratories) were used as purchased. All other chemicals were of reagent grade. Quartz-distilled water was used throughout.

---

* Abbreviations: Bovine α-chymotrypsin = CT, soybean trypsin inhibitor = STI, fluoresceinisothiocyanate = FITC
fluoresceinthiocarbamyl-chymotrypsin = CT-FTC

Methods

Labeling procedure

Bovine α-chymotrypsin (x moles) was labeled with nx moles of fluoresceinisothiocyanate isomer I in NaCl-phosphate buffer (0.145 M NaCl; 0.0075 M Na_2HPO_4; 0.0025 M NaH_2PO_4; adjusted to the desired pH). The reaction was carried out in absence of light and at a constant pH. The incubation solution was slowly stirred. Different modes of addition of FITC to the protein were used (all at once or continuously over the whole time of reaction). The reaction was stopped by rapidly freezing the solution.

Biospecific adsorption of the labeled protein to STI sepharose gel

STI sepharose gel (capacity 1,1 mg CT/ml wet gel) was equilibrated with buffer of pH 8.2 (0.2 M NaCl; 0,103 M TRIS; 0.05 M HCl). 0.1 ml of the STI gel was added to 1 ml of the sample solution in a test tube and shaken at room temperature over 5 min. After filtering on a nylon net, the gel with adsorbed active labeled protein was washed twice with buffer of pH 8,2 and twice with distilled water to remove the nonreacted FITC and the inactive protein (9). The bound labeled chymotrypsin was desorbed from STI gel changing the pH to 2 - 3 by use of a buffer of pH 2.3 (0,2 M NaCl, adjusted with HCl) or by application of the moist gel as spot on an isoelectric focusing gel near the anode as described in (8).

Analytical isoelectric focusing

Commercial polyacrylamide gels(PAG plate pH 3.5 - 9.5) were focused with a LKB multiphor at 10°C as described in the instruction manual. After establishing the pH gradient

(~ 2 h, 600 V, 5 mA) the samples were applied on a filter piece or STI sepharose gel near the anode. The focusing was performed at 5 mA with increasing voltage (600 - 1100 V) over 3 hours. The fluorescent bands on the gel were photographed. An UV light source was used for excitation.The pH gradient was estimated slicing the gel in pieces of 5 mm, leaching them in 1 ml of destilled water for at least 1 hour.

The protein zones were identified after staining with Coomassie brilliant blue R 250 and measured in a Gilford spectrophotometer 250 using a gel scan accessory.

Preparative isoelectric focusing

The gel was prepared with Sephadex G-75 superfine as described in (10). The Ampholines pH 5 - 7, 7 - 9 and 5 - 8 were used. The sample was applied near the anode and focused at 10°C on a LKB multiphor at 8 W over 16 hours. Each fluorescent zone was seperately scraped off. The pH of the gradient was measured directly in the gel. The proteins were eluted from the gel in small columns with 4 ml of the NaCl-Tris buffer of pH 8,2. Separation from Ampholines was achieved by gel filtration using a biogel P 6 column.

Absorption spectra were recorded using a Gilford spectrophotometer 250.

Fluorescence spectra were recorded at 25°C with a fluorimeter AMINCO SPF-500 using 1×10^{-6} M fluorescein in 1 M NaOH as an internal standard.

Determination of the enzyme activity

The activities of the labeled CT were determined with a pH-stat (Radiometer) at 25°C, using N-acetyl-L-tyrosine ethylester as substrate at pH 8.0 in 10 % methanol (v/v; 0.2 M NaCl and TrisxHCl, I = o.oo2) as described in (9).

Results

Optimal labeling conditions

The pH and temperature dependence of the labeling reaction between protein and FITC was studied. Some results are shown in fig. 1. They show that the rate of labeling measured as the amount of labeled protein increases with pH and temperature, and that the product distribution also depends on pH and temperature. No marked change in the fraction of labeled molecules or product distribution was observed when FITC was added to the reaction mixture at once, continuously or after a previous incubation. Thus, first or second order decay reactions of FITC that would influence the labeling yield are of marginal importance here. When n (given molar fluorochrome/ protein ratio) was increased > 3 and the incubation performed > 24 hrs at pH 8.5 a large fraction of the labeled protein was adsorbed to the gel in the biospecific adsorbent and could not be eluted with buffer. It had no fluorescence and was red. This is probably protein which has so many labels that intermolecular fluorescence quenching occurs as in concentrated solutions of fluorescein. This protein fraction had also lost its biological activity.

These results show directly that a high pH, as has been proposed previously (4), yields optimal labeling.

Analysis of isolated homogeneous labeled protein fractions

A sample labeled at the optimal labeling conditions and analyzed by biospecific sample adsorption and isoelectric

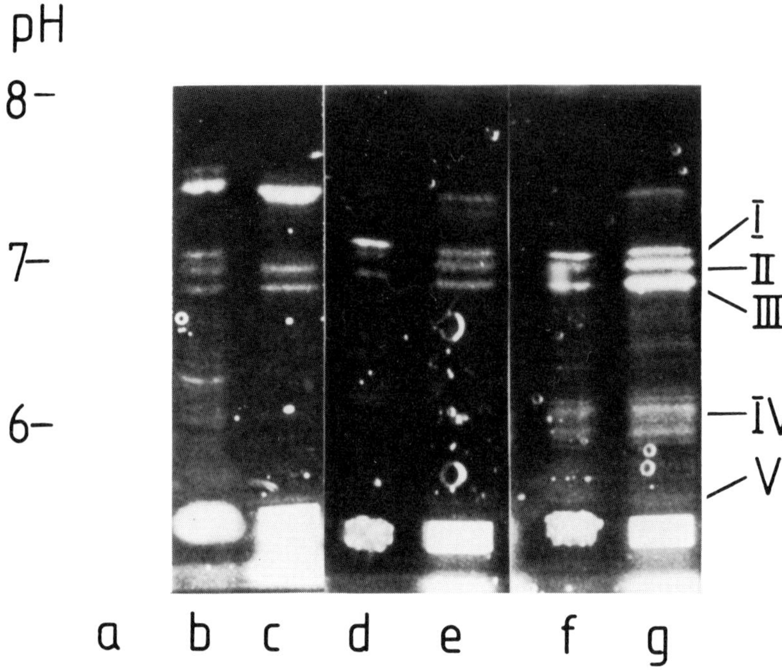

Fig.1. Analytical isoelectric focusing pattern of α-chymotrypsin labeled with fluoresceinisothiocyanate with and without biospecific sample application.
Lane a: Protein staining; lanes b - g: Fluorescence; initial concentrations: α-Chymotrypsin 85 μM, fluoresceinisothiocyanate 170 μM; labeling conditions: Buffer: 0.145 M NaCl, 0.0075 M Na_2HPO_4 and 0.0025 M NaH_2PO_4 adjusted to the desired pH; incubation time 30 min; lanes b,c: pH = 8.5; T = 4°C; lanes d,e: pH = 9.5; T = 4°C; lanes f,g: pH = 9.5; T = 25°C; lane a: α-Chymotrypsin; samples in lanes b,d,f were applied with the other without biospecific sample application. The bands analyzed further are numbered on the right hand side.

focusing was fractionated by preparative isoelectric focusing. The isolated fractions were separated from the compounds used to establish the pH gradient by gel filtration. The F/P ratio (fluorochrome molecules per protein conjugate) in these fractions was determined from the absorption spectrum assuming that the absorptivities are not perturbed in the conjugate. The adsorptivities published for fluorescein (11) and the proteins (12) were used in these calculations. The results are given in table I. As expected the isoelectric point of the labeled protein decreases with increasing F/P-ratios. Absorption and fluorescence spectra of protein conjugates with different F/P ratios are shown in fig. 2. The fluorescence intensity does not increase linearly with the maximum absorption when another fluorescein molecule is bound to the protein. This indicates that the quantum yield for fluorescence per bound fluorescein molecule depends on the labeling position and the F/P ratio.

The enzyme activities as a measure of the biological activity for the different protein conjugates are also given in table I. They show that the activity decreases with increasing number of bound fluorochrome molecules and depends on the labeling position.

Kinetics of labeling

From the F/P ratios given in table I for the bands in fig. 1, the fraction of conjugated protein with F/P = 1,2 ... was determined. This and the fraction of non-labeled protein is given as a function of time in fig. 3. From the initial slope of the curve for the formation of the conjugated protein with F/P = 1 the rates of the reactions

$$P_i + F \xrightarrow{k_i} P_i \ldots F \qquad (1),$$

Table 1.
Properties of different fluorescein labeled protein fractions isolated by isoelectric focusing.

substance	band no[x]	IP	F/P	relative biological activity		relative fluorescence quantum yield per fluorochrome	
				enzyme activity	interaction with STI	excitation at 284 nm	excitation at 492 nm
CT			0	1	-	-	-
$CT-(FTC)_1$	I	7.01	1	0.3	1	1.33	0.81
$CT-(FTC)_1$	II	6.88	1	0.3	0.3	1.28	0,57
$CT-(FTC)_1$	III	6.69	1	0.01	0.2	0.69	0.35
$CT-(FTC)_2$	IV	6.10	2	0.14	0.5	0,65	0,36
$CT-(FTC)_2$	V	5.05	2	0.01	0,2	0.60	0.42
$CT-(FTC)_3$	-		3[xx]	-	-	0,23	0.27
fluorescein						1	1

[x] see fig. 1

[xx] not purified

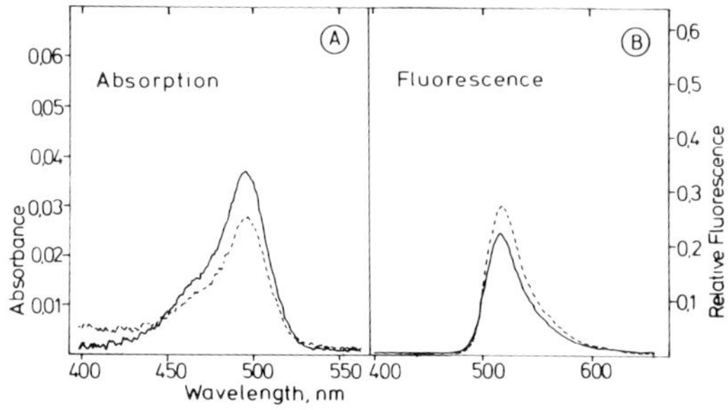

Fig.2. Visible absorption (A) and fluorescence (B) spectra after excitation at 492 nm of single- and double-labeled FTC-chymotrypsin. The spectra were measured for the conjugates in buffer of pH 8.2 (0.2 M NaCl, 0.103 M Tris, 0.05 M HCl) at 25°C; - - - - single labeled chymotrypsin (band II in fig. 1), ———— double labeled chymotrypsin (band IV in fig. 1).

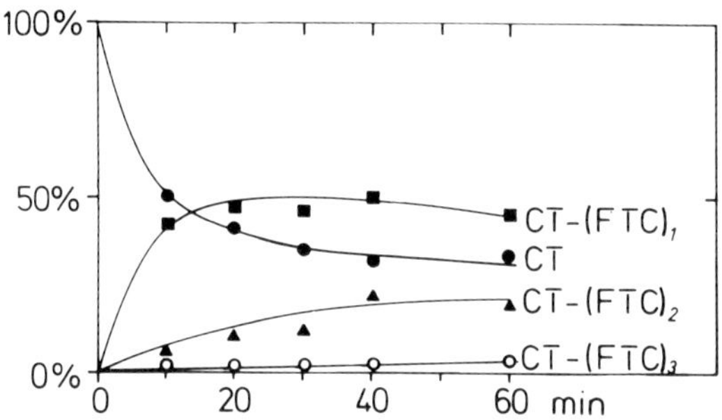

Fig.3. The kinetics of labeling α-chymotrypsin with fluoresceinisothiocyanate. Labeling conditions: 85 μM α-Chymotrypsin incubated with 170 μM fluoresceinisothiocyanate in buffer of pH 9.5 (0.145 M NaCl, 0.0075 M Na_2HPO_4 and 0.0025 M NaH_2PO_4) at 25°C. At different times samples were withdrawn and fractionated by isoelectric focusing. The relative amount of the different labeled protein conjugates was determined from protein staining.

where i denotes the labeling site i with the rate constant for labeling k_i, can be determined. The sum of the rate constants k_i was estimated from fig. 3 and found to be 6 M^{-1} x sec^{-1}. This sum can be used to estimate incubation times for labeling when the initial protein and fluorochrome content is known.

Direct determination of the biological activity and labels per protein molecule in homogeneously labeled protein fractions

This can be obtained directly from the isoelectric focusing pattern of the labeled protein with and without biospefic sample application. When the latter is performed so that the adsorbent is saturated the ratio of the amount of conjugated protein j, x_j, adsorbed by the biospecific adsorbent to adsorbed non-labeled protein, x_o, is (13)

$$\frac{x_j}{x_o} = \frac{K_j\ C_j}{K_o\ C_o} \qquad (2),$$

where C denotes total protein contents, K association constants for the species denoted by the subscript. The ratio (x_j/x_o) can be determined from the focusing pattern after biospecific sample application. The ratio (C_j/C_o) is determined from the corresponding pattern without biospecific sample application. Thus the ratios (K_j/K_o) for the differently labeled protein conjugates can be determined directly from the analytical isoelectric focusing pattern with and without biospecific sample appliation as shown in fig. 1. The ratios are given in table I, where they can be compared with the relative enzyme activities. The unlabeled protein is not pure and can therefore not be used as a reference in these calculations.

The number of labels per protein molecule can ideally also be determined from the analytical focusing pattern. The relative intensities (fluorescence with fluorochromes; counts per sec with nuclides) of two bands with labeled molecules are

$$\frac{I_1}{I_2} = \frac{(F/P)_1}{(F/P)_2} \cdot \frac{C_1}{C_2} \cdot \frac{\varphi_1}{\varphi_2} \qquad (3),$$

where the subscript indicates the band, φ the fluorescence quantum yield, (F/P) is the number of labels per protein molecules and C the protein content. The ratio (C_1/C_2) is determined from the protein stained focusing pattern. As the ratio (I_1/I_2) can be determined directly from the fluorescence or nuclide counts, the ratio of the number of labels per protein molecule can be obtained from eq. (3), provided that the fluorescence quantum yield is independent on the labeling position and the number of labels per molecule. For the fluorescein chromophore used here this was found not to apply. Therefore the number of labels per protein molecule cannot be determined directly by analyzing the focusing pattern.

Discussion

When the protein α-chymotrypsin is labeled with fluorescein-isothiocyanate several different protein conjugates are formed. The biological properties and the fluorescence yield per fluorochrome were found to differ markedly for the different protein conjugates. This applies also for the different conjugates with the same number of labels per protein molecule.

The procedure used here can be generally applied to analyze labeled protein conjugates and to prepare homogeneous conjugate fractions that are suitable to use in further studies.

Besides those, where the labeled proteins are used in specific binding assays, the homogeneous protein conjugates should be of interest also for other purposes. They can be used to study the microenvironment around the label on the protein surface and energy transfer as this influences the fluorescence properties (14). When these properties also are perturbed markedly in specific protein - protein (nucleic acid) interactions fluorochrome-labeled proteins could be used to study the kinetics of these interactions directly. Current techniques for such studies based on difference spectra in the UV spectrum are limited due to the small (∿ 1 %) difference spectra involved (13, 15).

The fluorescence quantum yield was found to decrease with the number of fluorochromes per protein molecule (table I). When this observation can be generalized this implies that a high label/protein ratio not necessarily increases the fluorescence per conjugate. Unfortunately, however, quantum yields are rarely published for fluorochrome-labeled proteins and then only for a mixture of conjugates (3,16). Therefore further studies on the fluorescence yields of homogeneous conjugate samples are necessary.

Although fluorochrome-labeled protein conjugates are used extensively for analytical purposes no general procedure has been established for the analysis of the protein conjugates with respect to labels/protein molecule and the biological activity of homogeneous conjugates (1-3). In practice mixtures of different protein conjugates are used that are prepared by seperating non-reacted fluorochrome from the incubation mixture. Properties as specific and non-specific interactions and signal intensity per label in the conjugate (fluorescence quantum yield) observed in analytical procedures differ for different conjugates. Thus the currently used analytical procedures will depend on the distribution of conjugates in the labeled sample.

In quantitative analytical studies it would therefore be valuable to use homogeneous fractions of a protein conjugate that has a high biological activity and fluorescence yield/ fluorochrome. The procedure developed here is suitable for this purpose.

The method to analyze protein conjugates based on biospecific sample application in the isoelectric focusing procedure and preparative isoelectric focusing should be possible to adapt also for other label- (nuclides, enzymes)-protein-systems provided a biospecific adsorbent is available. Then samples can be analyzed and homogeneous protein conjugates prepared that are suitable to use in the further quantitative studies.

References

1. Yalow, R.S.: Science 200, 1236-1245 (1978).
2. van Weemen, B.K., Schuurs, A.H.W.M.: FEBS Lett. 24, 77 (1972).
3. Knapp, W., Holubar, K.,Wick,G., eds: Immunofluorescence and Related Staining Techniques,North-Holland Publishing Company, Amsterdam 1978
4. Naim, R.C.: Fluorescent Protein Tracing, Third edition, E. u.S. Livingstone Ltd, Edinburgh-London 1969
5. Poste, G., Papahadjopoulos, D., Nicolson, G.L.: Proc. Natl.Acad.Sci. 72, 4430-4434 (1975).
6. Hirs, C.H.W., Timasheff, S.N., eds.:Methods in Enzymology, Academic Press, New York-London, Vol. XXV, 1972
7. Gabel, D., Kasche, V., Amneus, H., Lundqvist, H.: in: Hoffmann-Ostenhof, D., ed. :Affinity Chromatography, Pergamon Press, Oxford-New York 1978
8. Amneus, H., Näslund, L., Gabel, D., Kasche, V.: Biochem. Biophys.Res.Commun.90, 1313-1320 (1979).
9. Kasche, V., Amneus, H., Näslund, L., Gabel, D.: Biochim. Biophys.Acta 490, 1-18 (1977).
10. LKB Application Note: Preparative Flat-Bed Electrofocusing in a Granulated Gel with LKB 2117 Multiphor 198(1975).

11. Fasman, G.D., ed.: Handbook of Biochemistry and Molecular Biology, Proteins, CRC Press, Cleveland, Vol.II, 1976

12. Lindqvist, L.: Arkiv Kemi 16, 79-138 (1961).

13. Kasche, V.: Biochem.Biophys.Res.Commun. 38, 875-881 (1970).

14. Vaz, W.L.C., Nicksch, A., Jähnig, F.: Eur.J.Biochem. 83, 299-305 (1978).

15. Gennis, L.S., Cantor, C.R.: J.Biol.Chem. 251, 769-775 (1976).

16. Shechter, Y., Schlessinger, J., Jacobs, S., Chang, K.-J., Cuatrecasas, P.: Proc.Natl.Acad.Sci., 75, 2135-2139 (1978).

THE ABILITY OF THYROLIBERIN TO COORDINATE SOME TRANSITION METAL IONS

Grażyna Formicka-Kozłowska, Henryk Kosłowski
Institute of Chemistry, University of Wrocław, Joliot-Curie 14, 50-383 Wrocław, Poland

Gotfryd Kupryszewski
Institute of Chemistry, University of Gdańsk, Sobieskiego 18, 80-952 Gdańsk, Poland

Introduction

As a part of our studies on metal ion interactions with naturally occuring peptides (1-3) we have begun the investigation of the coordination ability of the thyrotropin-releasing factor (TRF) (4-6). Though no involvement of this hormone in the interactions with the metal ions in vivo is known up to date, its amino acid residue pattern (L-pyroglutamyl-L-histidyl L-prolinamide) seemed to promise quite rare coordination ability. There are no data available about the metal ion interaction with pyroglutamic acid,and histidine is very often the coordination site of the metal ion in many metalloproteins. This interesting peptide sequence has inspired us to examine the interactions of TRF with metal ions which are essential or beneficial to human organism.

Experimental

The TRF and Pyr-His (L-pyroglutamyl-L-histidine) was synthesized as described earlier (7). $Cu(ClO_4)_2 \cdot 6H_2O$, $Ni(ClO_4)_2 \cdot 6H_2O$ and $Co(ClO_4)_2 \cdot 6H_2O$ were used as obtained from Fluka AG, Buchs (Switzerland).

The 1:1 Cu(II):TRF, $[Cu^{+2}]$ = 0.01M, and 1:10 Co(II):TRF, $[Co^{+2}]$ = 0.01M, molar ratios were used for absorption and CD spectra experiments. All experiments for the CoII-TRF system were done in strictly oxygen-free atmosphere. Solutions with molar ratio of Ni(II):peptide of 1:5 were used for the absorption and CD spectra measurements with metal ion concentrations of 0.01M. Solutions of molar ratio 1:2 with $[Ni^{+2}]$ = 0.05M were prepared for ^{1}H NMR studies.

The magnetic moment in solution was established by the Evans method (8).

The absorption spectra were recorded on Beckman UV 5240 and Cary 14 spectrophotometers. Circular dichroism (CD) spectra were measured on an automatic recording spectropolarimeter JASCO J-20. All CD spectra are expressed in terms of $\Delta\varepsilon$ ($\varepsilon_l - \varepsilon_r$).

The proton magnetic resonance (^{1}H NMR) spectra were recorded on a Jeol JNM-PS-100 spectrometer, the electron paramagnetic resonance (EPR) spectra on a Jeol JES-ME-3X spectrometer at liquid nitrogen temperature. All studies except NMR were done in H_2O solution and the NMR spectra were recorded in $^{2}H_2O$ (D_2O). The histidine residue $\alpha CH \beta CH_2$ proton spectra were analyzed as ABC spectra on a JEC-6 computer.

Results and Discussion

^{1}H NMR Characterization of Metal-free Ligands

^{1}H NMR spectra of metal-free TRF have been reported previously (9,10). We have done a detailed analysis of the ABC histidine spectra at several pH values for TRF and for the Pyr-His peptides. The data are given in Table I. The notation of rotamers of the His residue is shown in Fig.1. The conformation of the histidine side chain in Pyr-His is largely

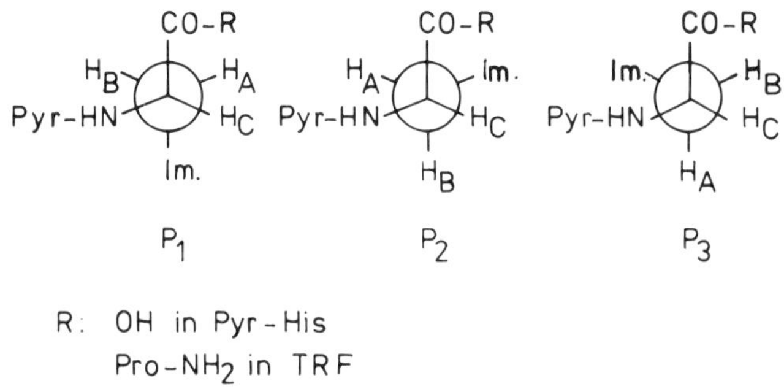

Fig.1. The Rotamer Notation for the Histidine Residue in TRF and Pyr-His Peptides.

Table I. The Coupling Constants and the Rotamer Populations for TRF and Pyr-His Metal-free Peptides.

Peptide	PH	Coupling Constants in Hz			Rotamer Populations		
		$\|J_{AB}\|$	J_{AC}	J_{BC}	p_1	p_2	p_3
TRF	0.5-4.0	16.0	8.8	5.2	0.56	0.24	0.20
	>6.0	16.0	8.0	7.5	0.49	0.45	0.06
Pyr-His	3.52	15.6	9.3	4.6	0.61	0.19	0.20
	6.52	15.6	9.7	4.6	0.65	0.19	0.16
	11.53	15.6	9.8	4.5	0.66	0.17	0.17

independent of pH. In the most stable rotamer the imidazole ring is in a trans position to the carboxyl group (isomer 1). The minor pH dependence of the histidine residue rotational isomerism was also found in the Gly-His dipeptide (11), although in the latter case the population of isomer 1 is smaller than that in Pyr-His. The increase of this rotamer population in the Pyr-His peptide derives from the destabilization of the rotamer 3 in which the possible close contact of the carboxyl group with the bulky pyroglutamic acid residue is very unfavorable.

Distribution of the histidine residue rotamers in solutions containing TRF at pH values up to 4 is close to that of Pyr-His (Table I). The increase of pH to about 6 (N3 deprotonation) causes a quite distinct increase in the population of rotamer 2. This suggests some interactions between imidazole and the C-terminal groups of the TRF molecule, eg. hydrogen bond formation between the N3 nitrogen and the NH_2 group of the Pro-NH_2 residue.

Copper(II) Complexes with TRF

The d-d transition energies and the EPR parameters for Cu(II)-TRF solutions are presented in Table II. The stepwise increase of the d-d transition energy with increasing pH suggests a coordination of the successive nitrogen donors. The selective ^1H NMR peptide line broadening when Cu(II) ions are added suggests that the metal ion begins the coordination at N3 imidazole nitrogen at pH about 4.

Table II. Absorption Spectra and EPR Data for the Cu(II):TRF 1:1 System.

pH	λ, nm	ε	$g_{\parallel}$	$A_{\parallel}$, G
5.72	758sh			
	617	28.1		
6.32	602	49.4	2.268	182.9
			2.223	190.1
6.80	599	61.6		
8.32	595	71.3		
9.50	592	73.3	2.222	188.0
10.68	585	75.6		
11.30	571	77.6	2.212	197.6
11.82	568	78.1	2.209	201.8

EPR spectra of Cu(II)-TRF solutions in the pH region 6-10 are well resolved (Fig.2) and show five-line superhyperfine

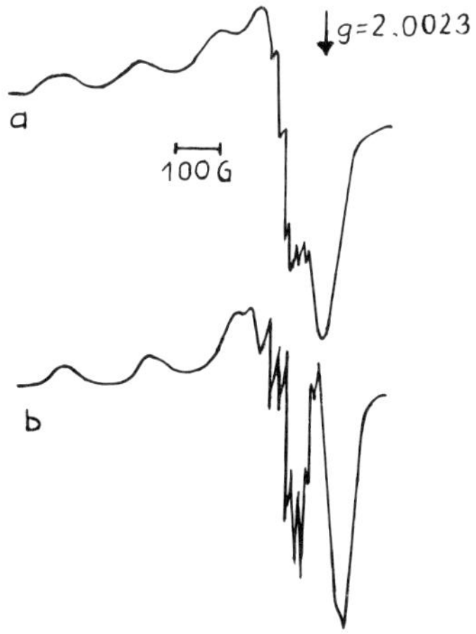

Fig.2. EPR Spectra of 1:1 Cu(II):TRF Frozen Solutions at pH=9.40 (a) and 11.40 (b).

structure deriving from two nitrogens coordinated with cupric ion. The d-d band position at 599 nm is also consistent with the coordination of two nitrogen donors to the Cu(II) ion, which are most probably N3 of imidazole and N^- of deprotonated peptide linkage.

When the pH increases above 10 the d-d band position shifts to 570 nm and a seven-line superhyperfine EPR structure appears. These results demonstrate the third nitrogen coordination to the cupric ion. Consideration of the molecular model suggests that the most probable donor set in the 1:1 Cu(II):TRF system is the N3 of imidazole, N^- of the peptide linkage and the N^- of the pyroglutamic acid.

The results for the Cu(II)Pyr-His system are similar to that of Cu(II)-TRF and they prove that none of the two proline residue nitrogens is involved in the binding of the metal ion.

The formation of Cu(II)TRF complexes proceeds as follows:

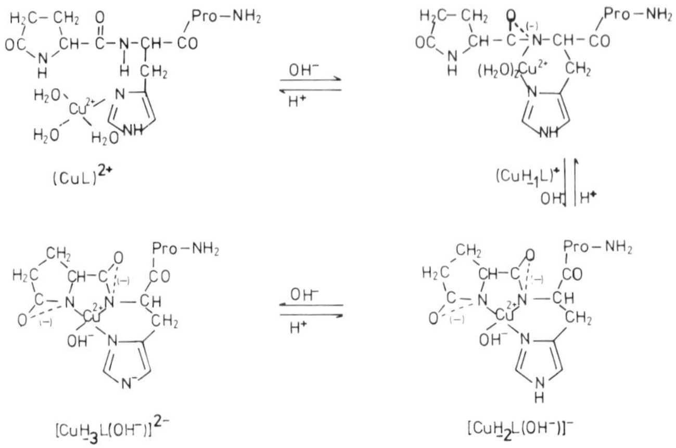

CD spectra of 1:1 Cu(II):TRF and Cu(II):Pyr-His systems are presented in Table III. The CD spectrum of metal-free TRF is characterized by two peaks: the positive peak at about 212 nm deriving from the $n \rightarrow \pi^*$ transition of the carbonyl chromophore, mainly of the prolinamide group (12,13) and the negative peak at 236 nm deriving from the cis proline isomer present to the extent of about 14 % in metal-free TRF (14). Both these peaks remain in the system with Cu(II) ion. This suggests a minor influence of the coordinated metal ion on the proline cis-trans isomer equilibrium, as well as on the proline carbonyl chromophore contribution to the CD sign. The enhancement of CD peaks observed in the Cu(II)-TRF complex is due to more restricted rotation in the metal chelates.

Two other UV features are found for Cu(II)-TRF solutions: The positive peak at 283 nm and the negative peak at 315 nm (Table III). They are slightly sensitive to pH. Both these bands derive from charge transfer transitions from peptide nitrogens to the copper ion. The comparison of the CD bands in this region of the spectra for the Cu(II)-TRF system with

Table III. Circular Dichroism of Cu(II) Complexes with TRF and Pyr-His.

Cu(II):TRF 1:1

pH 6.00 - 9.30		pH = 11.25		pH = 12.02	
λ,nm	Δε	λ,nm	Δε	λ,nm	Δε
588	+0.93	572	+1.09	572	+1.12
		475	-0.11	474	-0.09
405	+0.01			368	+0.05
328	-1.52	318	-0.73	315	-0.61
289	+1.35	283	+0.97	283	+0.96
230	-4.17	234	-3.86	235	-3.86
208	+4.70	210	+4.62	211	+4.09

TRF, Metal-free

pH = 7.10		pH = 10.20	
λ,nm	Δε	λ,nm	Δε
236	-0.13	236	-0.11
212	+1.94	212	+2.24

Cu(II):Pyr-His 1:1

pH 6.00 - 9.20		pH = 11.09		pH = 11.94	
λ,nm	Δε	λ,nm	Δε	λ,nm	Δε
587	+0.95	574	+1.15	572	+1.19
418	-0.005	470	-0.06	474	-0.10
331	-1.13	326	-0.75	319	-0.52
290	+1.04	287	+1.20	284	+1.14
233	-2.26	238	-3.01	238	-2.98
205	+6.93	205	+6.18	207	+5.80

Pyr-His, Metal-free

pH = 6.78		pH 10.20 - 12.00	
λ,nm	Δε	λ,nm	Δε
210	+3.71	210	+5.38

those of the copper(II) complexes with two other TRF analogues i.e. L-pyroglutamyl-L-tyrosyl-L-prolinamide and picolyl-L-histidyl-L-prolinamide, allowed us to describe these bands more exactly. The positive CD peak at about 280 nm derives from imidazole nitrogen to copper ion charge transfer transition, while the negative CD peak at about 320 nm derives from peptide linkage nitrogen to Cu(II) ion charge transfer transition.

In the visible region a positive peak at 588 nm is observed for the Cu(II) complex with two coordinated nitrogens. This peak derives from the $B_{1g} \rightarrow B_{2g}$ transition within the copper ion. For the Cu(II)-TRF complex with three coordinated nitrogens two bands are observed: at 572 nm and 475 nm which arise from $B_{1g} \rightarrow B_{2g}$ and $B_{1g} \rightarrow E_g$ d-d transitions. Small variations of the CD spectrum as the pH increases from 11 to above 12 (i.e. the appearence of the positive band at 368 nm) could be caused by the deprotonation of the imidazole N1 nitrogen. The Cu(II) ion coordinated to the N3 imidazole nitrogen can promote such a deprotonation at high pH (15,16).

The CD results for the Cu(II)-Pyr-His system are similar to those for Cu(II)-TRF, and they support the conclusion that the proline residue has only minor influence on the coordination and the symmetry in the Cu(II)-TRF complex.

Nickel(II) Complexes with TRF

Absorption spectra as well as the magnetic moment values of the solutions containing Ni(II) ions and TRF are presented in Table IV. These results show clearly that up to pH about 8 Ni(II) creates with thyroliberin only octahedral complexes. At pH range 8-10 the equilibrium between octahedral and diamagnetic square planar species is formed and the square planar complex is the only complex at pH above 11. Considering the d-d transition energy of the planar species (Table IV) one can distinguish two different square planar complexes. One of them created at pH region 8-10 with d-d transition energy equal to 440 nm and another one formed at pH above 11 with d-d transition at 460 nm. The decrease of the d-d transition energy of 1000 cm^{-1} may suggest some changes in the metal ion environment as the pH increases from 9 to 11. The CD spectra in the visible region (Fig.3) show quite univocally that two different square planar Ni(II)-TRF complexes are created at pH region 8 to 12. At pH below 10 one observes two CD peaks, a positive at 426 nm and a negative one at 494 nm. They are

Table IV. Absorption Spectra and Magnetic Moment Values for Ni(II)-TRF, 1:5 Molar Ratio Solutions.

pH	λ, nm	ε	μ, BM
7.53	1100	2.4	3.06
	709sh		
	645	4.2	
	383	11.2	
8.80	1020	10.0	2.67
	620sh	26.0	
	420sh	65.0	
9.51	1020	1.2	1.50
	600sh		
	446	77.0	
	303	2200.0	
11.20	458	93.0	0.00
	303	2600.0	
12.00	460	89.1	0.00
	303	2500.0	

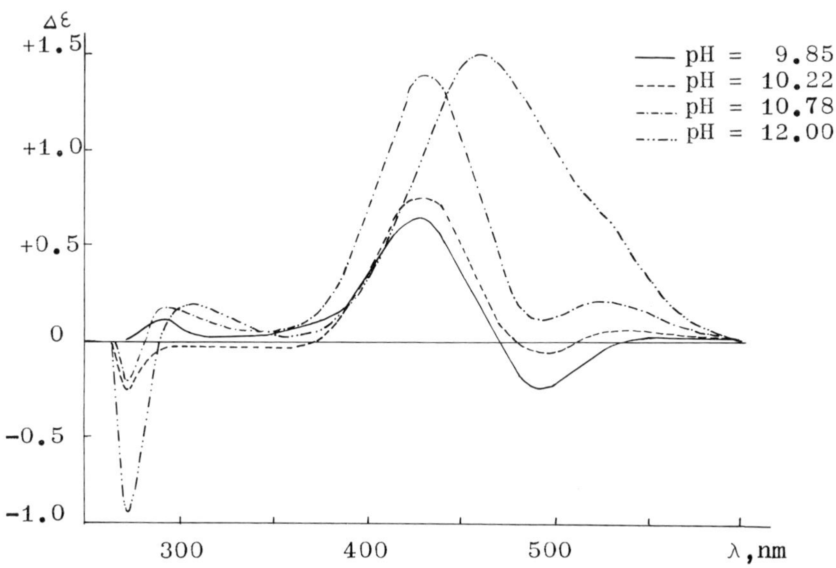

Fig. 3. The CD Spectra for Ni(II)-TRF, 1:5 Molar Ratio Solutions at Different pH.

assigned to $A_{1g} \rightarrow E_g$ (E) and $A_{1g} \rightarrow A_{2g}$ (A) d-d transitions. When the pH increases above 10 these two CD bands disappear slowly, while two new bands are created at 461 and 520 nm (Fig.3). Both of them have a positive Cotton effect.There are also two other CD extrema in the charge transfer region at 272 and 307 nm. Both latter bands derive most probably from the imidazole nitrogen → Ni(II) CT transition.

The CD spectra of both planar complexes suggest that when pH increases from 9 to 11 a variation in the conformation of the chelate rings takes place. The reason for this variation is most probably a deprotonation of a pyrrole nitrogen N1 (15-19) and the change of the Ni(II) – N3 bond geometry (6).

The absorption and CD spectra of the Ni(II)-Pyr-His system are the same as those of the Ni(II)-TRF solutions, as far as the d-d and charge transfer transitions are considered (6). Thus the former system was used for ^1H NMR studies to find more precisely the coordination sites of the metal ion in the peptide molecule. The chemical shift variations of the proton resonances of the Pyr-His dipeptide upon Ni(II) ion coordination in the diamagnetic square planar complex are very distinct (Table V).They clearly show that the coordination sites of Ni(II) in a planar complex are:N3 imidazole nitrogen, N^- of the peptide linkage between Pyr and His and N^- of the Pyr residue

Table V. The Upfield Chemical Shifts (in ppm) of ^1H NMR Resonances of Pyr-His Dipeptide Protons upon the Ni(II) Ion Coordination at pH = 11.5.

H2 Im.	0.58
H5 Im.	0.23
αCH His	0.46
βCH_2 His	0.16
αCH Pyr	0.55
βCH_2 Pyr	0.32
γCH_2 Pyr	0.18

as the most dramatic upfield chemical shifts are observed for H2 Im. , αCH His, αCH Pyr and βCH_2 Pyr (Table V). The resonances of H2 and H5 protons of imidazole in the Ni(II) complex at pH > 11 are strongly upfield shifted by 1.53 and 0.60 ppm compared to the metal-free ligand with N3 protonated. These large upfield shifts may provide quite strong support for deprotonation of the N1 nitrogen in the square planar complex at high pH region. A proposed coordination in the planar Ni(II) complexes with TRF is given below:

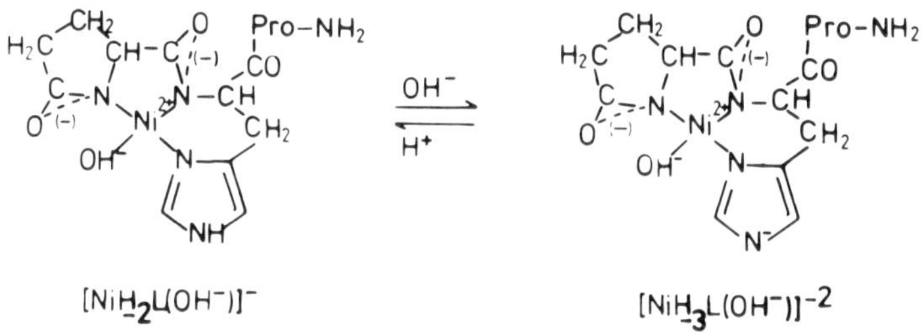

Cobalt(II) Complexes with TRF

The absorption spectra of the Co(II)-TRF solutions at high pH region, measured immediately after the preparation of the oxygen-free solution, show the d-d transition bands characteristic for the tetrahedral Co(II) complexes. Our primary ^1H NMR studies as well as the earlier investigations of Co(II)-His system (15,20) strongly suggest that in the tetrahedral Co(II)-TRF complex the coordination sites are: N3 imidazole nitrogen and the N^- donor of the peptide linkage.

The Co(II)-TRF solution kept under argon changes its colour and its absorption spectrum dramatically (Fig.4). Two hours after the solution preparation the spectrum consists of a single band at about 500 nm. Also the magnetic moment changes very considerably from 4.3 BM at the beginning (tetrahedral species) to 0 BM two hours later. The single band at 500 nm could suggest the low spin Co(II) complex of planar symmetry (25,27) or low spin Co(III) complex (21,24). The magnetic moment equal to 0 BM, however, strongly supports diamagnetic Co(III) species.

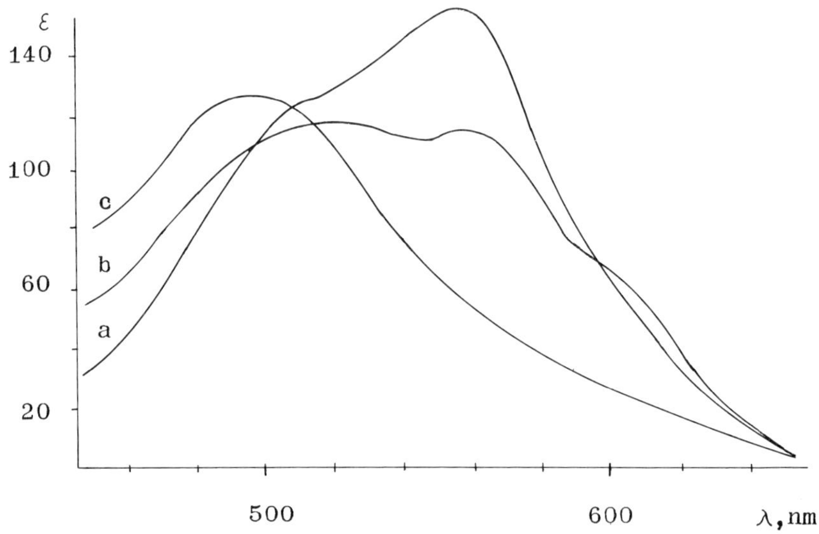

Fig.4. Absorption Spectra of 10:1 TRF:Co(II) Solution at pH= 11.5 Immediately after Preparation (a), one Hour (b) and two Hours (c) after the Preparation.

CD spectra in the visible region of the latter species consist of three peaks at 498, 343 and 400 (shoulder) nm. Such CD pattern could also support that the final species formed in the solutions containing Co(II) and TRF is the complex with the Co(III) ion (21,23,24). The precise mechanism of the Co(II) oxidation in the studied system could not be concluded from the performed studies. It is rather doubtful that Co(II)-TRF systems undergo oxygenation, as there are no bands in the absorption spectra which are very characteristic for the oxygenated Co(II) peptide complexes (21,22,26).

Conclusions

The results presented for TRF complexes with Cu(II), Ni(II) and Co(II) have clearly shown that the thyroliberin may act as effective coordinating agent with some metal ions already at physiological pH. The formed complexes seem to be quite stable (19) and their structural features are extremely interesting. Thus, the metal TRF system studies may be very useful for elucidation of the structural problems of the metal-containing biomolecules and on the other hand they may suggest the potential possibilities of the involvement of the metal ions in some biological activities of the thyroliberin.

Acknowledgements

Authors wish to thank Polish Academy of Sciences for the financial support, research grants No 10.4.2.01.5.2 and MR-I-9.03.

References

1. Jeżowska-Trzebiatowska, B., Formicka-Kozłowska, G., Kozłowski, H.: J.Inorg.Nucl.Chem. 39, 1265-1268 (1977). Ostern, M., Formicka-Kozłowska, G., Jeżowska-Trzebiatowska, B., Kozłowski, H.: Inorg. Nucl. Chem. Lett. 14, 351-357 (1978).
2. Jeżowska-Trzebiatowska, B., Formicka-Kozłowska, G., Kozłowski, H.: Bull. Acad. Polon. Sci., Ser. Sci. Chim. 24, 987-995 (1976).
3. Formicka-Kozłowska, G., Kozłowski, H., Jeżowska-Trzebiatowska, B.: Acta Biochim. Polon. 26, 239-248 (1979) and references therein.
4. Formicka-Kozłowska, G., Kozłowski, H., Kupryszewski, G., Przybylski, J., Jeżowska-Trzebiatowska, B.: Peptides,1978

Ed. Siemion, Z., Kupryszewski, G., Wrocław University Press, 496-498 (1979).

5. Formicka-Kozłowska, G., Kozłowski, H., Jeżowska-Trzebiatowska, B., Kupryszewski, G., Przybylski, J.: Inorg. Nucl. Chem. Lett. 15, 387-392 (1979).

6. Formicka-Kozłowska, G., Kozłowski, H., Kupryszewski, G.: Inorg. Chim. Acta (in press).

7. Maliński, E., Nauman, J., Przybylski, J., Ryżecka, M., Szafranek, J., Kupryszewski, G.: Pol. J. Med. Pharm. (in press).

8. Evans, D.F.: J. Chem. Soc. 2003-2005 (1959).

9. Fermandjian, S., Pradelles, D., Fromageot, P.: FEBS Lett. 28, 156-160 (1972).

10. Boilot, J.C., Clin, B., Bellocq, A.M., Lemanceau, B.: C.R. Acad.Sci. Paris 276, 217-220 (1973).

11. Kozłowski, H., Matczak-Jon, E.: Inorg. Chim. Acta 32, 143-148 (1979).

12. Voelter, W., Zech, K., Göbel, P., Oster, O., Attanasio, A.: Z. Naturforsch. 30b, 142-143 (1975).

13. Pradelles, P., Vicar, J., Morgat, J.L., Fermandjian, S., Blaha, K., Fromageot, P.: Collect. Czech. Chem. Commun. 42, 79-89 (1977).

14. Deslauriers, R., Garrigou-Lagrange, C., Bellocq, A.M., Smith, I.C.P.: FEBS Lett. 31, 59-66 (1973).

15. Sundberg, R.J., Martin, R.B.: Chem. Rev. 74, 471-517 (1974).

16. Kruck, T.P.A., Sarkar, B.: Can. J. Chem. 51, 3563-3571 (1973).

17. Bogges, R.K., Martin, R.B.: Inorg. Chem. 13, 1525-1527 (1974).

18. Brookes, G., Pettit, L.D.: J. Chem. Soc. Dalton Trans. 2112-2117 (1975).

19. The thermodynamic studies of the metal TRF systems are already in progress.

20. Morris, P.J., Martin, R.B.: J. Am. Chem. Soc. 92, 1543-1546 (1970).

21. Michailidis, M.S., Martin, R.B.: J. Am. Chem. Soc. 91, 4683-4689 (1969).

22. Harris, W.R., Bess, R.C., Martell, A.E., Ridgway, T.H.: J. Am. Chem. Soc. 99, 2958-2963 (1977).

23. Hawkins, C.J., Lawrance, G.A., Palmer, J.A.: Aust. J.Chem. 31, 2399-2414 (1978).

24. Morris, P.J., Martin, R.B.: Inorg. Chem. 10, 964-968 (1971).

25. Cotton, F.A., Holm, R.H.: J. Am. Chem. Soc. 82, 2979-

2983 (1960).

26. Gillard, R.D., Phipps, D.A.: J. Chem. Soc. A 1074-1082 (1971).

27. Hipp, C.J., Baker, W.A. Jr.: J. Am. Chem. Soc. 92, 792-798 (1970).

TELEVISION DENSITOMETRIC MICROASSAY OF PROTEINS IN THE NANO-GRAM RANGE ON CELLULOSE ACETATE MEMBRANES

Werner Giebel [+,++]

Klinik für Hals-, Nasen- und Ohrenheilkunde der Universität Tübingen, D-7400 Tübingen, Germany

Introduction

Analytical procedures to determine very small amounts of protein are sometimes desirable: For example in liquid chromatography to calculate the enzyme activity per protein content of individual fractions, or, as in our case, to check the purity of samples of biological fluids (perilymph, endolymph, or cerebrospinal fluid). If the protein content of a fluid is low (< 1 g/l) "microbiuret techniques" (1,2), absorption differences (3) or the fluorometry after reaction with fluorescamine (4,5,6) or orthophthalaldehyde (7,8) can be used. If, in addition, the volume of the sample is small (some microliters) most of these methods are not applicable. To overcome this difficulty the proteins were hydrolyzed in "oil wells" (9,10) followed by a reaction with orthophthalaldehyde (11). Another method capable of handling very dilute solutions of protein was by spotting the protein solution on a membrane filter and dansylating it with a cycloheptaamylose-dansyl chloride complex and subsequent scanning fluorometry (12). The drawback of fluorometric assays are the precautions which must be taken for samples containing interfering material. In order to avoid the disadvantages of fluorometry we tried to measure the density of protein-dye complexes of small spots applied to cellulose acetate foil (CAF). For staining Coomassie blue and amido black were tested despite the reported unsatisfactory sensitivity of the latter

[+,++]Dedication to G.Pfleiderer, Stuttgart and expr.to thanks to DFG see end of this article.

Structure and Activity of Natural Peptides

(13,14,15).

Material and methods

Application of the protein solutions was tested under different conditions. The aim of these studies was to produce spots small and intense enough for densitometric readings with volumes of 0.1 µl or less. Very small areas were obtained by denaturing the proteins during application. This was done with CAF soaked with sulfosalicylic acid (10 %). If fixed volume nanoliter capillaries are used to transfer the samples the membranes should be dry before application. For protein solutions with a content of more than 70 mg/l, normal dry CAF (Boskamp, Hersel, Germany) is used. The transfer and measurement of the fluid can be carried out by nanoliter capillaries made of metal (16) (Antech, Bad Dürkheim, Germany, Carl Zeiss, Oberkochen, Germany) or glass (Desaga, Heidelberg, Germany) as well as with microliter syringes (Hamilton, Darmstadt, Germany) which were used in this study.

Transparence of the CAF can easily be achieved by immersion in oil, as reported by Kohn (17). Since the protein-Coomassie blue complex is not stable (electrostatic and van der Waals bonds between dye and protein (18))the usual mixture of isopropanol and dioxan cannot be used because the complex breaks. The CAF was dried at room temperature between filter papers overnight. When shortening this procedure by drying at 105°C (2h) slight pressure should be applied to avoid wrinkling. The CAF becomes transparent by immersion in paraffin oil (DAB 6, Merck, Darmstadt, Germany).

Densitometry was carried out with a Quantimet 720 D (Cambridge Instrument Company, Dortmund, Germany). The usefulness of this television densitometer for exact quantitative densitometry was proved earlier (19,20). The preparation is put on the epidiascope and with the appropriate objective (F = 63 mm, f = 1.9) the whole CAF is visible on the television screen. With the aid of a light pen single spots can be selected and the integrated optical density (IOD) can be read digitally on the screen. In our case the television densitometer was directly connected with a programmed calculator (HP 9830, Hewlett - Packard , Böblingen, Germany) which calculated the protein concentration.

Maximal absorbance of the protein-dye complex was determined with a spectrophotometer (PMX III, Carl Zeiss, Oberkochen, Germany). For this assay a foil was soaked in a protein solution, fixed, stained, destained, made transparent in the usual way and put in the light path of the photometer. As the results show (Fig.1) the absorption spectra of Coomassie blue and amido black have their maxima at 590 nm and 620 nm respectively. A combination of two coloured glass filters (OG 570 and BG 380, Schott, Mainz, Germany) was used as broad band filter and put into the light path of the epidiascope between the light source and the object.

Standard curves were determined from a series of diluted bovine serum albumine (BSA) (Bio Mérieux, Nürtingen, Germany). The protein concentration of the dilutions was measured by means of absorbtion differences (3). From each solution (protein concentrations of 0.07, 0.14, 0.26, 0.52, 0.76, 1.01, 1.26, 1.57 and 1.99 g/l respectively) three spots were put on one CAF. Three of the membranes containing 27 spots each were stained with amido black and seven with Coomassie blue. The statistical evaluation is based on nine spots per concentration for amido black and 21 spots per concentration for Coomassie blue.

Statistics. For each concentration the mean valve($MV=\bar{x}$), the standard deviation (SD) and coefficient of variation (VC)

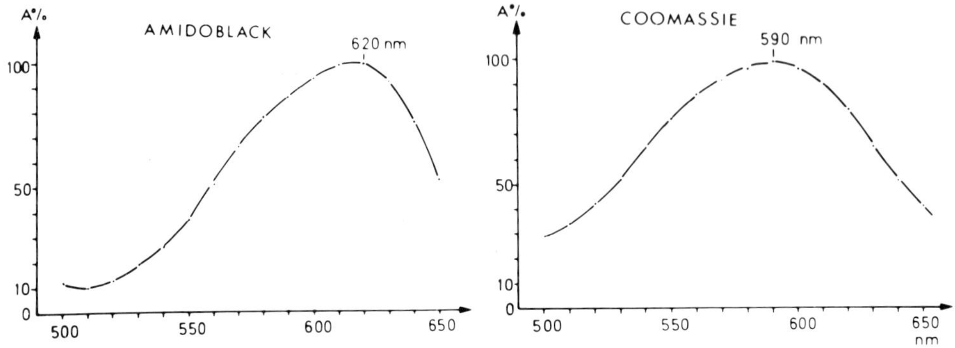

Fig. 1. Absorption spectrum of protein-dye complex on CAF after destaining. The wavelength is given in nm (x-axis) and the relative absorption in percent of the maximum (y-axis).

were calculated. The linear regression was determined with the following formula (21,22):

slope (S) $S = S_{xy} / S_x$

correlation coefficient (R) $R^2 = S^2_{xy} / S_x \, S_y$

intercept (I) $I = \bar{y} - S \cdot \bar{x}$

$\bar{x} = \Sigma x/n$, $S_x = \Sigma(x-\bar{x})^2$, $S_{xy} = \Sigma(x-\bar{x})(y-\bar{y})$.

Results

A series of protein solutions of which the protein content ranged from 72 mg/l to 1.99 g/l was spotted on the foils. Of each concentration 0.1 µl was applied three times. A picture of an original CAF stained with Coomassie blue is shown in Fig. 2 as well as the image of the same membrane on the television screen.

A statistical evaluation of the IOD read on the television densitometer is given in tables 1 and 2. If the proteins are stained with amido black the standard curve for the amount from 7 - 100 ng/spot is linear (Fig. 3). The coefficient of variation is above 10 % at lower concentrations (Tab.1). The linear regression (Tab. 1, Fig.3) was calculated for the range between 77.5 and 198.6 ng/0.1 µl in which the standard deviations are about 1 % of the mean values. The significance of the linearity is very high ($p < 0.001$, $N = 45$).

When staining with Coomassie blue the variation coefficient of the IOD reading lies between 3.8 % and 7.4 % in the protein concentration range from 14.1 to 198.6 ng/0.1 µl. Figure 3 shows that the standard curve is linear only up to about 100 ng/spot. The linear regression (Tab. 2, Fig.3) was therefore calculated for 7.2 to 77.5 ng/0.1 µl and found to be highly significant ($p < 0.001$, $N = 105$). Because the variation of approximately 5 % for single measurements is too large, the accuracy of the mean values of three spots of the same concentration on the same membrane was calculated. As the example shows (Ab. 3) the standard deviation of the mean values is 2.6 %.

Discussion

As the results show, the technique reported here, the application of small amounts (0.1 µl) of dilute protein solutions (<2 g/l) followed by staining and densitometric measurement, is sensitive and reproducible enough for our purpose,namely the control of purity of samples of inner ear fluids.

Compared to other methods using unhydrolized protein, the sensitivity of our technique is not less than that of the fluorometric results with fluorescamine (10 - 50 ng) (4,5) or orthophthalaldehyde (100 ng) (7). For protein hydrolysis and the reaction of amino acids with orthophthalaldehyde a lower limit of 3 ng has been reported (11). Proteins spotted on membrane filters and dansylated with a cycloheptaamylose-

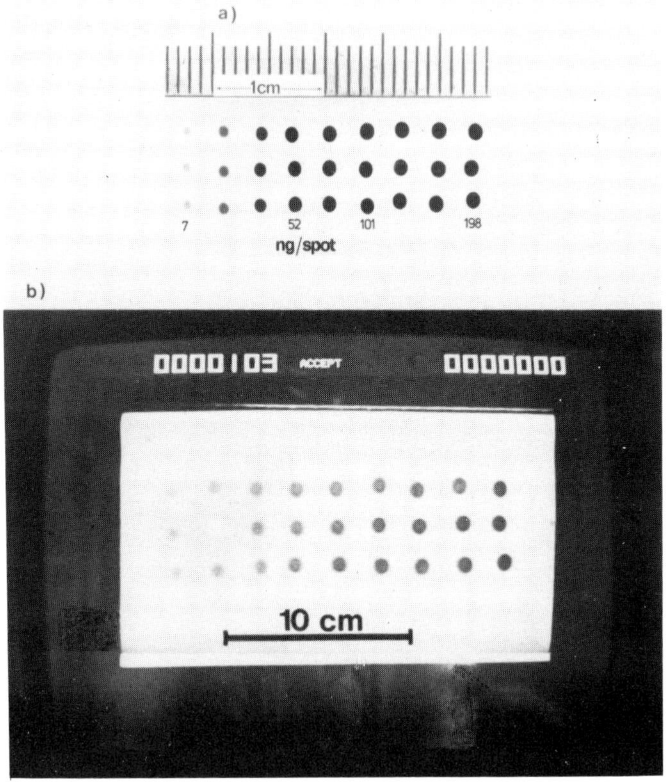

Fig. 2. Original CAF (a) with three spots (0.1 µl) for each concentration of a series of dilutes stained with Coomassie blue. The numbers give the amount of protein for each spot. The same membrane on the display of the television densitometer (Quantimet 720 D) (b): One spot containing 14.1 ng protein is accepted with the light pen. The integrated optical density (IOD 103) is given digitally on the screen.

dansyl chloride complex can be quantitatively evaluated in the range from 50 ng to 4 µg (12).

Staining with amido black or Coomassie blue is more reproducible than dansylation or the reaction with orthophthalaldehyde or fluorescamine. Moreover, the staining solutions are more stable (14) and easier to handle. The whole procedure can be carried out quicker. Interfering material can be washed out (14,15) in an additional step between fixation of the proteins onto the membrane and staining. Furthermore, no elaborate equipment is necessary because the densitometry should be possible with every conventional densitometer or photometer which has a device to scan normal CAF electropherograms.

Table 1. Staining with amido black. Measurement of the integrated optical density (IOD) of 9 spots for each concentration (n = 9).

protein (ng/0.1 µl)	IOD		
	MV	SD	%
7.2	7.9	1.4	17.9
14.1	15.2	1.9	12.7
26.3	26.6	1.2	4.5
52.3	52.4	1.3	2.5
77.5	77.5	1.4	1.8
101.2	101.2	0.8	0.9
126.3	126.4	0.8	0.7
156.6	156.5	1.5	1.0
198.6	198.5	1.8	0.9

Linear regression for 75 - 200 ng/0.1 µl
I: - 0.01 R: 0.999 $p < 0.001$
S: 0.99 N: 45

Table 2. Staining with Coomassie blue. Measurement of the integrated optical density (IOD) of 21 spots for each concentration (n = 21).

protein (ng/0.1 µl)	IOD		
	MV	SD	%
7.2	55.2	5.3	9.7
14.1	106.9	4.6	4.3
26.3	186.1	7.8	4.2
52.3	336.1	25.1	7.4
77.5	471.0	27.8	5.9
101.2	554.7	28.5	5.1
126.3	666.0	25.9	3.8
156.6	772.1	33.4	4.3
198.6	911.6	57.5	6.3

Linear regression for 7 - 75 ng/0.1 µl
I: 21.3 R: 0.992 $p < 0.001$
S: 5.88 N: 105

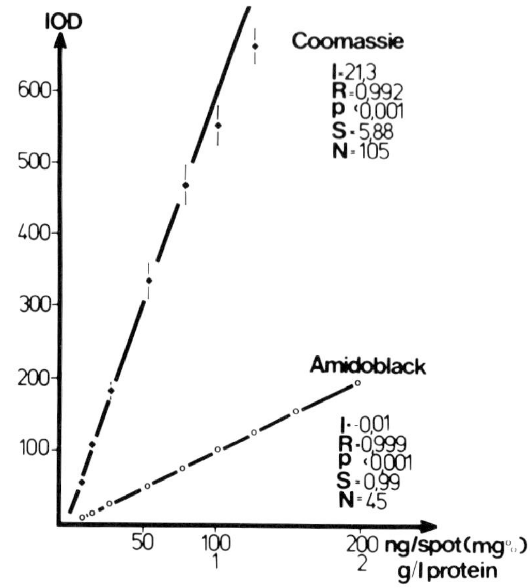

Fig. 3. Standard curves for the integrated optical density (IOD) of the protein-dye complex with Coomassie blue and amido black respectively. 0.1 µl of a series of protein solutions was spotted three times onto each membrane. The points plotted are the mean values of three membranes (9 spots) for amido black and of seven membranes (21 spots) for Coomassie blue if possible standard deviation is drawn. The lines drawn are the linear regressions calculated for 7 to 77 ng/spot for Coomassie blue and for 77 to 200 ng/spot for amido black. The statistical values are intercept (I), correlation coefficient (R), probability (p), slope (S) and number of values (N).

Amido black had already been used on membranes e.g. with extraction of the stained spot for the measurement of absorbance (13,14,15). A good linear response was obtained for 3 - 30 µg of BSA (15). With our method linearity is from 7 to 200 ng and the coefficient of variation is approximately 2 % for concentrations higher than 0.5 g/l (50 ng/spot).

With Coomassie blue the method is more sensitive. The IOD measured for the same concentration is about 6 times the value of amido black staining.(The slope of the standard curve is approximately 1 with amido black and approximately 6 with Coomassie blue). The coefficient of variation (between 5 and 10 %) is not as good as with amido black. But this variation seems to depend far more on the exactness of the sample measurement than on the staining. This conclusion can be drawn from the values given in Tab. 3. The range of the single measurements on one foil is usually bigger than the variation between different membranes. To achieve more confidence it is advisable to spot each unknown protein solution three times. Thus a coefficient of variation of about 2.5 % can be achieved.

In routine analysis, when the time of staining and destaining is not as exactly controlled as in a methodical investigation, we put three different protein solutions of known concentrations on each foil as internal standards in addition to the unknown samples.

If solutions with a protein content of lower than 70 mg/l are to be determined,"prefixed" foils can be used for application. On these CAF, soaked in sulfosalicylic acid and dried before use, volumes larger than 0.1 µl give small spots. Even when the fluid of a dilute protein solution gives an area of approximately 5 mm in diameter the protein is immediately precipitated during the application and gives a spot with a diameter below 1 mm. With this method samples with a concentration of 5 - 10 mg/l (15) can possibly be assayed in microliter amounts.

Morever, the sensitivity of this method may be increased by monochromatic light or small band filters. With the "very broad band filter" used in our investigation the densitometric reading was enhanced from about 90 with "white light" to about 190.

The colour yield for amido black is different for different proteins (15). For Coomassie blue an even greater scatter (21 %) was reported by Fazekas de St. Groth et al. (23), who additionally observed a deviation from Beer's law at high protein concentrations which is confirmed in our measu-

Table 3. Calculation of accuracy for triplicate determinations. (Protein concentration of solution 0.26 g/l, 26.3 ng/spot (0.1 µl).

	IOD	
foil no.	single measurement	mean value of one foil
1	191 [92]+)	
	187 [91]	
	193 [94]	190.33
2	182	
	182	
	164	176.0
3	200	
	182	
	176	186.0
4	186	
	184	
	200	190.0
5	193	
	188	
	181	187.33
6	182	
	185	
	191	186.0
7	188	
	186	
	188	187.33
MV	186.14	186.14
SD	7.84	4.8
%	4.2	2.6
SE	1.71	
%	0.9	
range	164 - 200	176 - 190

+) The values in brackets give the IOD reading if the whole spectrum of halogen lamps (white light) is used.

rements (Fig. 3). If an exact quantitative evaluation of different proteins is necessary, for Coomassie, the specific colour yield for each protein should be determined.

In conclusion, the advantages of the method reported herein are 1. high sensitivity, allowing the determination of 7 ng of protein at a concentration of 70 mg/l with a reproducibility of ± 2 % for amido black and ± 5-10 % for Coomassie blue on different membranes, 2. insensitivity to interference by nonproteinaceous compounds, 3. the possibility to analyse submicroliter samples. The manual efforts are less then in most other sensitive techniques because handling of test tubes, filter arrangements and cuvettes is avoided by carrying out the whole procedure on the cellulose acetate foil. The disadvantages of different colour yields dependent from the time of staining and destaining can be avoided by internal standards on each membrane.

References

1. Daughaday, W.H., Lowry, O.H., Rosenbrough, N.J., Fields, W.S.: J.Lab.clin.Med. 39, 663-665 (1952).

2. Itztaki, R.F., Gill, D.M.: Anal.Biochem. 9, 401-410 (1964).

3. Waddel, W.J.: J.Lab.clin.Med. 48, 311-314 (1956).

4. Böhlen, P., Stein, S., Dairman, W., Udenfriend, S.: Arch. Biochem.Biophys. 155, 213-220 (1973).

5. Böhlen, P., Stein, S., Imai, K., Udenfriend, S.: Anal. Biochem. 58, 559-562 (1974).

6. Taylor, S., Tappel, A.P.: Anal.Biochem. 56, 140-148 (1973).

7. Benson, J.R., Hare, P.E.: Proc.Nat.Acad.Sci. 72, 619-662 (1975).

8. Weidekamm, E., Wallach, D.F.H., Flückinger, R.: Anal.Biochem. 54, 102-114 (1973).

9. Matschinsky, F.M., Passonau, J.V., Lowry, O.H.: J.Histochem.Cytochem. 16, 29-39 (1968).

10. Lowry, O.H., Passonau, J.V.: A Flexible System of Enzymatic Analysis, Academic Press, New York 1972

11. Butcher, E.C., Lowry, O.H.: Anal.Biochem. 76, 502-523 (1976).

12. Kinoshita, T., Iinuma, F., Tsuji, A.: Anal.Biochem.66, 104-109 (1975).

13. Kuno, H., Kihara, H.K.: Nature 215, 974-975 (1967).

14. Nakao, T., Nakao, M., Nagai, F.: Anal.Biochem. 55, 358-367 (1973).

15. Schaffner, W., Weissman, C.: Anal.Biochem. 56, 502-514 (1973).

16. Kaiser, R.E.: Dosiertechnik in der HPLC, Institut für Chromatographie, Bad Dürkheim 1976

17. Kohn, J.: Clin.Chim.Acta 2, 297-303 (1957).

18. Maurer, H.R.: Disc Electrophoresis and Related Techniques of Polyacrylamid Gel Electrophoresis, Second revised and expanded edition, Walter de Gruyter, Berlin-New York 1971

19. Giebel, W.: Fortschritte der quantitativen Bildanalyse (Imanco GmbH.), 295-302 (1976).

20. Giebel, W., Striebel, H.M.: Mikroskopica Acta Suppl. 1, 133-149 (1977).

21. Kollegium Biomathematik NW.: Biomathematik für Mediziner, Second revised edition, Springer-Verlag, Berlin-Heidelberg-New York 1976

22. Diem, K., Lentner, C.: Documenta Geigy, Wissenschaftliche Tabellen, Seventh revised edition, J.R.Geigy A.G., Basel 1969

23. Fazekas de St. Groth, S., Webster, R.G., Datyner, A.: Biochim.Biophys.Acta 71, 377-391 (1963).

+ Dedicated to Professor Dr.G.Pfleiderer (director of the Biochemical Institute of Stuttgart University) on the occasion of his 60th birthday.

++ Supported by a grant of Deutsche Forschungsgemeinschaft (Gi 59/5).

A METHOD FOR THE DETERMINATION OF CORTICOTROPIN (ACTH) IN RAT PLASMA AND RAT PITUITARIES

Wolfgang Klingler
Wissenschaftliche Abteilung, Amersham Buchler,
D-3300 Braunschweig

Klaus Dieter Döhler, Chun Cheung Wong
Medizinische Hochschule, Abteilung für Klinische
Endokrinologie, D-3000 Hannover

Introduction

Twenty years ago R. Yalow and S. Berson (1) developed the first radioimmunoassay. Meanwhile a variety of substances, particularly steroidhormones and hormones from the pituitary gland, are routinely determined by this method. The measurement of corticotropin (ACTH) by radioimmunoassay, however, is still a problem. The reasons are manifold.

1. ACTH concentrations in blood are extremely low (pg/ml).
2. ACTH is metabolized rapidly in fresh blood (2). The concentration of exogenously applied ACTH to fresh blood decreases significantly within a few minutes and it completely disappears within a few hours. ACTH degradation is probably induced by enzymes localized on red blood cells.
3. Also in plasma, ACTH immunoreactivity decreases to some extent (2, 3).
4. Blood sampling by venous puncture is a stress situation and may stimulate ACTH secretion. Thus values may be falsified.

Structure and Activity of Natural Peptides

All these reasons are not only valid for ACTH determination in humans, but for rats as well.

Material and Methods

Preparation of pituitary and plasma samples:
To avoid stress-induced ACTH release rats should rapidly be decapitated with a minimum of preexperimental handling. Blood should be collected into heparinized (10 IU/ml) polystyrene tubes. To avoid hormonal degradation the blood should be centrifuged immediately and the plasma should be stored frozen until assay determination. As ACTH is adsorbed to glass very rapidly, the use of glass ware (i. e. syringes, tubes, pipettes etc.) at any step of the procedure -beginning with blood or pituitary collection and ending with the radioimmunoassay procedures- should be avoided.

Pituitaries were collected from the rats immediately after decapitation. They were homogenized in 3 ml of physiological saline containing 0.05 M phosphate and 1% bovine serum albumin (pH 7.6). After centrifugation the supernatant was stored frozen until assay determination.

Assay procedures:
Because of the low basal concentrations of ACTH in the plasma, extraction from plasma and subsequent concentration of the hormone is required. A plasma sample of 1 ml or more should be used. The measurement of ACTH in rat plasma or pituitary extracts can be divided into three steps:

Extraction
Evaporation of the extract
Radioimmunoassay.

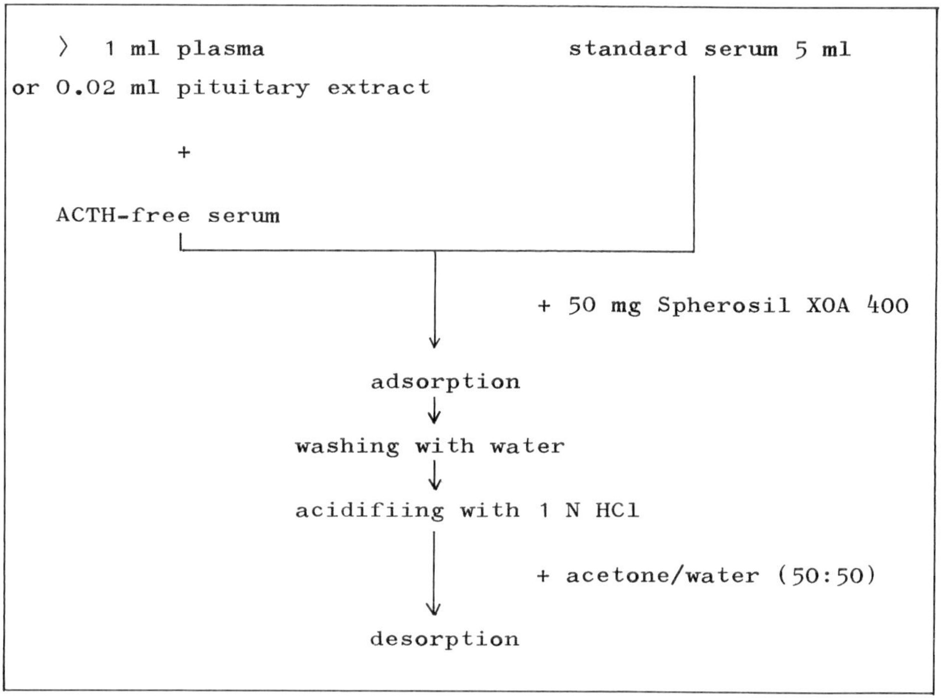

Fig. 1. Extraction procedure.

The extraction procedure (Fig. 1) is a slightly modified method of J. C. Ratcliffe and C. R. W. Edwards (4). At least 1 ml of rat plasma or 0.02 ml of pituitary extract solution is adjusted to 5 ml with ACTH-free human serum. The standard serum is prepared by adding 4 ng of ACTH to 5 ml ACTH-free human serum. 50 mg of Spherosil XOA 400 is given to each plasma/serum and this mixture is agitated on a rotary mixer for 30 minutes. ACTH is adsorbed to the Spherosil particles during this step. After centrifugation the supernatant liquids are discarded and the particles are washed with water to remove residues of plasma or serum. After acidifiing with 1 N HCl ACTH is desorbed from the particles with acetone/water (50:50).

The extraction yield is 50% on the average determined with radioactive tetracosapeptide (1-24 ACTH). Tab. 1 shows the extraction yield and the loss of ACTH during the procedure.

	A_1	A_2	A_3
Serum residue	21.05	18.34	21.87
Water	0.94	0.88	0.90
HCl	25.36	24.78	26.34
Acetone/Water	49.81	50.73	50.92
Spherosil (after desorption)	4.18	4.11	4.37
	101.34	98.84	104.35

Tab. 1. Extraction yield of ACTH (all data expressed in % of total activity).

Because of the ACTH loss during the acidifiing step we have tested the possibility to combine the acidifiing and desorption step by introducing a mixture of 1 N HCl/acetone (50:50). The results in Table 2 are promising.

	B_1	B_2	B_3
Serum residue	18.67	16.63	18.29
Water	0.91	0.57	0.87
HCl/acetone	76.27	78.57	77.77
Spherosil (after desorption)	6.10	5.90	6.19
	101.95	101.67	103.12

Tab. 2. Extraction yield of ACTH with shortened procedure (all data expressed in % of total activity).

The acetone/water or the HCl/acetone solutions are evaporated to dryness by immersing the tubes in a water bath at 55 ^{o}C and passing a gentle nitrogen stream on to the surface of the liquids. The residue of the standard serum is dissolved in 700 µl of a barbiturate buffer, the residues of the rat plasmas or pituitary extracts are dissolved in 300 μl of this buffer.

Five standard solutions for the radioimmunoassay are prepared by serial dilutions of the standard serum extract solution. A commercial radioimmunoassay (ACTH-Radioimmunoassay Kit, The Radiochemical Centre, Amersham, England) is used for the measurements. The flow scheme of this assay is shown in Fig. 2.

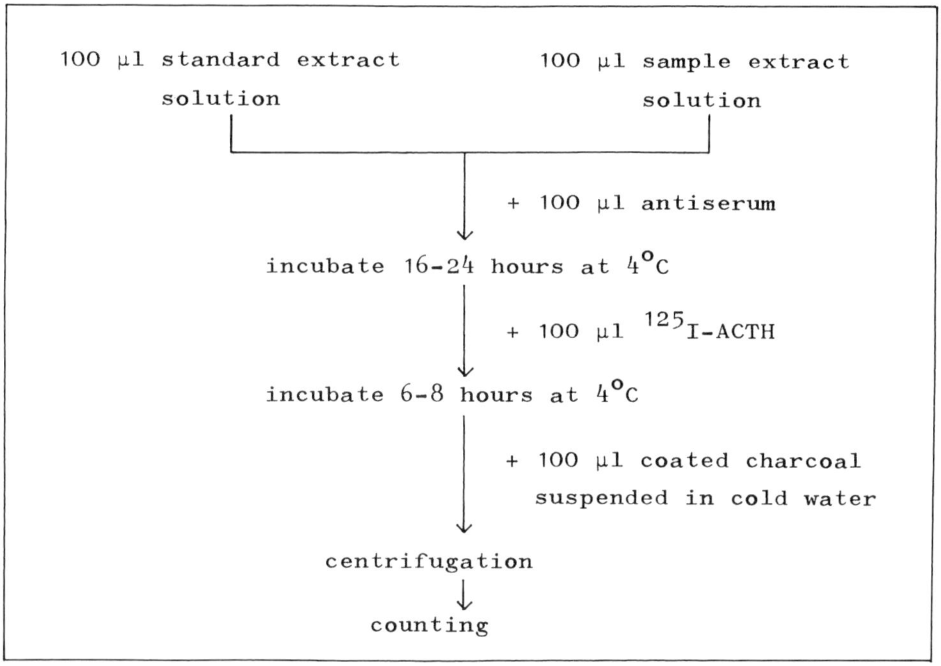

Fig. 2. Radioimmunoassay.

The radioactive antigen is synthetic 1-24 ACTH (tetracosapeptide) labelled with ^{125}I using the chloramine T method. The specific activity is 200 μCi/μg i.e. each molecule contains 0.3 iodine atoms.

The antibody reacts with the N-terminal amino acid sequence of the molecule (1-24). This sequence is identical in most mammalian animals. Fig. 3 shows the cross-reactivity of the antibody to ACTH and some parts of the molecule.

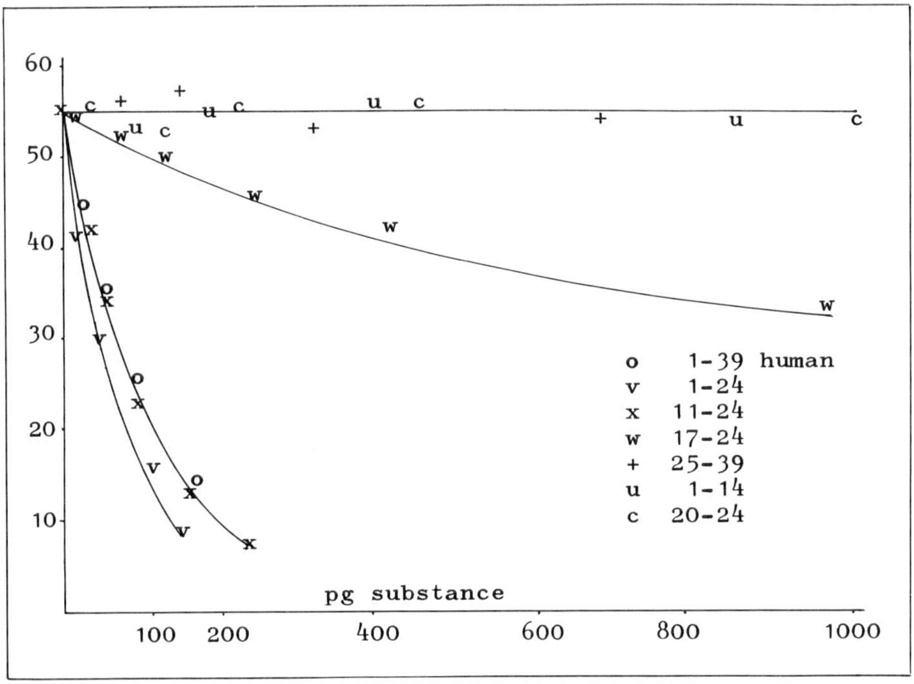

Fig. 3. Cross-reactivity of antibody to ACTH.

The ACTH concentration of rat plasma or pituitary gland is calculated by use of the formula:

$$x \text{ pg/ml} = c \times \frac{5}{V} \times 0.43$$

c = interpolated concentration from standard curve

V = volume of unknown plasma sample or volume of pituitary extract solution

0.43 = correction factor, because the standard serum extract is dissolved in 700 µl buffer and unknown sample extracts are dissolved in 300 µl buffer.

ACTH levels of three groups of rats were tested with this assay.

Group 1:
Adult male Sprague-Dawley rats were housed in an air-conditioned ($20^o \pm 1^oC$) and light-controlled (light period from 7 h to 19 h) animal room under specific pathogen-free conditions. They received food (Altromin 1320) and water ad libitum and remained undisturbed for several hours before the experiment was started.

At the time of experiment six animals were moved into an adjacent room, where they were decapitated immediately. Subsequently another six animals were moved into the same room but they were decapitated 20 minutes later. Blood was collected into heparinized plastic vials and was centrifuged immediately. The plasma was kept frozen until the performance of the assay.

Group 2:
Adult male Wistar rats were housed in an animal room under day-light conditions (light period from approximately 4 h to 20 h). They received food and water ad libitum. The animals could not be considered completely undisturbed, as a number of people had entered the room repeatedly before the experiment was started.

At the time of experiment eight rats were moved into an adjacent room. Four of them were decapitated immediately. The remaining four animals were disturbed repeatedly by opening and closing the lid of their cage every five minutes and by decapitating a number of rats right next to their cage. After a period of 20 minutes these four animals were also decapitated. Blood samples were collected as described for group 1.

Group 3:
Newborn male and female Wistar rats were raised in our air-conditioned ($20^o \pm 1^oC$) and light-controlled (light period from

7 h to 19 h) animal room. At 25 days of age they were separated from their mothers and received food (Altromin 1320) and water ad libitum. On day 35 of age ten males (55 to 85 g body weight) and ten females (65 to 80 g body weight) were decapitated rapidly without anaesthesia.

Their blood was collected as described for group 1 and their pituitary glands were removed and homogenized as described.

Results

Plasma ACTH levels in the unstressed Sprague-Dawley rats of group 1 ranged from 7 to 42 pg/ml (n = 6, $\bar{x}$ = 25.5 $\pm$ 5.9 (SEM) pg/ml). The animals decapitated 20 minutes later showed plasma ACTH concentrations from 64 to 125 pg/ml (n = 6, $\bar{x}$ = 84.8 $\pm$ 9.8 (SEM) pg/ml).

Plasma ACTH levels were higher in the Wistar rats of group 2. Immediately after removel from the animal room they ranged from 91 to 138 pg/ml (n = 4, $\bar{x}$ = 108.3 $\pm$ 10.3 (SEM) pg/ml). After repeated disturbance the animals exhibited values from 247 to 801 pg ACTH/ml (n = 4, $\bar{x}$ = 411 $\pm$ 130.5 (SEM) pg/ml).

The male pubertal rats of group 3 had plasma levels from 14 to 68 pg ACTH/ml (n = 9, $\bar{x}$ = 34.3 $\pm$ 1.9 (SEM) pg/ml). The tenth animal with a value of 419 pg ACTH/ml was not used for the calculations. The reason for this high value is unknown. Pituitary ACTH contents in these animals is 58.4 $\pm$ 3.61 (SEM) ng/mg pituitary gland.

The female pubertal rats of group 3 showed plasma levels from 16 to 66 pg ACTH/ml (n = 10, $\bar{x}$ = 45.9 $\pm$ 5.3 (SEM) pg/ml). The pituitary content is 34.5 $\pm$ 3.38 (SEM) ng/mg pituitary gland.

Discussion

These results demonstrate increased plasma ACTH levels in male Sprague-Dawley rats 20 minutes after cage transport and in male Wistar rats 20 minutes after repeated disturbance. The higher basal ACTH levels in the adult Wistar control rats, compared to the levels of adult Sprague-Dawley control rats, were probably due to the difference in housing conditions, rather than due to strain differences.

Plasma ACTH levels were similar in unstressed pubertal male and female Sprague-Dawley rats and they were in the same range as the levels of unstressed adult male Sprague-Dawley rats. Pituitary concentrations of ACTH were, however, significantly ($p < 0.001$) higher in males than in females.

References

1. Yalow, R.S., Berson, S.A.: J. Clin. Invest. 39, 1157-1175 (1960).

2. Besser, G.M., Orth, D.N., Nicholson, W.E., Bynny, R.L., Abe, K., Woodham, J.P.: J. Clin. Endocrinol. Metab. 32, 595-603 (1971).

3. Demura, H., West, C.D., Nugent, C.A., Nakagawa, K., Tyler, F.H.: J. Clin. Endocrinol. Metab. 26, 1297-1302 (1966).

4. Ratcliffe, J.G., Edwards, C.R.W.: In: Radioimmunoassay Methods, Kirkham, K.E., Hunter, W.M. (eds.), Churchill Livingstone, Edinburgh 502-512 (1971).

INVESTIGATION OF INTERACTIONS BETWEEN IMMOBILIZED LYSINE RESIDUES AND OLIGODEOXYRIBONUCLEOTIDES BY A COLUMN CHROMATOGRAPHIC PROCEDURE

Herbert Schott, Heiner Eckstein
Institut für Organische Chemie der Universität Tübingen
D-7400 Tübingen-1, Germany

Introduction

It is well-known that lysine-rich histones interact with A·T-rich DNA more strongly than with G·C-rich DNA (1-3) and their amino acid sequences contain several consecutive lysine residues (4). It is assumed, that the basic regions of histones are responsible for the histone-DNA binding (5,6). At this point the question arises as to the reason for this A·T specificity. It is conceivable that the preference of the A·T-rich DNA is due to the tertiary structure of the lysine-rich histones. It is, however, also possible that the lysine-rich regions of the histones resemble "active centres" which cause the A·T specificity. If the second assumption is correct, oligolysine peptides should interact preferentially with oligonucleotides rich in A and T rather than with corresponding C- and G-rich fragments. To settle this question, which is of considerable importance in respect of the DNA-protein recognition, we have investigated the lysine peptide-oligonucleotide interactions using a column chromatographic procedure.

In this chromatographic procedure defined lysine peptides are bound covalently to an insoluble support. Mixtures of oligonucleotides are then chromatographed on the resulting lysine peptide gels. The oligonucleotide-lysine peptide interaction can be determined directly from the retardation of the mobile oligonucleotides on the immobilized lysine peptides. The advantages of this method over physico-chemical methods lie in its simplicity and in the possibility of analysing multicomponent mixtures under simulated natural conditions. To investigate the lysine peptide-oligonucleotide interactions, lysine (gel I), pentalysine (gel II) and the synthetic undecapeptide Pro-Ala-Lys-Lys-Lys-Lys-Ala-Ala-Lys-Lys-Pro-OMe (7) are immobilized on an insoluble polyacrylic matrix (8). The undecapeptide corresponds to the region 17 - 27 of the lysine-rich histone from rabbit thymus (9).

Results and Discussion

The synthesis of gel I begins with the reaction of acrylic acid with the free α-amino group of the otherwise protected N^{ε}-butoxycarbonyl lysine methyl ester. The acrylamide of lysine thus obtained is transferred into the insoluble gel I by copolymerization with the cross-linking agent 1,4-tetramethylene dimethacrylate. After copolymerization, the protecting group in ε-position is removed, whereby the terminal carboxyl group remains protected.

To synthesize gel II an acryloyl lysine amide, which is protected in the ε-position of the lysine residue, is extended on its free carboxyl group in four condensation steps to the acrylamide of the penta-(N^{ε}-benzyloxycarbonyl)-lysine peptide. The protected peptide monomer is converted to an insoluble pentalysine peptide gel by cross-linking copolymerization with 1,4-tetramethylene dimethacrylate, followed by cleavage of the ε-amino protecting groups.

Gel III is obtained by covalent binding of the synthetic un-

The system of abbreviations is principally as has been suggested by the IUPAC-IUB commission, Eur.J.Biochem. 15, 203 (1970).

decapeptide, which carries only a free amino group, to a copolymer from acrylic acid and 1,4-tetramethylene dimethacrylate. After cleavage of the side-chain protecting groups, gel III is obtained. The terminal carboxyl group remains permanently protected as the methyl ester.

The structure of the gels I - III is shown in the scheme. All three gels represent different models of lysine-rich proteins. The lysine gel (gel I) mimics natural, lysine-rich proteins in which the lysine residues are incorporated statistically into the molecule. The structure of the lysine peptide gels (gel II and III) represent approximately lysine-rich histones, in which the lysine residues are not distributed randomly but are rather accumulated in certain regions of the molecule, as is known in the case of the H1 histone (9).

The oligopeptide-oligonucleotide interactions are determined chromatographically by the elution volumes with which the

Scheme: Structure of the gels I - III.

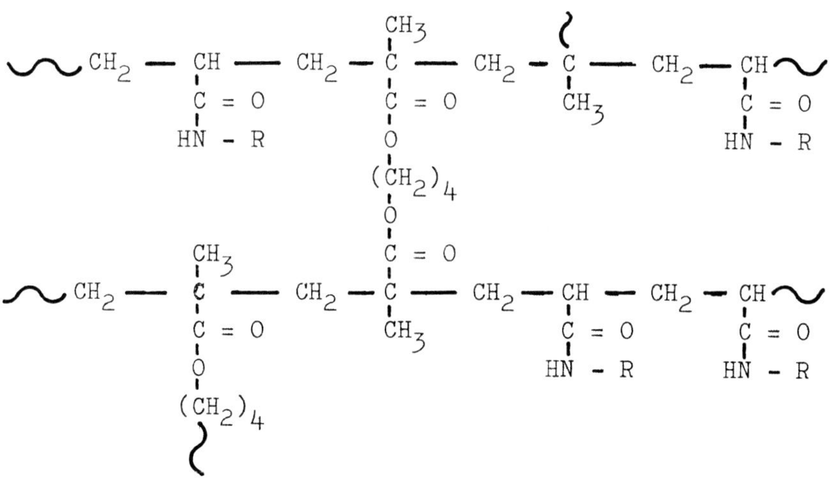

Gel I: R = Lys-OMe; Gel II: R = $(Lys)_5$;
Gel III: R = Pro-Ala-$(Lys)_4$-$(Ala)_2$-$(Lys)_2$-Pro-OMe

nucleotides are eluted from the column, either with increasing or with constant salt concentrations. The resulting elution profiles by chromatography of mixtures of oligodeoxythymidylic acids on the gels I - III show that the oligomers are retarded but only partially separated on gel I as can be seen in Fig. 1. On the gels II and III, however, the mixture is totally separated into its single components. The results demonstrate clearly that oligonucleotide-lysine peptide interactions do not originate from statistically distributed but rather from accumulated lysine residues.

Fig. 1. Chromatography of the mixture $p(dT)_{1-4}p$ a) on gel I ---- and on gel II ——, column size 23x2 cm, b) on gel III, column size 45x1 cm. The peak numbers correspond to the monomer units of the individual components. c) Chromatography of the mixture $(dT)_2$, $p(dT)_2$ and $p(dT)_2p$ on gel II, column size 23x2 cm. In all runs a linear NaCl gradient is used: 500 ml H_2O (mixing vessel), 500 ml 0.5 M NaCl, 25 mM Tris/HCl pH 7.5 (reservoir); elution temperature 25°C, flow rate a) and c) 50 ml/h, b) 25 ml/h.

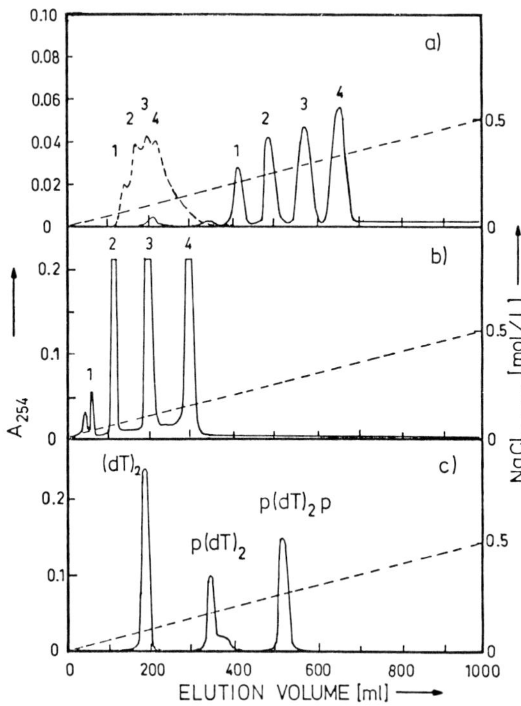

The interactions between oligonucleotides and the immobilized H1 histone fragment correspond about with those between oligonucleotides and the immobilized pentalysine peptide (Fig. 1). Therefore all the other separations are conducted only on gel II.

The strength of the interactions increases not only with increasing number of monomer units (see Fig. 1a,b) but also with increasing ratio of phosphate groups to monomer units.

With the following separations, a specific influence of the bases upon the interaction can be observed. The base specificity is demonstrated, for instance, by chromatography of mixtures of corresponding oligomers of the deoxyribothymidylic and deoxyriboadenylic acids on gel II (Fig. 2). The elution profiles of Fig. 2a - c show that the thymidylic acids pdT,

Fig. 2. Column chromatographic separation of oligomer mixtures of deoxyribothymidylic and deoxyriboadenylic acids on a pentalysine peptide gel (gel II). Corresponding pairs of the oligomers of the series $p(dT)_{1-4}$ and $p(dA)_{1-4}$ are chromatographed together. Thymidylic acids are eluted in peak 1 and adenylic acids in peak 2. A 1 M NaCl stock solution buffered with 0.05 M Tris/HCl to pH 7.5 is used and diluted to the concentration as is indicated by the broken line; elution temperature 25°C, column size 23x2 cm, flow rate 50 ml/h.

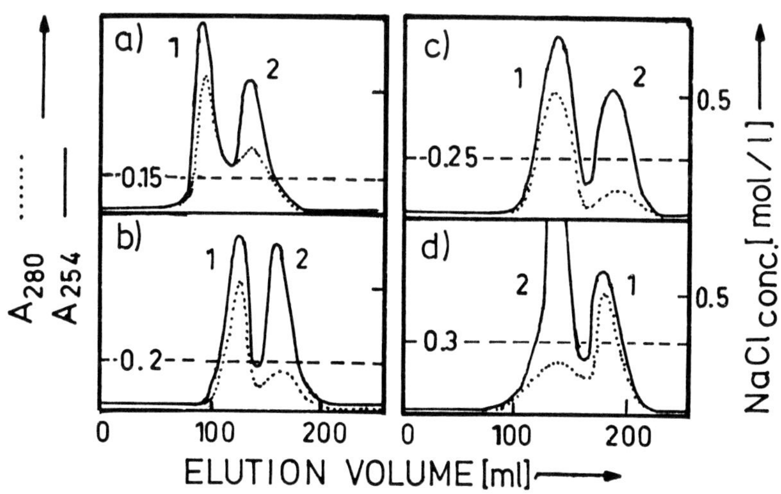

$p(dT)_2$ and $p(dT)_3$ interact with the immobilized pentalysine residues to a lesser extent than the corresponding adenylic acids pdA, $p(dA)_2$ and $p(dA)_3$. In all three runs the thymidylic acids are eluted prior to the adenylic acids. When going from the trimer to the tetramer, however, the retardation of the thymidylic acid increases discontinuously and thus the tetramer of the thymidylic acid is eluted after the tetramer of adenylic acid.

A very important result with regard to the histone-DNA recognition can be deducted from the elution profiles in Fig. 3. Chromatographic behavior of mixtures of pyrimidine nucleotides on gel II demonstrates clearly that the pentalysine peptide interacts with oligomers of deoxyribocytidylic acids to a far lesser extent than with corresponding oligomers of de-

Fig. 3. Chromatography of mixtures of pyrimidine nucleotides on gel II. The elution is conducted using the same stock solution as described in Fig. 2. The relative elution volume is determined as the quotient elution volume/bed volume. Column size 23x2 cm, elution temperature 25°C, flow rate 50 ml/h.

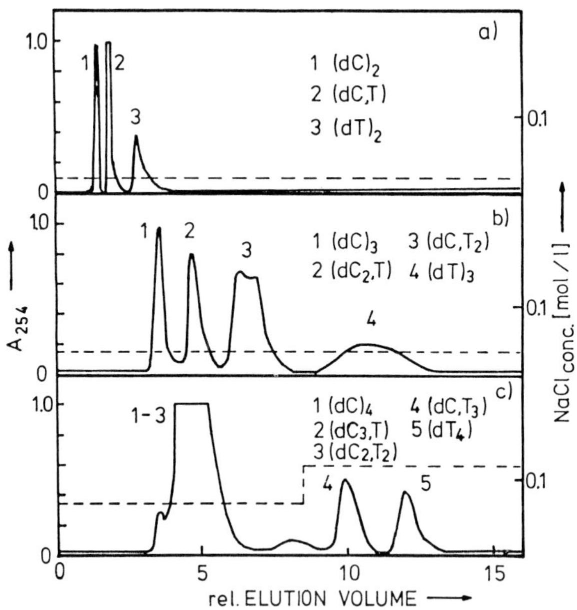

oxyribothymidylic acid. The strength of the pentalysine peptide-oligopyrimidine nucleotide interaction decreases, if thymidylic residues are exchanged by cytidylic residues.

The results of these separations are in accordance with the well-known observation that lysine-rich histones interact preferably with DNA rich in A·T. Moreover the separations demonstrate also, that the A and T specificity originates from the lysine-rich regions of histones. The interaction can occur on the level of oligonucleotide-oligopeptide and is not necessarily based on the tertiary structure of the lysine-rich histones.

Summarizing the results of these separations three conclusions can be drawn:

1. The strength of the lysine peptide-oligonucleotide interactions dependson the number of phosphate groups of the oligonucleotide.
2. The interactions are base specific: The adenylic and thymidylic acids show a higher affinity towards the pentalysine peptide than the corresponding cytidylic acids.
3. The interaction increases discontinuously when going from the trimers to the tetramers of thymidylic acids.

We explain the results with the following hypothesis: The interactions are based mainly on electrostatic forces between the positively charged ε-ammonium groups of the immobilized lysine peptide residues and the negatively charged phosphate groups of the oligonucleotides. This assumption explains the increase of the interaction with increasing phosphate residues.

Considering electrostatic forces exclusively, one can explain neither the base specificity nor the discontinuous increase of the interaction when going from the trimer to the tetramer of thymidylic acid. We therefore propose a so-called "conformation fit mechanism" with which we can give a reasonable explanation for these effects: the lysine peptide-oligonucleotide interaction is dominated by the arrangement of the ε-ammonium groups. Lysine residues, which are distributed statistically as for instance in gel I, give rise only to an unspecific adsorption of oligonucleotides. Selectivity of the lysine peptide-oligonucleotide interaction occurs only if the positively charged groups are arranged in a specific mode as is the case for instance in the lysine peptides of the gels II and III. Oppositely charged partners can approach only to such an extent as is permitted by their conformations. The conformation of oligonucleotides depends strongly upon the nature of the constituent nucleobases. It therefore can be assumed that the electrostatic forces which principally exist between the ε-ammonium groups of lysine residues and the phosphate groups of the nucleotides can be exploited to differing degrees. In case of an optimal fit of the conformation of both partners, the electrostatic forces can be exploited maximally. Moreover, these forces will be enhanced by additional physical forces always existing between groups in very close contact (e.g. from surface charge densities, hydrophobic and van der Waals' forces etc.). These principally unspecific forces intensify the electrostatic interaction resulting in the discontinuous increase of the lysine peptide-oligonucleotide interaction in the case of the tetrameric deoxyribothymidylic acid. With the aid of such a "conformation fit mechanism", nature can create highly specific interactions by modulation of unspecific forces.

According to our new and earlier published results (10, 11) we propose the following mechanism of the interaction between lysine-rich histones and DNA: most of the uncharged amino acids in a histone molecule are probably involved in the formation of a certain tertiary structure in which lysine residues are exposed on the protein surface (12). These regions may act in a similar way to a receptor, causing histone molecules to associate with any negatively charged region of the DNA. After formation of a rather loose dynamic contact between the histone and the DNA in the first step, the histone molecule can now in a second step brush along the DNA and search for optimal regions, using its "receptor". Along the DNA, which at first appears relatively featureless, there are regions which, as a consequence of the sequence of the nucleobases, differ slightly both in their negative partial charges and in their conformation. This enables a histone molecule to find all those regions of the DNA amongst the DNA's numerous fine structured areas which fit best to the conformation of its receptor. At these DNA segments, which have to be, according to our results, rich in A and T, the possible additional forces become now manifest and the histone adheres to this region.

References

1. Sawecka, J., Kornacka, L., Malec, J.: Experientia 34, 309 - 310 (1978).
2. Li, H.J., Leffak, I.M., Maciewicz, R.A., Levy, D.: Stud. Biophys. 67, 97-98 (1978).
3. Brongoszewska, M., Malarska, K., Opacka, J. Plucienniczak, A.: Stud. Biophys. 69, 79-80 (1978).
4. Huntley, G.H., Dixon, G.H.: J. Biol. Chem. 247, 4916-4919 (1972).
5. Bradbury, E.M., Chapman, G.E., Danby, S.E., Hartman, P.G., Riches, P.L.: Eur. J. Biochem. 57, 521-528 (1975).
6. Palan, J., Climent, F., Avilés, F.J., Morros, A., Soliva, M.: Biochim. Biophys. Acta 476, 108-121 (1977).
7. Eckstein, H., Bayer, E.: in preparation.
8. Eckstein, H., Schott, H.: Makromol. Chem., in press.
9. Rall, S.C., Cole, R.D.: J. Biol. Chem. 246, 7175-7190 (1971).
10. Schott, H., Eckstein, H.: Hoppe-Seyler's Z. Physiol. Chem. 360, 1689-1692 (1979).
11. Schott, H., Eckstein, H.: Eur. J. Biochem., in press.
12. Chapman, G.E., Hartman, P.G., Cary, P.D., Bradbury, E.M., Lee, D.R.: Eur. J. Biochem. 86, 35-44 (1978).

IV. PEPTIDE SYNTHESES

TOTALLY SYNTHETIC THYMOSIN-α_1, A LYMPHOCYTES STIMULATING POLYPEPTIDE OF THE THYMUS.

Christian Birr, Ulrich Stollenwerk

Max-Planck-Institute for Medical Research, Jahnstr. 29
D-6900 Heidelberg 1, F.R.G

Ortfried Brodner, Hans Georg Manke

Medical Clinics of the University, Bergheimer Str. 58
D-6900 Heidelberg 1, F.R.G

Introduction

The potential of polypeptides from the thymus for helpful pharmaceuticals against immune deficiency diseases in man is best demonstrated by the fact that cell-free protein extracts from calf thymus were found to suppress with varying success immune deficiency effects like diminished rejection of skin grafts, increasing susceptibility to infections, accelerated aging and enhanced probability of tumor formation (1). Nowadays, several clinical applications of protein extracts from calf thymus in patients with leucemia, other types of cancer and particularly small lung cell cancer are already reported to show curative effects (2)(3). One of the acidic components of the protein extract from thymus has been separated and analyzed in its sequence in 1977 by A.L.Goldstein and others (4). This polypeptide of 28 amino acid residues was named thymosin-α_1.

Having a long-standing fascination in the effects of thymus polypeptides on the differentiation and maturation of pre-T-lymphocytes into immune-competent T-cells, the publication of the sequence of thymosin-α_1 kindled our eager interest to search for a synthesis of this molecule and to investigate its biological activities and the potential of the synthetic fragments.

Structure and Activity of Natural Peptides

Methods

In the recent past we had experienced (5) the particular usefulness of excess N-(3,5-dimethoxybenzyl-α,α-dimethyl-oxycarbonyl)-amino acids (Ddz-amino acids)(6) in the mixed anhydride method. In our modernized variant of this famous classical Wieland method of peptide synthesis there is no longer the need for intermediate work-up. Our procedure with Ddz-amino acids results in analytically pure peptides ready for further peptide elongations within 24 hours.

Because of the lability of the temporary Ddz-protector to 1-5% trifluoroacetic acid , residues like glutamyl and aspartyl as well as seryl and threonyl in the fragments were advantageously protected on side functions by the tert.butyl group. This enhanced the lipophilicity of the thymosin-α_1 fragments we had synthesized by the method mentioned to such an extent that all building blocks for further fragment condensations were found soluble in dichloromethane. This is worth mentioning since the 28 amino acids of thymosin-α_1 include 6 glutamyl and 3 each of aspartyl, seryl and threonyl residues. As we also experienced, the N-terminal acetyl moiety of thymosin-α_1 very much decreases the lipophilic solubility of intermedate fragments. The synthetic units are shown in figure 1. From all mixed anhydride syntheses of the Ddz-protected fragments of thymosin-α_1 one can calculate an average yield per stepwise N-terminal peptide elongation of 83 %. As indicated in figure 1 this clearly demonstrates the superiority of our modernized method of peptide synthesis in solution.

We therefore like to explain our decision not to perform the thymosin-α_1 synthesis on polymer support, despite having recently demonstrated some success with the Merrifield method both in the total synthesis of the MCD-peptide (mast cell degranulating) (7) and in the synthesis of fully active insulin (8). First, by the thymosin-α_1 synthesis in solution we wanted to demonstrate the ease and efficiency of the application of excess Ddz-amino acid mixed anhydrides in solution on the preparation of a biologically relevant polypeptide mainly consisting of trifunctional amino acids. Second, in the quick search for an optimal route for the synthesis of an undescribed polypeptide from fragments of unknown physicochemical characteristics, we found it less time-consuming first to gather the necessary knowledge on the preparative details from an efficient method of synthesis in solution. Third, we are aiming at the transfer of this knowledge into a repeated synthesis of thymosin-α_1 on polymer phase utilizing the identical set of protective groups and fragments in a polystyrene gel phase (9) as explored in solution.

	FULLY PROTECTED THYMOSIN-α_1 FRAGMENTS	YIELD/ STEP [%]
I	DDZ-SER(BU^t)-ASP(OBU^t)-ALA-ALA-VAL-ASP(OBU^t)-OBZL	83
II	DDZ-THR(BU^t)-SER(BU^t)-SER(BU^t)-GLU(OBU^t)-ILE-THR(BU^t)-OBZL	86
III	DDZ-THR(BU^t)-LYS(Z)-ASP(OBU^t)-LEU-LYS(Z)-GLU(OBU^t)-LYS(Z)-OBZL	83
IV	DDZ-LYS(Z)-GLU(OBU^t)-VAL-VAL-GLU(OBU^t)-OME	88
V	DDZ-GLU(OBU^t)-ALA-GLU(OBU^t)-ASN(MBH)-OBZL	76
		83$^{a)}$

a) overall average yield/step of peptide synthesis

Figure 1. Ddz-peptide esters of thymosin-α_1 fragments prepared from Ddz-amino acid mixed anhydrides.

For the preparation of useful fragments we had cut the sequence of thymosin-α_1 into five portions. As can be seen from figure 1 the cuts were placed between identical or similar, polar amino acid residues with side functions protected by the lipophilic tert-butyl group. The guiding idea was to introduce a lipophilic interplay of those "fatty coated" polar side functions. In a polar solvent like dimethylformamide the formation of fatty clusters might hold the fragments close together to favour the fragment condensation reaction with dicyclohexylcarbodiimide/ 1-hydroxybenzotriazole (10). Though we changed the order of the fragment condensations during our search for an optimal synthesis, all cut positions remained unchanged. This indicates the versatility of the above mentioned concept of fatty cluster formation for the efficient condensation of the fragments.

Results

In our first trial for the total synthesis of thymosin-α_1 we were confronted with insolubility problems related to the N-terminal acetyl-hexapeptide. On elongation of that peptide by the 7-12 fragment (61% yield) the acetyl-dodecapeptide resulted, which was hardly useful in the final fragment condensation because of its very poor solubility (yield of fully protected thymosin-α_1 raw material 17%). Since the fragment condensations in the C-terminal region of the target molecule proceeded quite satisfactorily, our further trial for the total synthesis was concentrated on alterations in the (1-12) fragments. To overcome the insolubility problem caused by the acetyl-hexapeptide (1-6) we at first prepared the Ddz-(1-6) methyl ester. However, on the saponification with aqueous sodium hydroxide in dioxane/methanol we had to detect an α-β rearrangement of the aspartyl-alanyl peptide bond in position 2-3 by its attack on the β-tert.-butyl ester moiety at the aspartyl residue. Therefore, we finally synthesized Ddz-(1-6) benzyl ester and introduced the acetyl residue prior to the hydrogenolytic cleavage of the benzyl ester. The insolubility problem was solved by condensation of the fragment II (7-12) not with Ac-(1-6) but with the lipophilic bigger C-terminal (13-28) benzyl ester as presented in figure 2. Starting from fragment V (25-28) benzyl ester, the total synthesis of thymosin-α_1 successfully was performed by stepwise condensation with the N-terminally next fragment. In these syntheses the head components always were used in a two-fold amount.

From the fully protected end product thymosin-α_1 was released in 90% overall yield from three steps of deprotection. From model studies we had experienced first to cleave the benzyl ester and benzyloxycarbonyl groups by catalytic hydrogenolysis in 2,2,2-trifluoroethanol with Pd/C as catalyst, second to cleave mainly the tert-butyl esters and third, in the presence of anisole to cleave tert-butyl ethers and the C-terminal 4,4'-dimethoxybenzhydryl (Mbh) residue with concentrated trifluoroacetic acid. This procedure with stepwise increased acidity of the cleavage medium resulted in thymosin-α_1 of a very satisfying quality. It is worth mentioning that our 90% endproduct yield can be compared with a 9.2% yield from another trial for the total synthesis of thymosin-α_1 (11), where S.S. Wang and coworkers on deprotection of their entirely benzyl-protected endproduct with liquid hydrogen fluoride destroyed the main portion of the molecule.

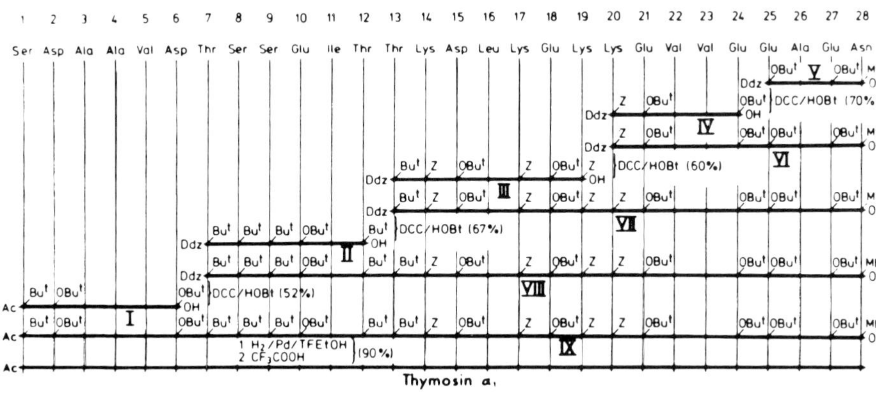

Figure 2. Scheme of the successful total synthesis of thymosin-α_1 (12) by stepwise fragment condensation in solution starting from the C-terminus. Two equivalents of each head component with dicyclohexylcarbodiimide/1-hydroxybenzotriazole were utilized for condensation.

The above mentioned yield of our totally synthetic thymosin-α_1 (90%) was obtained after chromatographic purification on BIOGEL P-6 in water containing 10% of 2,2,2-trifluoroethanol and 1% acetic acid (v/v).

Amino acid analyses after total hydrolysis in 6n HCl,110°C,24-72h:

	Asp	Thr	Ser	Glu	Ala	Val	Ile	Leu	Lys
theoretically	4	3	3	6	3	3	1	1	4
fully protected	3.65	2.76	2.45	6.05	2.23	2.51	1.00	1.09	4.08
free thymosin-α_1	3.54	2.70	2.58	6.16	3.13	2.46	0.96	1.00	4.08
electrofocussing	4.00	2.80	2.58	5.31	2.96	2.71	0.95	1.00	4.54

Figure 3. Amino acid analyses of the final preparations of totally synthetic thymosin-α_1.

On electrofocussing of our synthetic thymosin-α_1 in comparison with thymosin fraction no.5 [a standardized mixed preparation of polypeptides from calf thymus (4)] our material showed one acidic band at about pI 3.9, which only faintly stained with Coomassi Blue (figure 4). This faint band was also present in the mixture of natural thymus polypeptides. Goldstein (4), however, had reported pI 4.2 for thymosin-α_1, which was not in agreement with our synthetic product. We therefore isolated the strongly staining band at pI 4.2 of the electrofocussed fraction no.5 from natural sources and performed amino acid analyses in comparison with the data presented in figure 3. This resulted that the band at pI 4.2 is an artificial ampholine complex of about 10% polypeptide content with an analysis completely different from thymosin-α_1. We therefore like to stress that thymosin-α_1 under the conditions reported (4) shows pI 3.9 and only faintly stains with Coomassi Blue (13).

In the CD spectrum, as presented in figure 5, our synthetic thymosin-α_1 shows some indications for a partly helical character of the molecule, which also can be estimated by the method of Chou and Fasman (14). The calculation results two β-bends in the region 7-12 and 14-17, β-pleated sheet structure at 11-13 and a helical portion between 17-27, as shown also in figure 5 together with some ORD data.

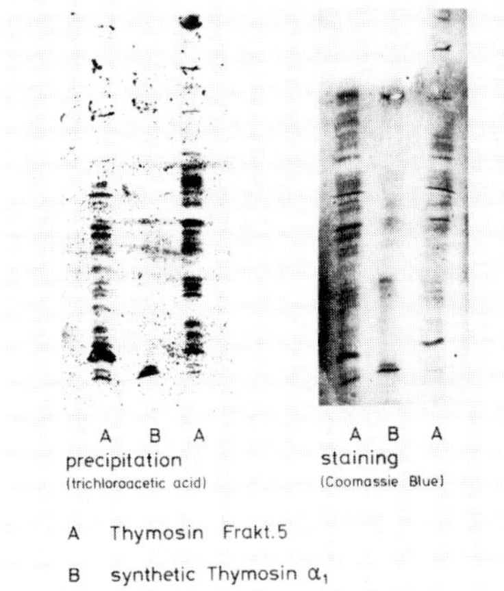

Figure 4. Isoelectric focussing at pH 3.5-5.5 (narrow range LKB plate) of synthetic thymosin-α_1 compared to natural thymus polypeptides (thymosin fraction no.5).

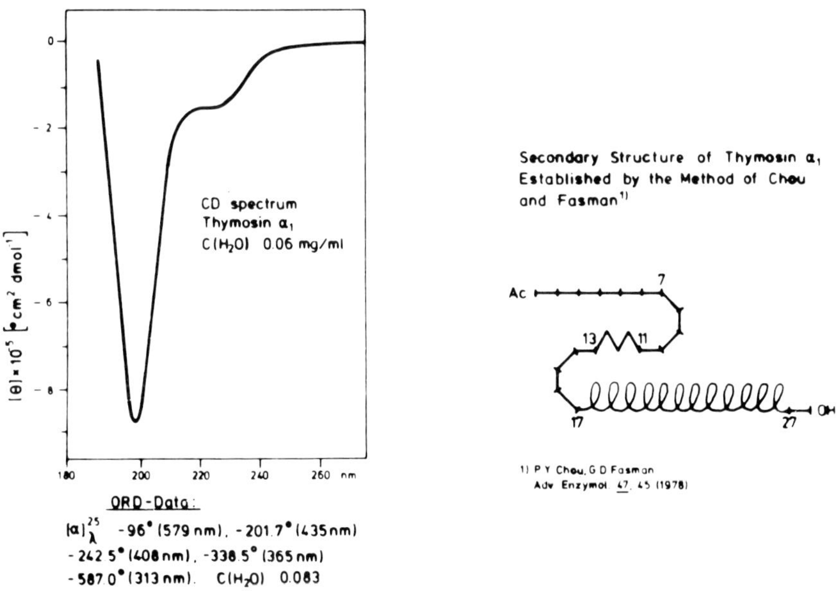

Figure 5. The CD spectrum, a calculatory estimation of the secondary structure and ORD data of thymosin-α_1.

Biological activity

The immune stimulatory potential of the synthetic thymosin-α_1 was compared to a highly purified natural thymus polypeptide, prepared by one of us (H.G.M.) from calf thymus. This latter preparation consists only of one immune stimulatory active polypeptide, not identical to thymosin-α_1, and another inert polypeptide. With these materials the E-rosettes assay was performed on human peripheral lymphocytes. The RNA-polymerase of the cells was inhibited by α-amanitine (15).

normal cell population: 64%[a] E-rosettes
α-amanitine inhibited : 25%

thymus polypeptide[b]		synthetic thymosin-α_1[c]	
μg/culture	[%][d]	μg/culture	[%][d]
6.25	10	5.00	33
3.12	67	2.50	57
1.25	100	1.00	64
0.62	62	0.50	100
0.31	6	0.25	39

a) absolute values b) natural c) analyt. pure d) rel. amount of E-rosettes restored on stimulation of α-amanitine inhibited peripheral human lymphocytes by thymus peptides.

Figure 6. Comparative data of synthetic thymosin-α_1 and of a natural thymus polypeptide in the E-rosettes assay with α-amanitine inhibition.

As presented in figure 6, totally synthetic thymosin-α_1 was found three times more potent than the natural thymus peptide. In 100 μl of a suspension containing 5 x 10^5 cells, only 1.6 nmoles (0.5 μg) of synthetic thymosin-α_1 completely restores the capacity to form E-rosettes of lymphocytes, which previously were blocked by α-amanitine. This stimulatory potency indicates that thymosin-α_1 might reactivate the inhibited protein biosynthesis on different stages of the differentiation of surface markers. Mature lymphocytes might carry the fully developed sheep erythrocytes recognizing protein sites on the cell surface. Therefore, their contribution to the E-rosettes formation cannot be suppressed by the α-amanitine inhibition. In lymphocytes pre-mature in this respect, however, thymosin-α_1 seems to cause the release of ribonucleic acid from a pool, which cannot be inhibited by α-amanitine, resulting in an enhancement of the differentiantion, as found by one of us (H.G.M.).

Human peripheral lymphocytes. Names of donors S., We., and Wa. 1µg thymosin-α_1 and α-amanitine/ml cell suspension. 120h incubation.
r: responder cell, s: stimulator cell (mitomycin-blocked)

cell pairs		S_r/Wa_s	Wa_r/S_s	S_r/We_s	We_r/S_s	We_r/Wa_s	Wa_r/We_s
[%]	a)	20	27	4	44	27	45
	b)	86	47	37	0	1550	51

a) Stimulation by thymosin-α_1 of non-inhibited cultures
b) Cultures inhibited by α-amanitine prior to thymosin-α_1 stimulation

Figure 7. Allogenic one-way mixed lymphocytes cultures stimulated by synthetic thymosin-α_1.

A most surprising effect of our synthetic thymosin-α_1 was observed in the mixed lymphocytes culture, as presented in figure 7. In several cell cultures thymosin-α_1 stimulated the responder cells on an average of 28% over the stimulatory effect of the mitomycin-blocked lymphocytes. This effect on the average was doubled if the responder cells previously were inhibited with α-amanitine. Also a thymosin-α_1 sample from A.L.Goldstein showed this activity in our assay, though it was found inactive in another type of mixed lymphocytes culture (4).

This remarkable stimulation of the proliferation of T-lymphocytes by the totally synthetic thymosin-α_1 indicates the potential of our material for suitable pharmaceuticals against immune deficiency diseases in man. In cases of congenital and deficiencies due to aging, but also on virus infections and during the genesis of different types of cancer, the recognition and the defense of defect and malignant cells by fully differentiated immune competent T-lymphocytes needs the stimulation by a powerful drug, which in addition to its stimulatory potency does not affect the weakened body negatively. This is the chance for a future immune therapy with the natural occuring synthetic thymosin-α_1. As mentioned in the introduction, several applications already of heterogeneous polypeptide extracts from calf thymus showed curative effects on immune deficiency diseases. Recently it was demonstrated (3) that patients with small lung cell cancer on additional therapy with thymosin fraction no.5 can survive a group on conventional chemotherapy.

References

1. Bach, J.-F., Carnaud, C.: Progr.Allergy 21, 342 (1976).
2. Chretien, P. B., Lipson, S. D., Makuch, R., Kenady, D. E., Cohen, M. H., Minna, J. D.: Cancer Treat. Rep. 62, 1787-1790 (1978).
3. Lipson, S. D., Chretien, P. B., Makuch, R., Kenady, D. E., Cohen, M. H.: Cancer 43, 863-870 (1979).
4. Goldstein, A. L., Low, T. L. K., McAdoo, M., McClure, J., Thurman, G. B., Rossio, J., Lai, C.-Y., Chang, D., Wang, S.-S., Harvey, C., Ramel, A. H., Meienhofer, J.: Proc. Natl. Acad. Sci. USA 74, 725-729 (1977).
5. Birr, Chr., Nassal, M., Pipkorn, R.: Int. J. Pept. Protein Res. 13, 287-295 (1979).
6. Birr, Chr.: Liebigs Ann. Chem. 763, 162-172 (1972).
7. Birr, Chr., Wengert-Müller, M.: Angew. Chem. 91, 156-157 (1979); Angew. Chem. Int. Ed. Engl. 18, 147-148 (1979).
8. Birr, Chr., Pipkorn, R.: Angew. Chem. 91, 571-573 (1979); Angew. Chem. Int. Ed. Engl. 18, 536-538 (1979).
9. Birr, Chr.: Reactivity and Structure, Concepts in Organic Chemistry, Vol. 8, Aspects of the Merrifield Peptide Synthesis, Springer, Berlin-Heidelberg-New York, 1978.

10. König, W., Geiger, R.: Chem. Ber. 103, 788-805 (1970).

11. Wang, S.-S., Kulesha, J. D., Winter, D. P.: J. Amer. Chem. Soc. 101, 253-254 (1979).

12. Birr, Chr., Stollenwerk, U.: Angew. Chem. 91, 422-423 (1979); Angew. Chem. Int. Ed. Engl. 18, 394-395 (1979).

13. Birring, T., Baxandall, J.: Anal. Biochem. 85, 1 (1978).

14. Chou, P. Y., Fasman, G. D.: Adv. Enzymol. 47, 45-148 (1978).

15. Manke, H. G.: J. Immunol., in press (1980).

VASOPRESSIN ANALOGS FOR AFFINITY LABELING - SYNTHESIS AND ACTIVITIES ON BOVINE KIDNEY PLASMA MEMBRANES AND FROG SKIN

Falk Fahrenholz, Karl-Heinz Thierauch, Peter Crause
Max-Planck-Institut für Biophysik
D-6000 Frankfurt, Germany

Anders Johnson
Institute of Biological Chemistry A, University of Copenhagen
DK-2100 Copenhagen, Denmark

Introduction

Interaction of vasopressin (antidiuretic hormone) with its receptors on the outer surface of cells in different target organs leads to different molecular events during stimulus-response coupling: In plasma membranes derived from the kidney inner medulla of mammalians (1) as well as in amphibian urinary bladder (2) and skin (3) vasopressin activates adenylate cyclase, whereas in hepatocytes the initial consequences of hormone receptor interactions are an increase in phosphatidylinositol turnover and in the Ca^{2+}-uptake (4).

For the analysis of signal transfer through membranes isolation of the functional components is of great importance. However, the first step in this direction, the solubilization of membrane proteins, leads to difficulties: Roy et al. (5) found that treatment of plasma membranes, derived from pig kidney, by nonionic detergents diminishes the binding affinity of vasopressin receptors for the hormone and disrupts the coupling between the receptor and adenylate cyclase. The specific labeling of receptors in plasma membranes prior to their solubilization would greatly facilitate, both, the iso-

Structure and Activity of Natural Peptides

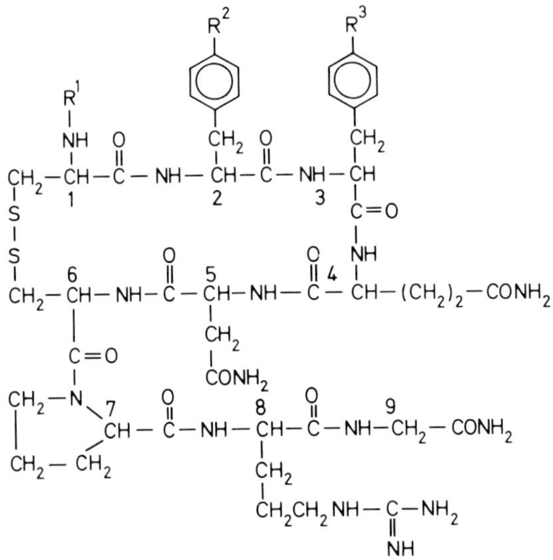

1	R^1 = H	, R^2 = OH	, R^3 = H	AVP
2	R^1 = H	, R^2 = NH_2	, R^3 = H	[Phe(p-NH_2)^2] AVP
2a	R^1 = H	, R^2 = N_3	, R^3 = H	[Phe(p-N_3)^2] AVP
2b	R^1 = Dns,	R^2 = N_3	, R^3 = H	N^α-Dns-[Phe(p-N_3)^2] AVP
2c	R^1 = H	, R^2 = $NHCOCH_2Br$,	R^3 = H	[Phe(p-$NHCOCH_2Br$)^2] AVP
3	R^1 = H	, R^2 = H	, R^3 = NH_2	[Phe^2, Phe(p-NH_2)^3] AVP
3a	R^1 = H	, R^2 = H	, R^3 = N_3	[Phe^2, Phe(p-N_3)^3] AVP
3b	R^1 = H	, R^2 = H	, R^3 = $NHCOCH_2Br$	[Phe^2, Phe(p-$NHCOCH_2Br$)^3] AVP

Fig. 1. Structural formulas of arginine vasopressin and its synthesized analogs.

lation of the receptors and the investigation of the molecular mechanisms underlying hormone action. For this purpose, five analogs of 8-arginine-vasopressin (AVP) were prepared which can react with other molecules, thereby rendering covalent incorporation into their receptors possible (Fig. 1). Tyrosine (position 2 in AVP) or phenylalanine (position 3) were replaced either by p-azido-L-phylalanine which can be activated by light (6) or by p-(bromoacetylamino)-L-phenylalanine which reacts preferentially with sulfhydryl groups In one analog the α-amino group of the N-terminal cysteine was additionally substituted by a dansyl group for fluorescence studies. In derivatives with a reactive group in position 3, tyrosine in position 2 was replaced by phenylalanine to avoid an intramolecular reaction with the phenolic hydroxyl group. The analogs were tested for their intrinsic activity and potency to activate adenylate cyclase in membrane preparations from the inner medulla of bovine kidney and for their action on frog skin epithelium.

Results

The vasopressin derivatives with reactive groups were prepared via precursors containing p-aminophenylalanine in either position 2 or 3 (7). For the incorporation of p-aminophenylalanine into the peptide chain, we used its derivatives with the tert.-butyloxycarbonyl or benzyloxycarbonyl group for permanent blocking of the p-amino function and with selectively cleavable protecting groups for the α-amino or α-carboxy group. The p-amino function could be protected via the copper complex or by specific acylation in aqueous solution at pH 4.6 (Fig. 2) as described by Fahrenholz and Thierauch (8).

Fig. 3 outlines the synthesis of AVP analogs with reactive groups in position 2. The protected nonapeptide 10 was prepared from fragments which correspond to the sequences 6-9, 3-5, and 1-2 in AVP. Their coupling by means of 1-hydroxybenzotriazole/N,N'-dicyclohexylcarbodiimide proceeded without racemisation as was demonstrated by enzymatic methods. From the protected nonapeptide 10 the amino and guanido protecting groups were removed by trifluoromethanesulfonic acid (9) yielding the S-protected derivative 10a. Its cyclization by means of iodine led to [Phe(p-NH_2)^2]AVP (2). In a similar way, by using Nps-Phe(p-NH-Boc)-ONSu for incorporation in

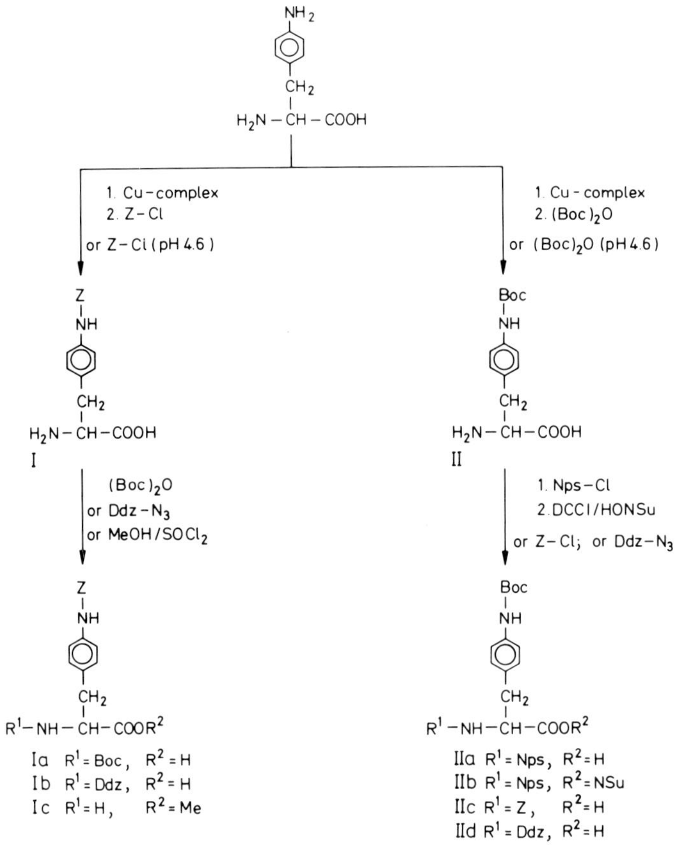

Fig. 2. Synthesis of protected derivatives of p-amino-L-phenylalanine.

position 3, [Phe^2, $Phe(p\text{-}NH_2)^3$]AVP (3) was prepared.

The ring closure by disulfide formation in derivatives 2 and 3 was confirmed by their circular dichroism (CD)- and ^1H-NMR spectra as well as by Edman degradation. The CD- spectra of 2 and 3 in aqueous solution of pH 2 (Fig. 4) contain in the

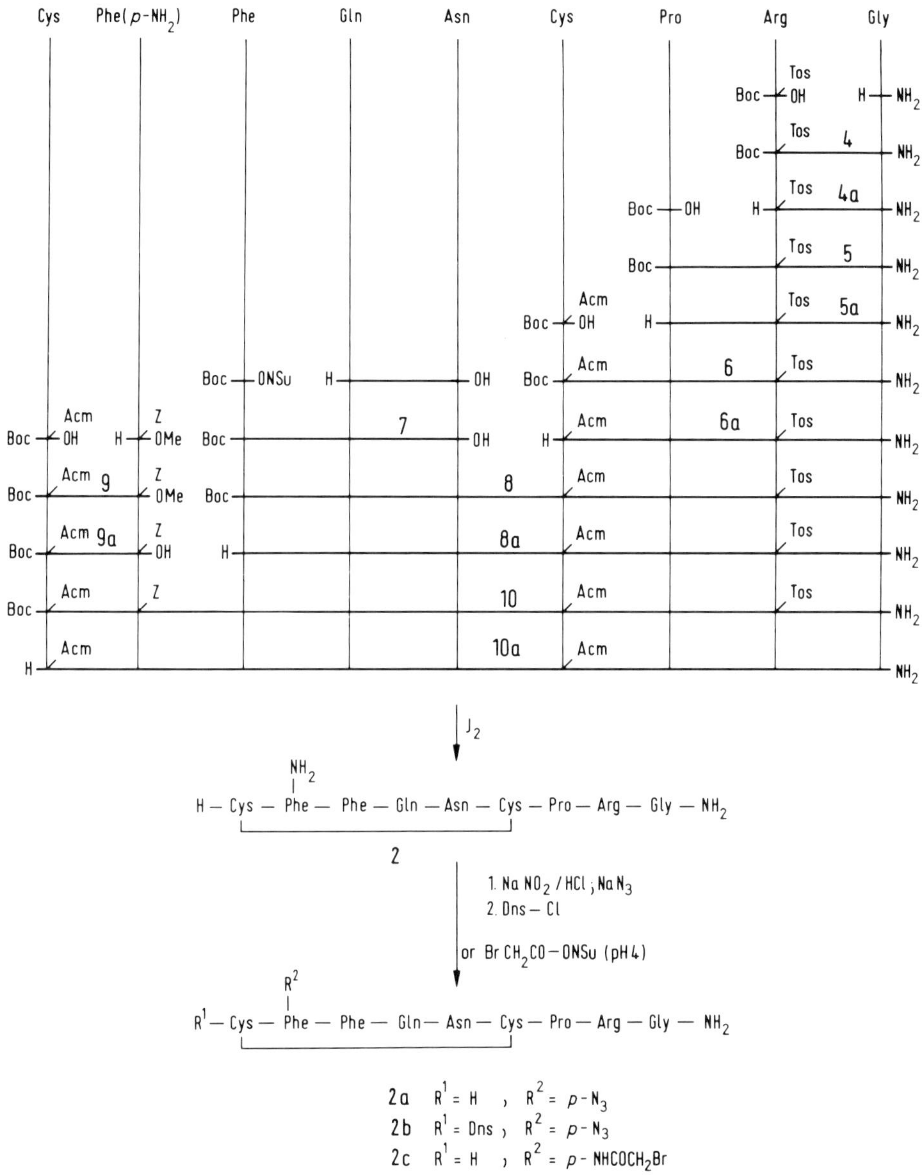

Fig. 3. Synthesis of vasopressin analogs with reactive groups in position 2.

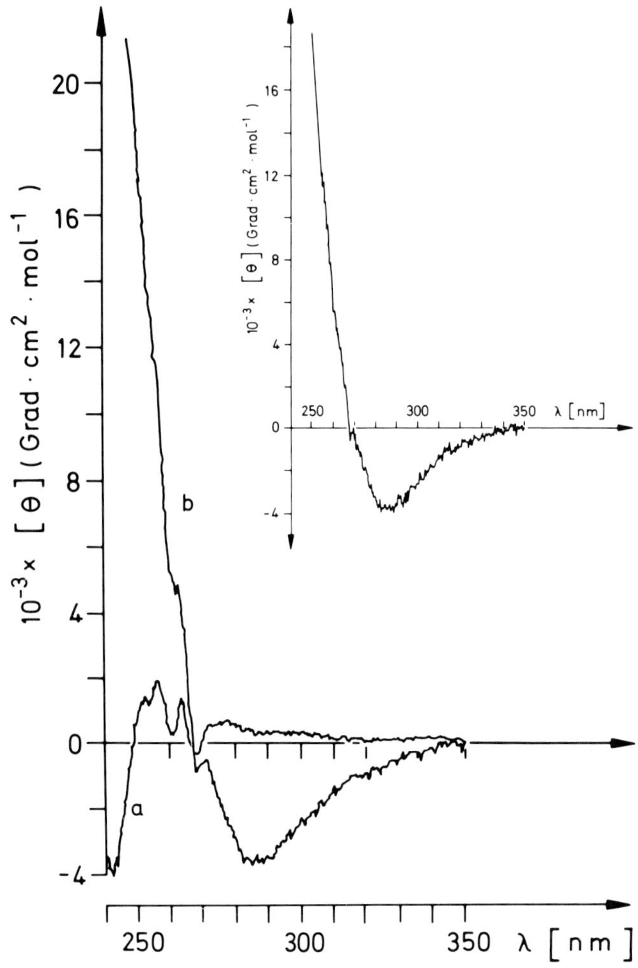

Fig. 4. Circular dichroism in the long wavelength region of (a) S-protected nonapeptide and (b) cyclic nonapeptide in an aqueous solution of pH 2.0. Inset: difference spectrum (b) - (a).

long wavelength region a negative band, whose minimum lies between 280-300 nm. This can be ascribed to electron transitions of the disulfide group. Its presence has been demonstrated in derivatives of vasopressin (10) and oxytocin (11) not containing tyrosine. This band is absent in the linear S-protected nonapeptides. The ^{1}H-NMR spectrum of [Phe $(p-NH_2)^2$]-vasopressin (2) in Me_2SO-d_6 shows in contrast to its linear precursor 10a the same pattern of peptide NH (and C^{α}-H) resonances as arginine vasopressin (12), thereby demonstrating that the conformation of the peptide backbone in 2 is similar to that in arginine vasopressin.

The p-aminophenylalanine analogs 2 and 3 were converted in high yields into the p-azidophenylalanine derivatives [Phe $(p-N_3)^2$]-vasopressin (2a) (Fig.3) and [Phe^2, Phe$(p-N_3)^3$]-vasopressin (3a) (Scheme 2) by diazotation in 0.1N HCl and reaction of the diazonium chloride with sodium azide. The disulfide bridge and α-amino group are stable under these reaction conditions (8). For fluorescence studies, the α-amino group of the N-terminal cysteine has been additionally substituted by reacting 2a with dansyl chloride which led to N^{α}-Dns-[Phe$(p-N_3)^2$]-vasopressin (2b). Reaction of the p-aminophenylalanine analogs 2 and 3 with bromoacetyl-N-hydroxysuccinimide ester in sodium formate buffer at pH 4.0 gave the p-bromoacetylamino-derivatives [Phe$(p-NHCOCH_2Br)^2$]-vasopressin (2c) and [Phe^2,Phe$(p-NHCOCH_2Br)^3$]-vasopressin (3b) with minor amounts of $N^{\alpha}N^{P}$-di-bromoacetyl-derivatives as revealed by electrophoresis at pH 1.6. Because of the low pKa value of 4.25 for the p-amino group, it undergoes acylation at low pH-values in preference to the α-amino group (8).

The intrinsic activities (percentage of the maximal velocities relative to arginine-vasopressin) and apparent affinities (expressed by the hormone concentration for halfmaximal stimulation $[H]_{0.5}$) of the described analogs in stimulating adenylate cyclase in a membrane preparation derived from the inner medulla of bovine kidney are shown in Fig. 5. Analogs modified only in position 2 show either the same (2c) intrinsic activity as arginine-vasopressin or a higher activity (2a),

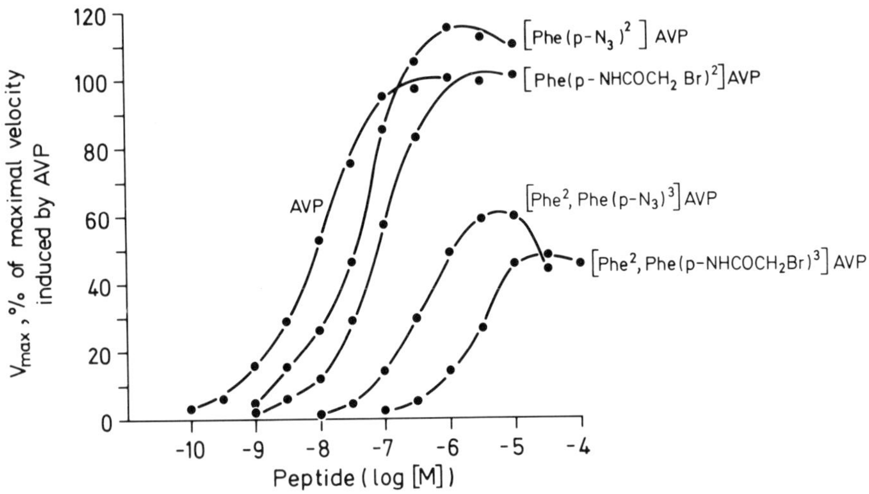

Compound	No	$\frac{V_{analogue}}{V_{AVP}} \cdot 100$	$[H]_{0.5}$ [M]	n_H
AVP	1	100	$8.8 \pm 1.5 \cdot 10^{-9}$	0.84 ± 0.10
[Phe $(p\text{-}N_3)^2$] AVP	2a	114.9 ± 4.6	$3.8 \pm 0.7 \cdot 10^{-8}$	0.95 ± 0.01
[Phe $(p\text{-}NHCOCH_2Br)^2$] AVP	2c	101 ± 3.4	$7.9 \pm 0.4 \cdot 10^{-8}$	1.01 ± 0.09
N^{α}-Dns-[Phe $(p\text{-}N_3)^2$] AVP	2b	64.9 ± 3.3	$6.1 \pm 0.8 \cdot 10^{-7}$	1.21 ± 0.07
[Phe^2, Phe $(p\text{-}N_3)^3$] AVP	3a	59.4 ± 5.0	$3.0 \pm 0.3 \cdot 10^{-7}$	1.14 ± 0.1
[Phe^2, Phe $(p\text{-}NHCOCH_2Br)^3$] AVP	3b	48.1 ± 2.5	$2.5 \pm 0.3 \cdot 10^{-6}$	1.01 ± 0.12

Fig. 5. Activation of adenylate cyclase by arginine-vasopressin and its analogs. Membrane preparations from the inner medulla of bovine kidney were incubated with various initial concentrations of hormones under steady state conditions as described by Fahrenholz et al (7). The table shows the following parameters: Maximal velocity V, hormone concentration for half maximal stimulation $[H]_{0.5}$ and apparent H ill coefficients n_H.

while the concentrations for half maximal stimulation are about 4 to 10 times higher than those for vasopressin. The N^{α} -dansyl derivative 2b and the analogs with reactive groups in position 3, 3a and 3b show considerably reduced intrinsic activities and apparent affinities for stimulating adenylate cyclase.

The frog skin is a convenient tissue for testing the physiological activity of different agents. The mammalian neurohypophyseal hormones vasopressin and oxytocin as well as the amphibian hormone arginine-vasotocin (AVT) increase the active sodium transport (natriferic effect) and osmotic water flow (hydroosmotic effect) across the isolated frog skin (13). The stimulation of active sodium transport by AVT, AVP, [Phe$(p-N_3)^2$]-AVP (2a) and [Phe^2, Phe$(p-N_3)^3$]-AVP (3a) was tested on skins of Rana temporaria (Fig. 6) by measurement of the short circuit current (SCC) (14). The analog with the photolabile group in position 2 (2a) was only a partial agonist whose affinity was about 7 times less than that of AVP, while the derivatives with p-azidophenylalanine in position 3, 3a gave almost the same maximal response as AVT and AVP. Its affinity was reduced by a factor 50 in contrast to AVP.

The hydroosmotic activity of [Phe$(p-N_3)^2$]-AVP (2a) was determined on Rana esculenta by simultaneous measurements of osmotic water flow (OFW) and SCC as described by Johnsen and Nielsen (15) (Table 1).

[Phe$(p-N_3)^2$]-AVP M	% stimulation OWF	% stimulation SCC
$1.6 \cdot 10^{-6}$	35 ± 11	64 ± 2
$4.9 \cdot 10^{-6}$	38 ± 10	72 ± 4

Table 1. Normalized hydroosmotic (OWF) and natriferic (SCC) response to two doses of [Phe$(p-N_3)^2$] AVP. Calculations were performed as described in Fig. 6. Response to $4 \cdot 10^{-8}$ M AVT = 100%. Values are mean $\pm$ SEM, n=4.

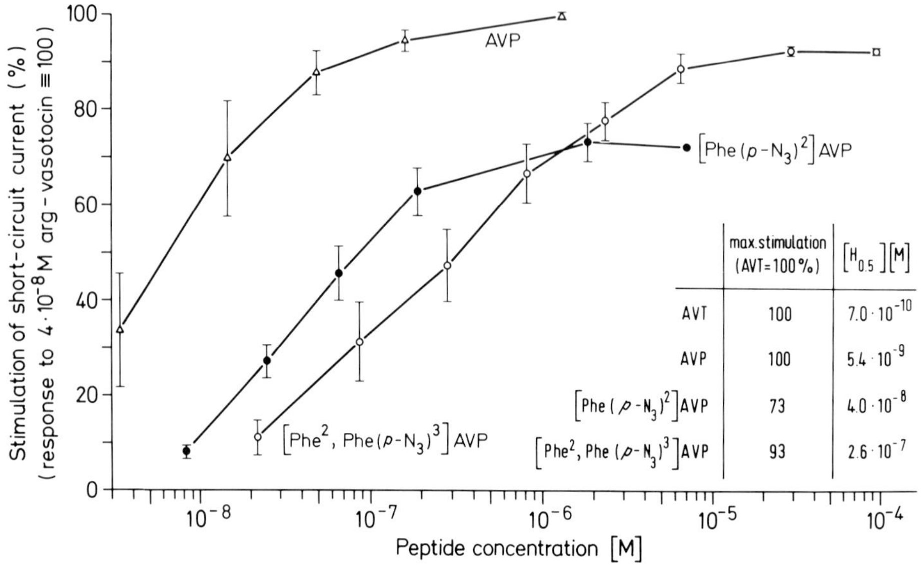

Fig. 6. Normalized (cumulative) response to increasing doses of hormone. Each experiment was performed on paired halves, one remaining unstimulated as control to compensate for time-dependent variation of SCC. After each addition of hormone the SCC was allowed to reach the new steady state level before the next addition. At start of experiment a=b. % stimulation of SCC = (a - b)·100/c, where a = SCC of experimental half, b = SCC of control half, c = a - b after addition of 4 · 10^{-8} M AVT at the end of the experiment. Results are given as mean ± SEM, n=4 for AVP and [Phe(p-N_3)^2]-AVP (<u>2a</u>) (n=2 for 7 · 10^{-6} M <u>2a</u>) and n=5 for [Phe^2,Phe(p-N_3)^3]-AVP.

Both responses are maximal at 4.9 · 10^{-6} M concentration of the hormone analog. However, while the maximal natriferic response is more than 70%, the maximal hydroosmotic response attains only about 40% of the maximal stimulation obtained with AVT. This discrepancy seems to be in accord with the separation of the OWF and SCC responses to hormonal stimulation reported by several investigators (16).

Finally, the stimulation of adenylate cyclase in isolated frog skin epethelia (<u>Rana temporaria</u>) by AVT, AVP and

[Phe(p-N_3)^2]-AVP was compared. AVP (10^{-6} M) produced 21.1 ± 3.7% (n=4) of the maximal rise in cAMP level induced by AVT ($4 \cdot 10^{-8}$ M). [Phe(p-N_3)^2]-AVP, however, showed only a moderate intrinsic activity (Table 2) as compared to $4 \cdot 10^{-8}$ M AVT, which increased the level to 17.6 ± 2.4 pmol/mg dry weight (n=8). At a concentration of $1.5 \cdot 10^{-9}$ M, the amphibian hormone AVT produced a similar small increase in cAMP level (from 1.29 ± 0.11 to 1.7 5 ± 0.4 pmol/mg dry weight n=3) as high doses of [Phe(p-N_3)^2]-AVP (2a) and it induced 76 ± 8% (n=4) of the maximal stimulation of SCC. This value is in the

[Phe(p-N_3)^2]-AVP (M)	0	$1.6 \cdot 10^{-8}$	$8 \cdot 10^{-8}$	$4 \cdot 10^{-7}$	$2 \cdot 10^{-6}$	10^{-5}
pmol cAMP / mg dry weight	1.58 ±0.16	1.57 ±0.24	1.71 ±0.21	2.07 ±0.08	1.81 ±0.17	2.26 ±0.19
n	6	6	6	6	5	4

Table 2. Stimulation of cAMP level by [Phe(p-N_3)^2]-AVP. Frog skins were divided into four pieces and the epithelia were separated from each skin piece. The epithelial pieces were incubated 5 minutes with the different hormone concentration and cAMP was measured as described by Johnsen and Nielsen (3). Values are mean ± SEM.

same order of magnitude as the increase of SCC induced by a maximal stimulating dose of compound 2a. (Table 1) Apparently the lack of full natriferic and hydroosmotic activity of [Phe(p-N_3)^2]-AVP is due to its low intrinsic activity in stimulating the amphibian adenylate cyclase.

Discussion

The pathway described for the synthesis of p-azidophenylalanine analogs of vasopressin - via the p-aminophenylalanine derivatives with protected p-amino group and introduction of the azido function in a final step of the synthesis - avoids the incorporation of the light-sensitive p-azidophenylalanine into the peptide chain at an early stage of the synthesis or the catalytic reduction of p-nitrophenylalanine in the presence of S-protected cysteine or of the disulfide bridge. The analogs can be labeled in two ways: firstly, dansylation of the α-amino group yields a photoreactive derivative (2b) with an environment-sensitive fluorophore. It retains substantial intrinsic activity and affinity for stimulating vasopressin sensitive adenylate cyclase. Secondly, it is possible to iodinate p-aminophenylalanine peptides and to exchange iodine catalytically against tritium (17,18). In the case of neurohypophyseal hormones iodine-tritium exchange in Tyr^2 followed by purification (to remove desulfurated by-products) gives labeled compounds with high specific radioactivity (17-26 Ci/mmol) (19).

The high intrinsic activities and affinities of the vasopressin analogs 2a and 2c modified only in position 2, in stimulating adenylate cyclase derived from the inner medulla of bovine kidney, are in accordance with the finding, that $[Tyr(OMe)^2]$-AVP (like derivative 2a) shows higher maximal stimulation of adenylate cyclase than arginine-vasopressin (20), and that the hydroxyl group of tyrosine is not essential for high antidiuretic activity. In the biologically active conformation proposed by Walter et al. (21) on the basis of NMR studies and structure activity relations, the tyrosine side chain points away from the 20-membered ring and is not a part of the active center of the hormone. According to this model, phenylalanine in position 3 is situated in the corner of a β-turn formed by the resudues 2-5 and should be mainly involved in binding and recognition. Our results concerning the reduced intrinsic activities and affinities of analogs 3a and 3b modified in the side chain of phenylalanine, suggest that the steric size in the 3 position plays a significant role not only in binding, but also in hormonal stimulation.

In contrast to the activation of adenylate cyclase derived from mammalian kidney, [$Phe(p-N_3)^2$]-AVP is only a partial agonist when tested on frog skin epithelium. A reduced intrinsic activity has also been reported for the action of [$Tyr(OMe)^2$]-oxytocin in toad bladder (22). Apparently, the steric size of the substituents of the aromatic amino acid in position 2 is of great importance for maximal hormonal stimulation in amphibians.

While N^{α}-(bromoacetyl)oxytocin is an irreversible antagonist in the activation of neurohypophyseal hormone-specific adenylate cyclase (23), the p-bromoacetyl derivatives 2c and 3b are agonists. It has to be shown whether this activation is irreversible. In conclusion, due to their high affinities for their receptors, the analogs of AVP with reactive groups in position 2 and 3 may be useful for the specific labeling of vasopressin receptors in biological membranes and for studies of covalent hormone receptor complexes.

References

1. Dousa, T., Hechter, O., Schwartz, J.L., Walter, R.: Proc. Nat.Acad.Sci. U.S.A. 68, 1693-1697 (1971).
2. Bär, H.-P., Hechter, O., Schwartz, I.L., Walter, R.: Proc. Nat. Scad. Sci. U.S.A. 67, 7-12 (1970).
3. Johnsen, A.H., Nielsen, R.: Acta physiol. scand. 102, 281-289 (1978).
4. Billah, M.M., Michell, R.H.: Biochem.Soc.Trans. 6, 1033-1035 (1978).
5. Roy, C., Rajerison, R., Bockaert, J., Jard, S.: J.Biol. Chem. 250, 7885-7893 (1975).
6. Wieland, T., Dungen, A., Birr, C.: Liebigs Ann. Chem. 752, 109-114 (1971).
7. Fahrenholz, F., Thierauch, K.-H., Crause, P.: Hoppe-Seyler's Z. Physiol. Chem., in press (1980).
8. Fahrenholz, F., Thierauch, K.-H.: Int. J. Peptide Protein Res., in press (1980).

9. Yajima, H., Kitagawa, K., Segawa, T., Nakama, M., Kataoka, K.: Chem.Pharmac.Bull. 23 (1975).

10. Fric, I., Kodicek, M., Flegel, M., Zaoral, M.: Eur.J.Biochem. 56, 493-502 (1975).

11. Beychok, S., Breslow, E.: J.Biol.Chem. 243, 151-154 (1968).

12. Walter, R., Ballardin, A., Schwartz, I.L., Gibbons, W.A., Wyssbrod, H.R.: Proc.Nat.Acad.Sci. U.S.A. 71, 4528-4532 (1974).

13. Handler, J.S., Orloff, J. in:Handbook of Physiology, section 8, (Orloff, J., Berliner, R.W. ed.), American Physiological Society, Washington D.C., 791-814 (1973).

14. Ussing, H.H., Zehran, K.: Acta physiol.scand. 23, 110-127 (1951).

15. Johnsen, A.H., Nielsen, R.: Acta physiol.scand., in press (1980).

16. Andreoli, T.E., Schafer, J.A.: Annual Review of Physiology 38, 451-500 (1976).

17. Eberle, A., Schwyzer, R.: Helv.Chim.Acta 59, 2421-2431 (1976).

18. Fischli, W., Caviezel, M., Eberle, A., Escher, E., Schwyzer, R.: Helv.Chim.Acta 59, 878-879 (1976).

19. Flouret, G., Terada, S., Yang, F., Nakahara, T., Hechter, O.: Biochem. 16, 2119-2124 (1977).

20. Butlen, D., Guillon, G., Rajerison, M.R., Jard, S., Sawyer, W.H., Mannings, M.: Mol.Pharmacol. 14, 1006-1017 (1978).

21. Walter, R., Smith, C.W., Mehta, P.K., Boonjarern, S., Arruda, J.A., Kurtzman, N.A. in: Disturbances in Body Fluid Osmolality (Andreoli, T.E., Grantham, J.J., Rector, F.C. ed.), American Physiological Society, Bethesda, Maryland, 1-36 (1977).

22. Kirchberger, M.A., Hechter, O., Walter, R., Schwartz, I.L.: Biochem.Biophys.Acta 500, 246-255 (1977).

23. Walter, R., Schwartz, J.L., Hechter, O., Dousa, T., Hoffmann, P.J.: Endocrinology 91, 213-222 (1972).

Acknowledgement

A.H. Johnsen was supported by a grant from The Danish Natural Science Research Council.

SYNTHESIS OF MCD-PEPTIDE AND APAMIN BY LIQUID PHASE FRAGMENT CONDENSATION

Peter Hartter
Physiologisch-chemisches Institut der Universität
Hoppe-Seyler-Str. 1, D-7400 Tübingen, Germany

Many animal venoms are composed of proteins, peptides and low molecular weight substances. Whereas proteins can be considered as allergic and spreading factors, the peptides from snake and bee venoms are cytotoxic and neurotoxic substances. Melittins, apamin, tertiapin, secapin and the mast cell degranulating peptide (MCD-peptide) are basic peptides from bee venom (1-3).

<u>Melittin</u>

Gly-Ile-Gly-Ala-Val-Leu-Lys-Val(8)-Leu-Thr-Thr-Gly-Leu-Pro-Ala-Leu-Ile-Ser-
Trp-Ile(20)-Lys-Arg-Lys-Arg-Gln-Gln(26)

<u>Melittin F:</u> Corresponding to melittin: Pos. 8 - 26

<u>Secapin</u>

Tyr(1)-Ile-Ile-Asp-Val(5)-Pro-Pro-Arg-Cys(9)-Pro-Pro-Gly-Ser-Lys-Phe(15)-Ile-Lys-Asn-
Arg-Cys(20)-Arg-Val-Pro-Val(24)

<u>MCD-peptide</u>

Ile-Lys-Cys-Asn-Cys(5)-Lys-Arg-His-Val-Ile-Lys-Pro-His-Ile-Cys(15)-Arg-Lys-Ile-
Cys(19)-Gly-Lys-Asn(22)-NH_2

<u>Apamin</u>

Cys(1)-Asn-Cys(3)-Lys-Ala-Pro-Glu-Thr-Ala-Leu-Cys(11)-Ala-Arg-Arg-Cys(15)-Gln-Gln-His(18)-NH_2

<u>Figure 1.</u> Basic peptides from bee venom.

Structure and Activity of Natural Peptides

About half of the dry weight of the venom is a family of closely related peptides, the melittins with pronounced surfactant properties and direct hemolytic activity (4). In contrast the content of the other basic peptides is about 1 - 2 %.

The neurotoxic apamin has a highly specific mode of action exclusively at the central nervous system. It produces on intraventricular application a long-lasting spinal spasticity at a dose of 50 ng/mouse (5).

MCD-peptide got its name from the ability to degranulate peritoneal mast cells from rats (3,6). The inhibition of inflammation was found by Hanson et al. (7) some years later. It is a potent anti-inflammatory agent and it suppresses the development of adjuvant arthritis (Billingham et al.) (8). A third biological activity of MCD-peptide was reported recently by Habermann (9). On intraventricular or intrachordal injection, the peptide produces local poisoning symptoms as convulsions and respiratory arrest in mice. The poisoning is different from those following apamin application (5).

Although there are structural similarities between MCD-peptide and apamin (see Fig. 2), apamin exerts no histamin-liberating activity.

Both peptides have two intramolecular disulfide bridges, a high ratio of basic amino acids and comparable chain lengths as well as a homologous amino acid sequence (MCD pos.: 3 - 6 and apamin pos.: 1 - 4; Cys-Asn-Cys-Lys). Considering these structural similarities the different and specific biological activities of both peptides are surprising. This raises the question whether these effects are caused by distinct amino acid sequences.
To examine this problem, the synthesis of derivatives with various amino acid sequences and subsequent examination of their biological action is a practical method. Furthermore sequence fragments should be examined for equivalent biological activi-

ties.

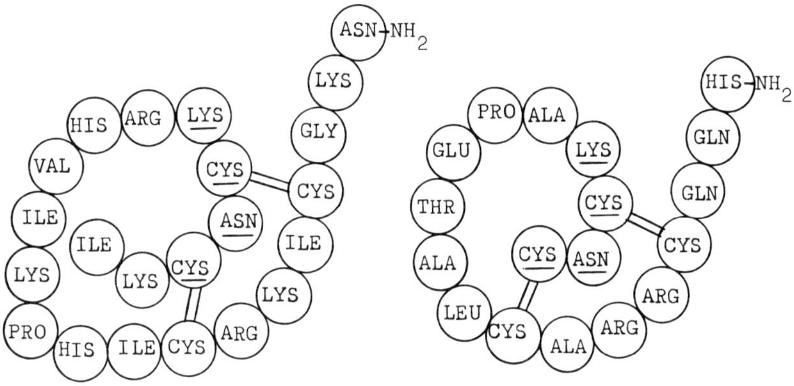

Figure 2. Structure of

MCD-peptide, Apamin.

Synthesis of both peptides (MCD and apamin) press the same problems such as disulfide bridges (10), the chain length and protection of the functional groups.
Primarily, the natural sequences of MCD-peptide (11) and apamin (12) should be synthesized in order to compare the biological activities of sequence variation products with those of the synthetic natural products.
For the synthesis of these basic bee venom peptides, the liquid phase method from Bayer et al. (13) was adapted to the liquid phase fragment condensation. Thus the advantages of the classic methods in peptide synthesis are combined with those of reactions on the soluble polymer support polyethylene glycol.

1 5 15
Ile-Lys-Cys-Asn-Cys-Lys-Arg-His-Val-Ile-Lys-Pro-His-Ile-Cys-Arg-
V IV III II

19 22
Lys-Ile-Cys-Gly-Lys-Asn-NH_2
I

MCD(17-21)	Boc-Lys(Z)-Ile-Cys(SiPr)-Gly-Lys(Z)	I
MCD(12-16)	Boc-Pro-His(Trt)-Ile-Cys(Trt)-Arg(Tos)	II
MCD(8-11)	Boc-His(Trt)-Val-Ile-Lys(Z)	III
MCD(5-7)	Boc-Cys(SiPr)-Lys(Z)-Arg(Tos)	IV
MCD(1-4)	Boc-Ile-Lys(Z)-Cys(Trt)-Asn(Mbh)	V

Figure 3. Synthetic fragments for preparation of MCD-peptide.

1 3 11
Cys-Asn-Cys-Lys-Ala-Pro-Glu-Thr-Ala-Leu-Cys-Ala-Arg-Arg-
IV III II

15
Cys-Gln-Gln-His-NH_2
I

APA(17-15)	Boc-Cys(SiPr)-Gln(Mbh)-Gln(Mbh)	I_A
APA(14-10)	Boc-Leu-Cys(Trt)-Ala-Arg(Tos)-Arg(Tos)	II_A
APA(9-5)	Boc-Ala-Pro-Glu(γBzl)-Thr-Ala	III_A
APA(4-1)	Boc-Cys(SiPr)-Asn-Cys(SiPr)-Lys(Z)	IV_A

Figure 4. Synthetic fragments for preparation of apamin .

In detail the protected MCD fragments (Fig. 3) and apamin fragments (Fig. 4) were synthesized via classic methods (14, 15,16). This was followed by fragment condensation steps, using DCC/1-hydroxybenzotriazole as coupling reagent, on the polyethylene glycol carrier. The polymer carrier was previously prepared by esterification of polyethylene glycol (MG 10 000) with asparagine or histidine, the C-terminal amino acids of MCD-peptide and apamin respectively (17). On this support the peptides I - V (Fig. 3) and I_A - IV_A (Fig. 4) are successively coupled in a reaction cycle as shown in Fig. 5.

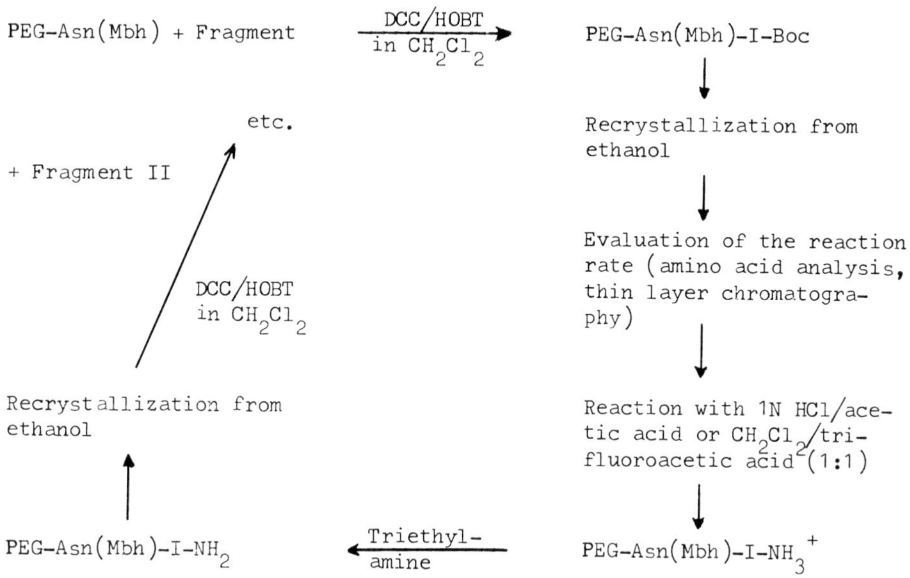

Figure 5. Liquid phase fragment condensation: Reaction cycle during the condensation step.

The completely protected MCD-peptide or apamin-polyethylene glycol product is than dissolved in HBr/trifluoroacetic acid to split off most of the protecting groups. Ammonolysis in dioxane/ammonia splitsto the Asn-amide and His-amide derivatives (see Fig. 1). To remove the tosyl groups from arginine the crude product is dissolved in liquid HF. Oxidative sulfitolysis of the disulfide bridges leads to the MCD-$(SSO_3)_4$ and apamin-$(SSO_3)_4$ "Bunte"-salts which are purified by gel filtration (Sephadex G50, Biogel P_6, P_4 and P_2) and ion-exchange chromatography (SP-Sephadex) (17).

After reduction of the $-SSO_3$ groups with mercaptoethanol, cautious air oxidation (10) and chromatography on SP-Sephadex, Biogel and Sephadex G50 superfine, peptides of good amino acid analysis and biological activities can be obtained (see Tab.1).

Biological activities of	
MCD-peptide:	
1. Histamine liberation from peritoneal mast cells	50 %
2. Blood pressure decrease in rat	50 %
3. Lysis of human erythrocytes	negative
4. Liberation of serotonin from thrombocytes	negative
5. Carrageenan edema: Inflammation inhibition	40 %
Apamin:	
Neurotoxicity on intraperitoneal application	45 %

The synthetic peptides show good agreement in UV , CD and amino acid analysis with the natural products.

Tab. 1.

The liquid phase fragment condensation method has two essential advantages:

1. Cutting the natural sequences into fragments with three to five amino acid residues, the coupling steps at the soluble polymer support are reduced to five (MCD) or four (apamin). The fivefold excess of fragments, contained in the reaction mixture (to get high reaction rates in the fragment condensation step) are easily removed by recrystallizing the PEG

peptide coupling product from ethanol, methanol, or dimethylformamide, solvents in which these fragments are soluble (18). The quantitative separation of these small peptides from the reaction product can be controlled by thin-layer chromatography.

2. The molecular weights of the deprotected peptide fragments range from four to seven hundred and most of the peptide fragments contain one or two basic amino acids in their sequence. If the finished reaction products contain other peptides than the natural sequence these by-products can readily be distinguished by their molecular weights and Ip values. This enables an effective purification by a careful ion-exchange chromatography and gel filtration.

Nevertheless, the biological activities of the purified, synthetic products do not surpass 50 % activity of the natural peptide. This loss in activity is due to possible side reactions of the asparagine residues (19), racemization (see Tab. 2) (20) and partial damage of amino acids during the liquid HF cleavage step.

D-amino acids	nat. MCD-peptide		PEG synth. product		Merrifield product	
	A	B	A	B	A	B
Ile	1,5	17,5	2,4	25,2	1,2	12,4
Asp	3,6	13,1	15,3	18,3	14,8	18,2
Lys	0,7	3,9	4,2	5,5	2,0	3,6
Pro	1,2	1,2	2,0	1,9	1,8	2,7

A: Disulfide peptides; B: MCD-$(SSO_3)_4$ derivative

Tab. 2. Determination of % D-amino acids in several MCD-peptides.

In summary, these results demonstrate that combination of the liquid phase method with classic methods for peptide synthesis is a valuable strategy for the preparation of basic peptides.

It is advantageous if three to pentapeptides with at last one functional side group are coupled on the polyethylene glycol support. These fragments require a good solubility in methanol, ethanol or dimethylformamide. With regard to the basicity and molecular weight of the fragments an effective purification of the synthetic products by ion-exchange chromatography and gel filtration is possible.

Apamin and MCD-peptide, synthesized by the liquid phase fragment condensation method, yield better amino acid compositions as well as higher biological activities than the products synthesized by the Merrifield method.

References

1. Habermann, E.: Science 177, 314-322 (1972).
2. Gauldie, J., Hanson, J.M., Rumjanek, F.D., Shipolini, R.A., Vernon, C.A.: Europ. J. Biochem. 61, 369-376 (1976).
3. Gauldie, J., Hanson, J.M., Shipolini, R.A., Vernon, C.A.: Europ. J. Biochem. 83, 405-410 (1978).
4. Kreil, G.: Monatsh. Chemie 96, 2061-2063 (1965).
5. Habermann, E., Reiz, K.-G.: Biochem. Z. 341, 451-466 (1965).
6. Haux, P.: Hoppe-Seyler's Z. Physiol. Chem. 350, 536-546 (1969).
7. Hanson, J.M., Morley, J., Soria-Herrera, C.: Brit. J. Pharmacol. 50, 383-392 (1974).
8. Billingham, M.E.J., Morley, J., Hanson, J.M., Shipolini, R.A., Vernon, C.A.: Nature 245, 163-164 (1973).
9. Habermann, E.: Naunyn-Schmiedebergs Arch. Pharmacol. Exp. Pathol. 300, 189-191 (1977).
10. Hartter, P., Weber, U.: Hoppe-Seyler's Z. Physiol. Chem. 356, 693-699 (1975).
11. Birr, C., Wengert-Müller, M.: Angew. Chem. 91, 156-157 (1979).

12. Granier, C., Pedroso Muller, E., van Rietschoten, J.: Europ. J. Biochem. 82, 293-299 (1978).

13. Mutter, M., Hagenmaier, H., Bayer, E.: Angew. Chem. 83, 883-884 (1971).

14. Hartter, P.: Hoppe-Seyler's Z. Physiol. Chem. 357, 1683-93 (1976).

15. Hartter, P.: Hoppe-Seyler's Z. Physiol. Chem. 358, 331-337 (1977).

16. Hartter, P.: Hoppe-Seyler's Z. Physiol. Chem. 361 - in press - (1980).

17. Hartter, P.: Hoppe-Seyler's Z. Physiol. Chem. 361 - in press - (1980).

18. Bayer, E., Mutter, M.: Chem. Ber. 107, 1344-52 (1974).

19. Gattner, H.G., Schmitt, E.W.: Hoppe-Seyler's Z. Physiol. Chem. 358, 105-113 (1977).

20. Woiwode, W., Frank, H., Graeme, J., Nicholson, J., Bayer, E.: Chem. Ber. 111, 1711-18 (1978).

SYNTHETIC ROUTES TO ENKEPHALIN

Hans Horn, Wolfgang Voelter

Institut für Organische Chemie der Universität Tübingen,
Auf der Morgenstelle 18, 7400 Tübingen, Germany

Nadarajah Vethaviyasar

Department of Chemistry, University of Calabar, P.M.B. 1115,
Calabar, Nigeria

Introduction

In 1975 Hughes et al. (1) isolated from the brains of pigs a substance which generated, both, in vitro and in vivo assay systems morphine -like effects. The identification of the unknown compound by amino acid analysis, mass spectrometry and sequence determination revealed the fact that the compound causing morphine-like effects consisted of two closely related pentapeptides with the sequences L-Tyr-Gly-Gly-L-Phe-L-Met-OH and L-Tyr-Gly-Gly-L-Phe-L-Leu-OH. Because these substances are produced mainly in the limbic system at the core of the brain where pain reception takes place, one thought that these peptides could be used for therapy of patients suffering from dependance of morphine and heroine. Unfortunately these expectations could not be answered up to now.

Results and discussion

Shortly after knowing the peptides' sequence one started with the syntheses of the compounds. One of the first synthesis was performed by Voelter et al. using "classical" methods.

Structure and Activity of Natural Peptides

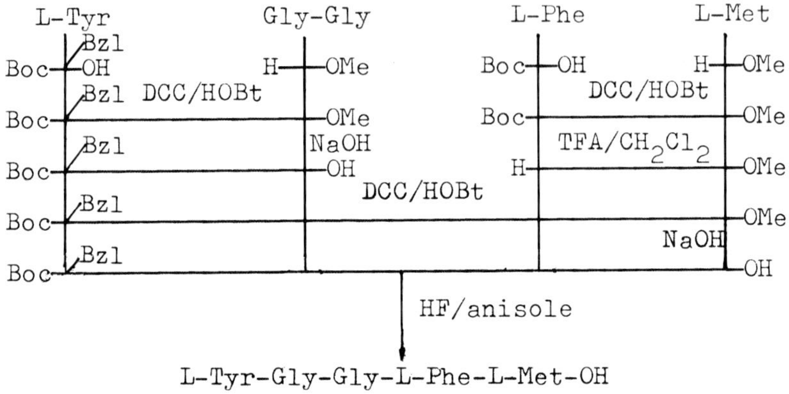

Fig.1. Synthesis of Met-enkephalin by Voelter at al.(2).

Boc-L-Tyr(Bzl)-OH and the dipeptide ester Gly-Gly-OMe are coupled by 1-hydroxybenzotriazole and dicyclohexylcarbodiimide to give the tripeptide Boc-L-Tyr(Bzl)-Gly-Gly-OMe. From Boc-L-Phe and L-Met-OMe the protected dipeptide ester Boc-L-Phe-L-Met-OMe is received by the same coupling step. After deprotection with trifluoroacetic acid in methylene chloride Boc-L-Phe-L-Met-OMe is coupled with the saponified tripeptide to give Boc-L-Tyr(Bzl)-Gly-Gly-L-Phe-L-Met-OMe. Alkaline hydrolysis and total deprotection with hydrogen fluoride yields the pentapeptide L-Tyr-Gly-Gly-L-Phe-L-Met-OH. However, the chosen way is not satisfactory as the desired pentapeptide forms in low yields and purification from the by-products is difficult. Removal of the benzyl ether group from the L-Tyr residue by liquid hydrogen fluoride results in partial destruction of the molecule as can be seen by thin-layer chromatography. For these reasons the synthetic concept was changed (3).

In figures 2 and 3 two possible synthetic ways are given which yield Met-enkephalin in satisfactory yields (3). Boc-L-Phe-OH and the sodium salt of L-methionine are coupled by the N-hydroxysuccinimido ester to give Boc-L-Phe-L-

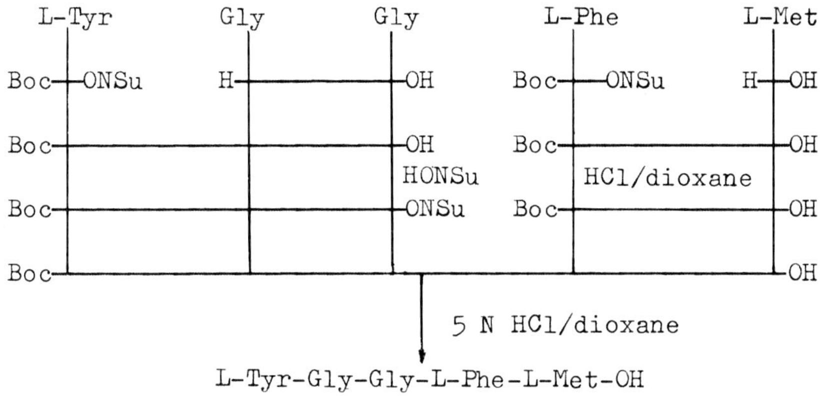

Fig.2. Synthesis for Met-enkephalin using hydroxysuccinimido esters (3).

Met-OH. This dipeptide is deprotected in a solution of hydrogen chloride in peroxide-free dioxane to give L-Phe-L-Met-OH·HCl. The reaction of Boc-L-Tyr-OH with Gly-Gly yields the tripeptide Boc-L-Tyr-Gly-Gly-OH. This protected compound is converted into the N-hydroxysuccinimido ester and coupled with the dipeptide to give the pentapeptide Boc-L-Tyr-Gly-Gly-L-Phe-L-Met-OH. After cleavage of the Boc-Tyr bond with hydrogen chloride in dioxane the crude product is received in a yield of 80 %. After recrystallisation from methanol/ether a chromatographically homogeneous white solid of the pentapeptide is received. In figure 3 another synthetic way is shown, which, after fragment condensation of L-tyrosine with the tetrapeptide, yields Met-enkephalin. The last coupling step with dicyclohexylcarbodiimide supplies the protected pentapeptide in high yields because activations of single amino acids by 1-hydroxybenzotriazole can be performed to a greater extent than those of tri-or tetrapeptides. This way is also suitable for syntheses of derivatives in which the N-terminal amino acid L-tyrosine is replaced by other amino acids. As the phenolic hydroxyl group is not blocked by a protecting group it is possible to cleave off the amino protecting group under very mild conditions by hydrogen chloride in peroxide-free dioxane.

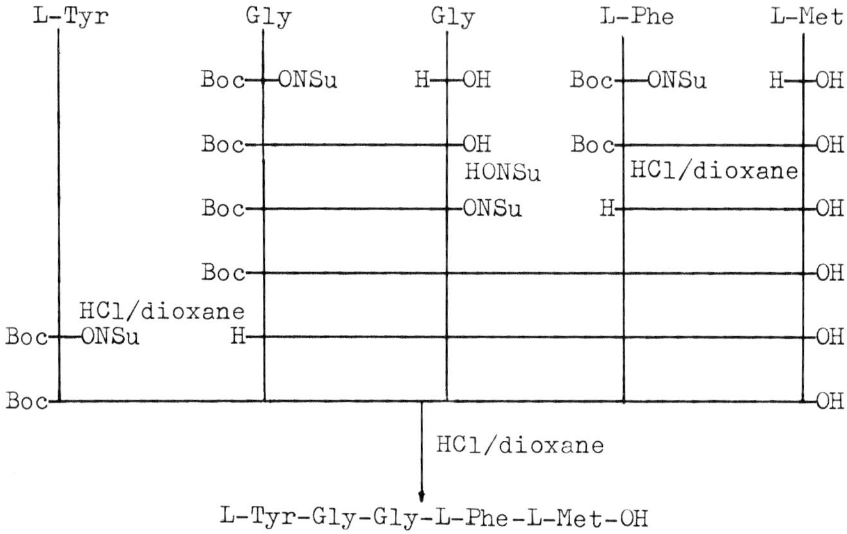

Fig.3. Synthesis of Met-enkephalin by Horn (3).

peroxide-free dioxane.

Another synthesis for enkephalins is described by Berger and Jakubke (4) (Fig.4). These authors apply fragment condensation for the NH_2-protected tetrapeptide with the C-terminal amino acid. By this method Leu- as well as Met-enkephalin can be synthesized from the identical protected tetrapeptide. Furthermore, variations in position 5 of the peptide can easily be realized to get derivatives of the native products.

As mentioned above, the alkaline hydrolysis of a pentapeptide ester may cause the formation of a great deal of by-products. Therefore, a carboxyl-protecting group which can be easily removed under mild conditions should be used. Quite recently, peptide chemists used tert-butyl esters. This ester can be saponified with hydrogen chloride in dioxane and with trifluoroacetic acid. Figure 5 demonstrates the synthesis of Met-enkephalin by the use of C-terminal tert-butyl ester protection.

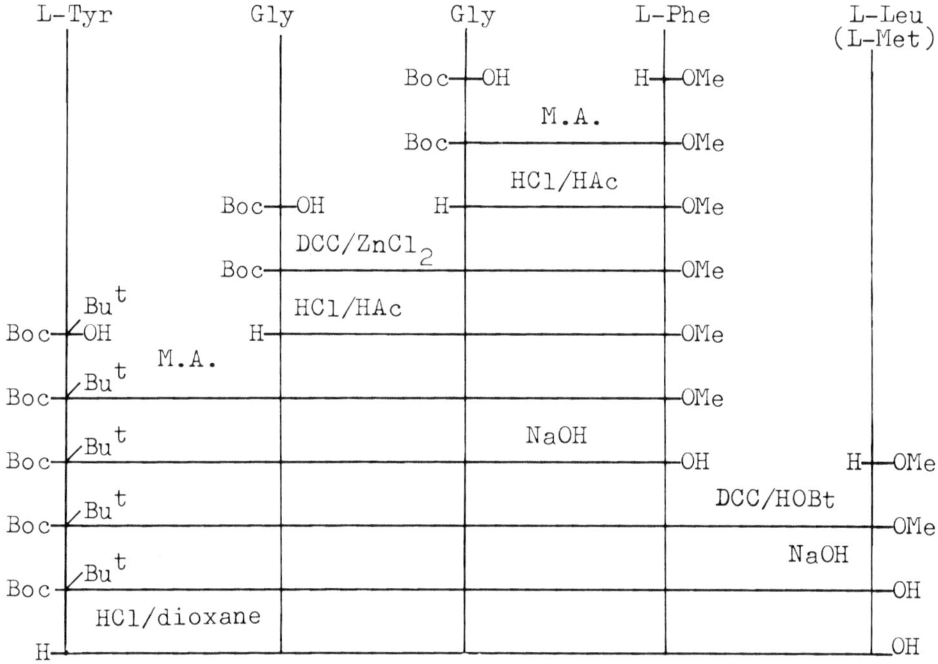

Fig.4. Synthesis of Leu-resp. Met-enkephalin by Berger and Jakubke (4).

Using the tert-butyl ester groups for L-methionine the α-amino group of L-phenylalanine has to be protected by a group which is not sensitive to acids, e.g. the phthalyl group. The commonly used carbobenzoxy residue is not suitable in this case because the presence of methionine poisons the catalyst so that no hydrogenolytic cleavage of the protecting group can be made. The phthalyl residue is removed from the dipeptide by stirring with hydrazine in methanol. Another new synthesis of Met-enkephalin is given by Wang et al. (5) (Fig. 6). The strategy uses amino protection by the Boc group and carboxyl protection as the benzyl ester. The benzyl ester is synthesized from the cesium salt of the carboxylic acid and

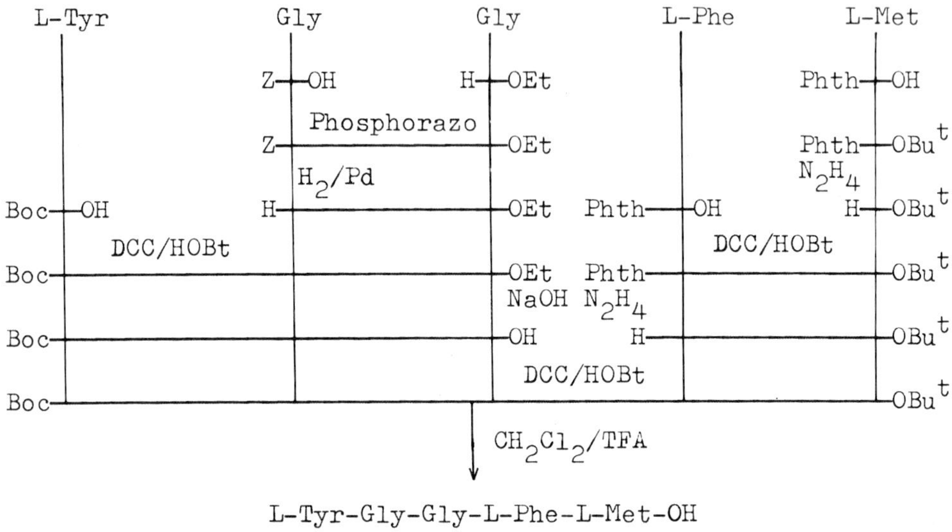

Fig.5. Synthesis of Met-enkephalin using C-terminal tert-butyl ester protection (3).

benzyl bromide.The reaction of Boc-L-Phe-L-Met-OH to Boc-L-Phe-L-Met-OBzl is performed in high yields. This ester is subsequently allowed to react with the tripeptide azide Z-L-Tyr(Z)-Gly-Gly-N_3 and the pentapeptide is received as a crystalline pure compound. No indication for thioether alkylation at the methionine residue during esterification is found. Leu-enkephalin is received by the same synthetic route but the carbobenzoxy and benzyl protecting groups can be split off by hydrogenation.

In a further synthesis of Leu-enkephalin the carbobenzoxy group is used for the protection of the α-amino group of tyrosine (fig.7 (3)).

Bower et al. (6) synthesized Met- and Leu-enkephalin by the of N-hydroxysuccinimide method. The side chain of the tyrosine residue is protected by the tert-butyl group. For α-amino protection Boc and Z groups are used. For details of the syntheses see figs. 8 and 9.

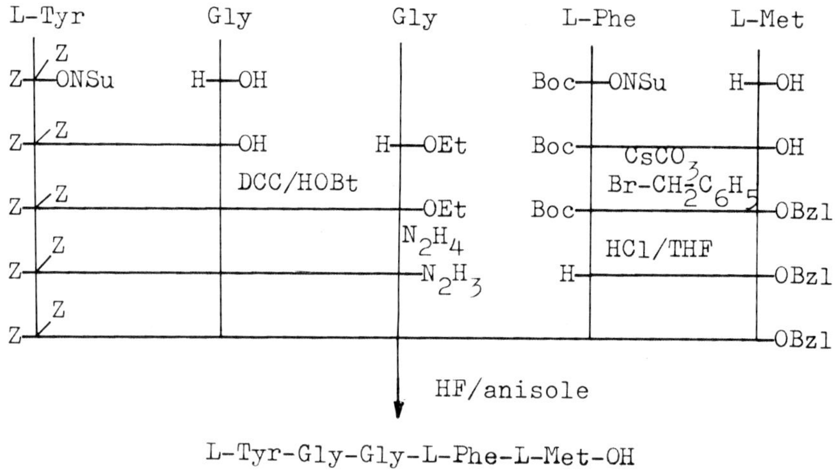

Fig. 6. Synthesis of Met-enkephalin by Wang et al. (5).

For Met- and Leu-enkephalin also solid phase and liquid phase peptide syntheses were described. To give an example, each, the synthetic routes of figs. 10 and 11 are discussed briefly.

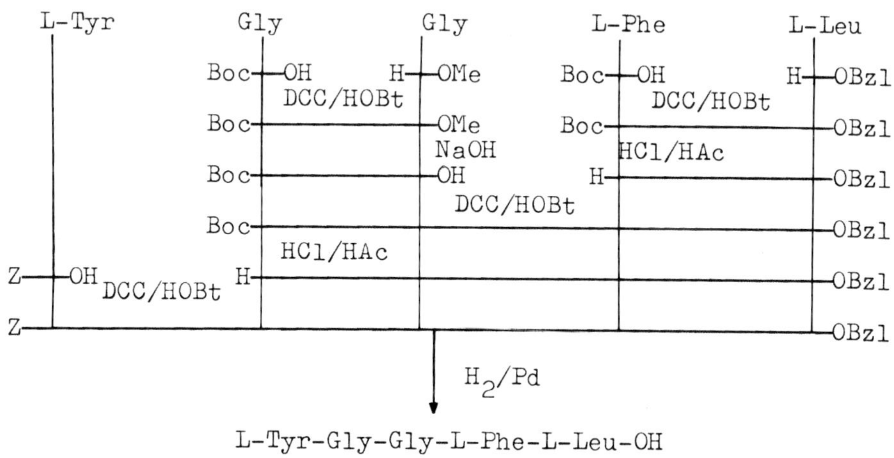

Fig. 7. Synthesis of leucine-enkephalin (3).

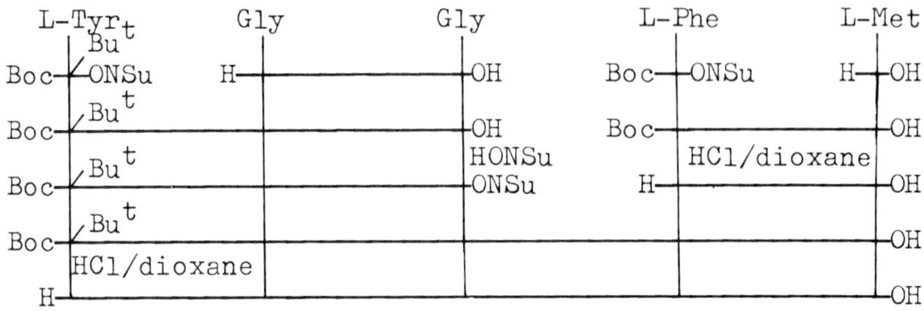

Fig.8. Synthesis of Met-enkephalin by Bower et al. (6).

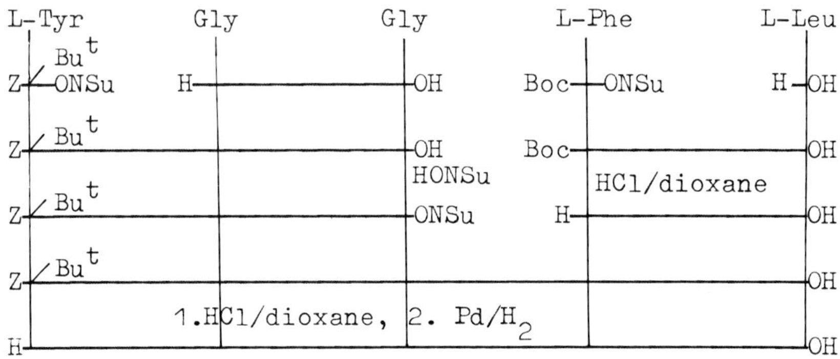

Fig. 9. Synthesis of Leu-enkephalin by Bower et al. (6).

Terenius et al. (7) use exclusively Boc amino acids for their synthesis (Fig.10). The side chain of tyrosine is protected by the 2,6-dichlorobenzyl group. A chloromethylated copolymer serves for the attachment of the C-terminal amino acid. 50 % Trifluoroacetic acid is used for the removal of Boc groups. Each coupling step is performed by the DCC method. After complete synthesis, the peptide is split off the resin with liquid hydrogen fluoride in the presence of anisole. One advantage of the liquid phase compared to the solid phase synthesis is that larger quantities can be produced. Jaureguiberry et al. (8) obtained Leu- and Met-enkephalin by this method (fig. 11) starting with Gly-Gly-L-Phe-OH,

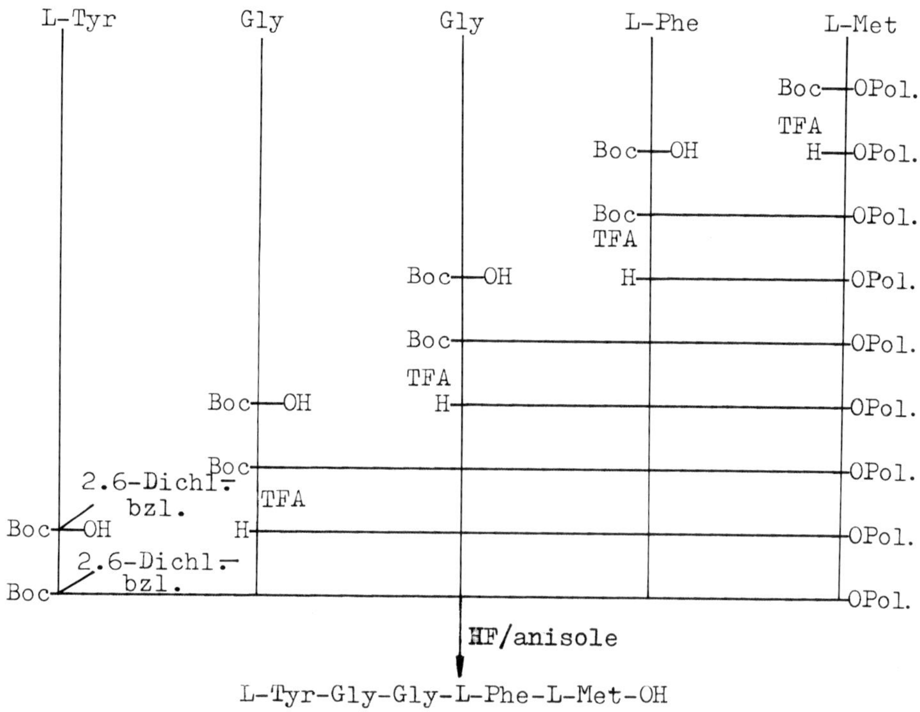

Fig. 10. Solid phase peptide synthesis of enkephalin by Terenius et al. (7).

Boc-L-Tyr-OH and L-Met-OMe. After protection of Gly-Gly-L-Phe-OH by the Boc group, Boc-Gly-Gly-Phe-OH is coupled by DCC with Met-OMe. The Boc group is cleaved off by trifluoroacetic acid from Boc-Gly-Gly-L-Phe-L-Met-OMe and the resulting tetrapeptide coupled by DCC with Boc-Tyr-OH. The desired enkephalin is obtained by saponification of the ester and splitting off the Boc group by trifluoroacetic acid.

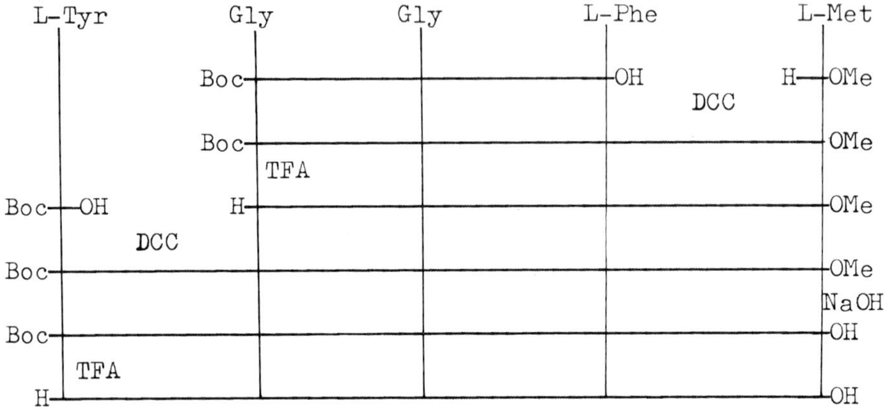

Fig.11. Liquid phase synthesis of enkephalin by Jaureguiberry et al. (8).

References

1. Hughes, J., Smith, T.W., Kosterlitz, H.W., Fothergill, L. A., Morgan, B.A., Morris, H.R.: Nature 258, 577-579 (1975).
2. Voelter, W., Bürvenich, C., Horn, H., Kalbacher, H., Pietrzik, E.: Angew.Chem. 88, 332-333 (1976).
3. Horn, H.: Thesis, Tübingen University, in preparation, 1980
4. Berger, E., Jakubke, H.-D.: Pharmazie 34, 349-350 (1979).
5. Wang, S.-S., Gisin, B.F., Winter, D.P., Makofske, R., Kulesha, I.D., Tzougraki, C., Meienhofer, J.: J.Org.Chem. 42, 1286-1290 (1977).
6. Bower, J.D., Guest, K.P., Morgan, B.A.: J.Chem.Soc. Perkin I, 2488-2491 (1976).
7. Terenius, L., Wahlström, A., Lindeberg, G., Karlsson, S., Ragnarsson, U : Biochem. Biophys.Res.Commun. 71, 175-179 (1976).
8. Garbay-Jaureguiberry, C., Roques, B.P., Oberlin, R., Anteunis, M., Lala, A.K.: Biochem.Biophys.Res.Commun. 71, 558-565 (1976).

1-(1-ADAMANTYL)-1-METHYLETHOXYCARBONYL (ADPOC): A NEW USEFUL AMINO PROTECTING GROUP +

Hubert Kalbacher and Wolfgang Voelter
Institut für Organische Chemie der Universität Tübingen, D-7400 Tübingen, Auf der Morgenstelle 18, Germany

Introduction

Although the well-known amino protecting groups of the urethane type (tert-butyl and benzyloxycarbonyl) fulfill many requirements of straightforward peptide synthesis (1), it is often desirable to remove NH_2 groups under very mild acidolytic conditions so that an optimal protection of acid-labile residues can be achieved in syntheses involving repetitive coupling and deblocking steps (solid phase method). Especially tryptophan-containing peptides are destroyed or alkylated by removing the protecting groups under strong acidolytic conditions (2). The protecting groups Bpoc (3), Ddz (4), Ppoc (5), Azoc (6) and TmZ (7) accomodate these requirements; hoever, only the first two find general use. In both cases the storage of the protected amino acids is not unproblematical (8) owing to autocatalytic decomposition and photolability. Also the Bpoc phenyl carbonate, frequently used for introduction of the Bpoc group, tends to decompose even at low temperatures (9). Bpoc-Trp-OH can be synthesized in only low yields.

In an investigation of amino-protecting groups designed to increase the acidlability, we found that the new 1-(1-adamantyl)-1-methylethoxycarbonyl(Adpoc)group has the following properties (10,11) :

+Dedicated to Prof.Dr.H.Auterhoff,Tübingen,on the occasion of his 65th birthday.

Structure and Activity of Natural Peptides

1. The Adpoc residue can be easily incorporated into amino acids and peptides by means of the crystalline and stable Adpoc phenylcarbonate (Adpoc-OPh), Adpoc fluoride (Adpoc-F) or Adpoc-oxiimino-2-phenylacetonitrile (Adpoc-ON), the formations of which are shown in Fig.2.
2. The Adpoc group can be removed in mild acidolytic medium, preferably 3 % trifluoroacetic acid in dichloromethane, within 4-5 minutes at $0^{o}C$ or 1-2 minutes at room temperature. In a mixture of acetic acid/83 % formic acid/water (7:1:2) deprotection of the Adpoc group is complete within 100 minutes ($40^{o}C$). Under these conditions the Adpoc group is removed 10^{3} times faster than the Boc group; consequently it is possible to cleave the Adpoc group selectively in peptides with protecting groups of the tert -butyl type and also with benzyloxycarbonyl, benzyl ether, and benzyl ester protecting groups.
3. The protecting group is completely stable towards alkali and hydrogenolysis.
4. The lipophilic character of the Adpoc group leads to excellent solubility of higher peptides.
5. Adpoc amino acids are crystalline compounds and stable for months at room temperature.
6. Adpoc amino acids are stable in the presence of sunlight.

Results

Synthesis of Adpoc compounds. An intermediate in the synthesis of Adpoc amino acids is the tertiary alkohol, 2-(1-adamantyl)-2-propanol. For its preparation 1-adamantanecarboxylic acid is converted by reaction with phosphorus pentachloride/ethanol into the ethyl ester, which, on reaction with Grignard reagent (from methyl iodide and magnesium), affords the alcohol in over 90 % yield.

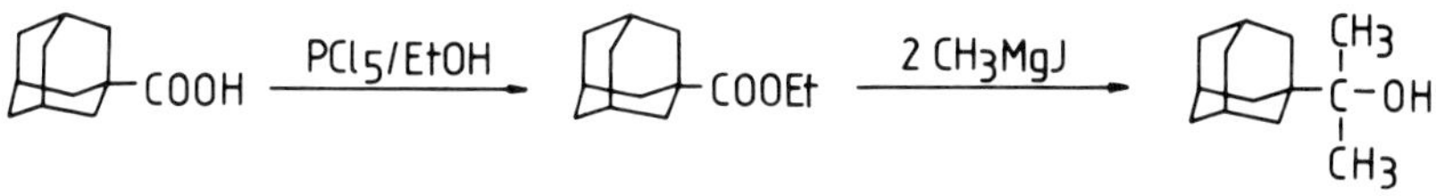

Fig.1. Synthesis of 2-(1-adamantyl)-2-propanol.

Incorporation of the Adpoc group can easily be performed by reagents according to Fig. 2.

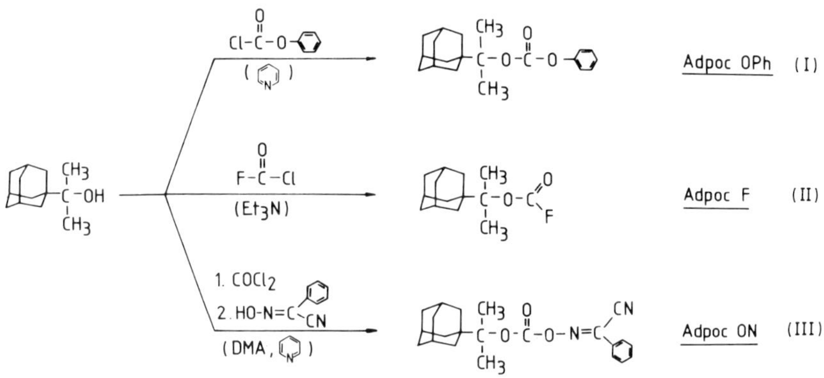

Fig.2. Synthesis of reagents for the incorporation of the Adpoc group.

Usually the benzyltrimethylammonium(Triton B) salts of amino acids are reacted with reagents I or II in DMF. Although the incorporation of protecting groups via phenylcarbonates is generally used, this method proved to be disadvantageous for several reasons in our case (11). However, Adpoc fluoride, a very reactive and stable compound, allows the introduction of the Adpoc group into amino acids under mild conditions (-10°C, reaction time only few minutes) and in high yields. The reagent is stable at -30°C in solution for several months. Alternatively, amino acids can be acylated in an aqueous system (water/dioxane, equimolar addition of triethylamine) with the crystalline (mp. 100-101°C) 2[1-(1-adamantyl)-1-

methylethoxycarbonyl-oxiimino]-2-phenylacetonitrile(Adpoc-ON) within one hour at room temperature in excellent yields.

In table 1 the physical data of some Adpoc derivates are collected. Also the yields of pure isolated material are given.

Table 1. Yields and physical data of selected Adpoc derivatives.

Compound	Yield [%]	M.P. [°C]	$[\alpha]_D^{20}$ [°]
Adpoc-Ala.DCHA	83	159-160	+6.2(c=0.7, EtOH)
Adpoc-Arg(NO_2)-OH .1/2 H_2O	88	206	-3.6(c=0.8, DMF)
Adpoc-Asp(Bu^t)-OH	73	178	+10.3(c=0.8, MeOH)
Z-Arg($(Adpoc)_2$)-OH	69	158-160	+15.2(c=1.2, $CHCl_3$)
Adpoc-Gln-OH	74	98	-2.9(c=0.8, EtOH)
Adpoc-Gly-OH	78	181	
Adpoc-His(Adpoc)-OH	87	113	+9.2(c=0.5, DMF)
Adpoc-Lys(Z)-OH	83	75	-4.5(c=0.9, MeOH)
Adpoc-Met.DCHA	74	150	-22.8(c=0.8, MeOH)
Adpoc-Leu-OH	89	176	-12.8(c=1, MeOH)
Adpoc-Phe.DCHA	77	124	+21.8(c=1.2, MeOH)
Adpoc-Ser(Bu^t)-OH	81	188-189	+15.7(c=0.8, MeOH)
Adpoc-Thr(Bzl)-OH	71	50-52	+19.8(c=1, MeOH)
Adpoc-Thr(Bzl).DCHA	94	130-131	+12.9(c=1, MeOH)
Adpoc-Trp-OH	86	116	-6.8(c=1, MeOH)
Adpoc-Val.DCHA	79	105	+7.8(c=0.9, MeOH)

Quantitative analysis of Adpoc deprotection by HPLC. The acidolytic cleavage of the Adpoc group can be easily followed by high-performance liquid chromatography (HPLC) as shown in figures 3 and 4, using Adpoc-Trp-OH as an example.

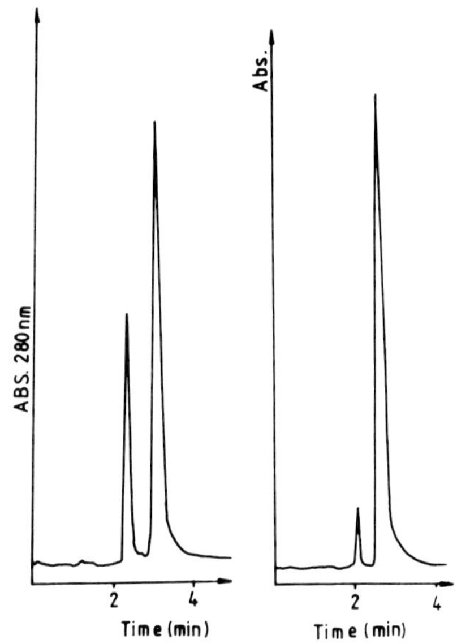

Fig. 3. HPLC chromatogram of the acidolytic cleavage of Adpoc-Trp-OH; left: 20 sec, right: 40 sec after treatment with 3 % CF_3COOH/CH_2Cl_2 at 25°C. Column: RP-8 (25 x 0.4 cm, 10 µm particle size); eluent: Methanol (90 %)/water (adjusted to pH 5.2 with H_2SO_4); flow rate: 2.5 ml/min; pressure at column inlet: 120 atm; UV detection: 280 nm.

Deprotection of N^{α}-Adpoc in the presence of a N^{ε}-Boc group. Investigations of Sieber and Iselin showed that the Boc group in N^{ε}-position has more than double the stability than in N^{α}-position. In the deprotection reagent acetic acid/formic acid (83 %)/water (7:1:2) it is possible to get selective cleavage of the Adpoc group in presence of N^{ε}-Boc groups (see Fig.5).

In table 2, data are given for the half-life times, times for a 99.9 % deprotection and the velocity constants k_1 in various acidolytic solvents.

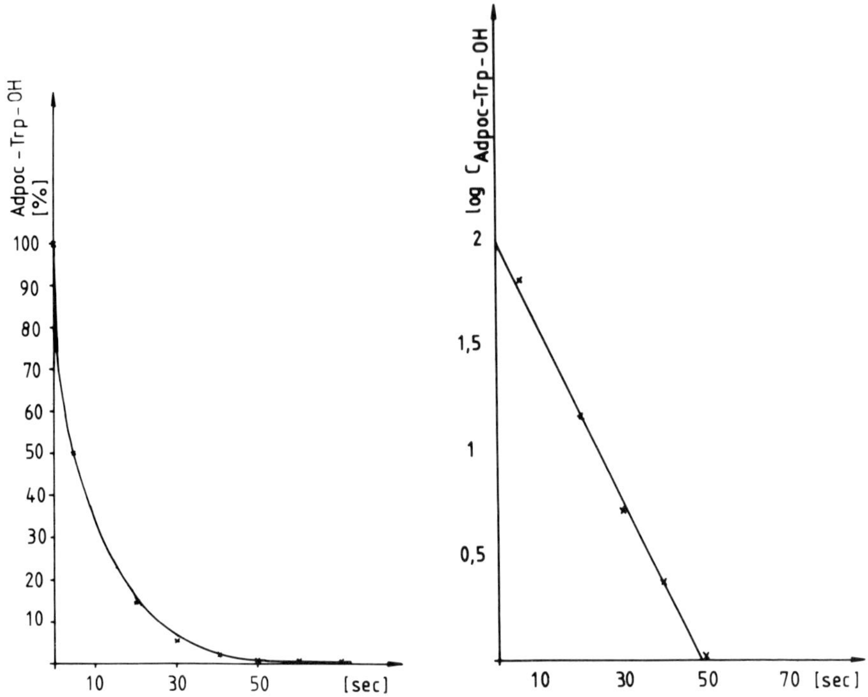

Fig. 4. Cleavage of Adpoc from Adpoc-Trp-OH in the presence of 3 % TFA/CH_2Cl_2 at 25°C. Left: Decrease of Adpoc-Trp-OH in [%] and right: In log C as a function of time.

Table 2. Rates of deprotection of Adpoc-Trp-OH using various acids.

Nr.	Acidolytic solvent	Deprotection		k_1 (min^{-1})	k_{rel} (Boc=1)
		50 %(min)	99.9 %(min)		
1	3 % TFA/CH_2Cl_2, 25°C	0.18	1.78	3.87	
2	3 % TFA/CH_2Cl_2, 0°C	1.1	11.7	0.63	
3a	Acetic acid/ 83 % (w/w) formic acid/ water, 7:1:2 (v/v), 40°C	10.8	107	$6.4.20^{-2}$	

Table 2 (continued)

Nr.	Acidolytic solvent	Deprotection 50 %(min)	Deprotection 99.9 % (min)	$k_1(min^{-1})$	k_{rel} (Boc=1)
3a	but 25°C	35.5	354	$1.95.10^{-2}$	590
4	Acetic acid 80 % (v/v) 40°C	34	338	$2.00.10^{-2}$	
For comparison cleavage in solvent 3b					
Bpoc-Pro-OH, 25°C		10.5		$6.5.10^{-2}$	0.30
N^{ε}-Boc-Lys-OH, 25°C		36 d		$1.3.10^{-5}$	1470
Boc-Gly-OEt, 25°C		14 d		$3.3.10^{-5}$	

The following relations for the reaction rate constants are found for the acidolytic solvent acetic acid/formic acid (83 %)/water (7:1:2):

N^{α}-Boc: Adpoc : Bpoc = 1:600 - 1000 : 1800 - 2700

N^{ε}-Boc: Adpoc : Bpoc = 1:1500: 4900 - 6800

Stability of Adpoc amino acids. The indole part of tryptophan is very sensitive to oxidation, even in the absence of oxygen. As main products of the photooxydation, ß-(3-oxindolyl)-alanine and N-formylkynurenine are idendified (12). Therefore, a one mmolar solution of tryptophan and Adpoc-tryptophan in 80 % methanol is allowed to stand for 3 months under normal light conditions. In this time, the tryptophan containing solution becames more and more dark yellow, while the Adpoc-Trp-OH solution remains completely colourless. More than three side products can be detected by HPLC. These changes are also visible in the UV. However, no spectroscopic changes are observed for on Adpoc-Trp-OH and a freshly prepared Trp solution, while the tryptophan solution, 3 months of age, shows quite a different absorption spectrum (Fig. 6).

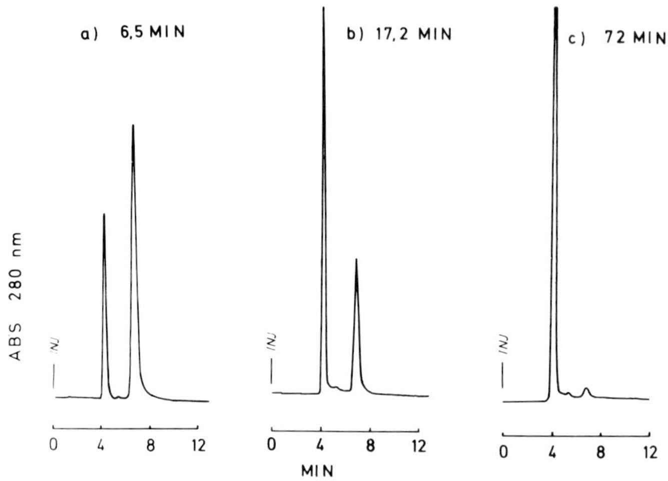

Fig. 5. HPLC chromatogram of the cleavage of Adpoc-Trp-Lys-(Boc)-OH after treatment with acetic acid/formic acid (83 %)/water (7:1:2) at 40 °C. The yield of liberated H-Trp-Lys(Boc)-OH is for a) 36.48 %; b) 65.99 %; c) 98.63 %. No deprotection of the N^{ε}-Boc group is observed. For chromatographic conditions see Fig. 4.

Application of the Adpoc group. The application of the Adpoc group for peptide synthesis is demonstrated by the syntheses of TRH and somatostatin.

TRH. The basic character and acylation of the imidazole ring of histidine, the low solubility of intermediates and the hygroscopic character of TRH are difficulties which have to be overcome during its synthesis. According to figure 7 TRH can be synthesized in a convenient manner.

Adpoc-His(Adpoc)-OH, obtained by reacting histidine with Adpoc-F in more than 80 % yield, was coupled with prolinamide by dicyclohexylcarbodiimide/1-hydroxybenzotriazole (DCC/HOBt). The lipophilic character of this dipeptide allows conventional washing operations and it can easily be purified by crystalli-

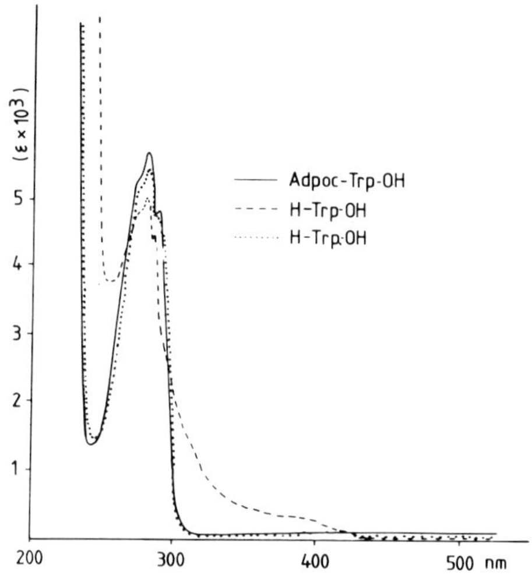

Fig.6. UV spectra of a H-Trp-OH (.....) solution and of Adpoc-Trp-OH (———) and H-Trp-OH (-----) solutionsafter exposure to normal light for 3 months (in 80 % methanol).

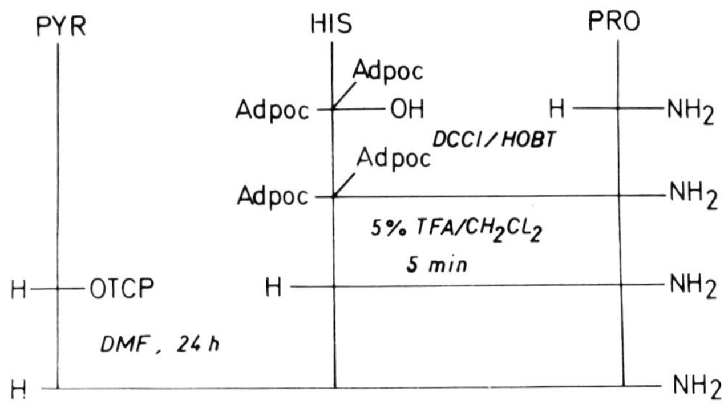

Fig. 7. Synthesis of TRH using the Adpoc group.

zation. The deprotection of both Adpoc groups is complete by treatment with CF_3COOH/CH_2Cl_2 (5 : 95 %) at room temperature within 4-5 minutes. Further reaction with Pyr-OTCP yields directly TRH.

Somatostatin. The successful synthesis of a protected C-terminal heptapeptide with the natural sequence of somatostatin using Adpoc protection for the tryptophan moiety demonstrates that the Adpoc residue can be selectively cleaved in a solvent of low acidity: $CH_3COOH/HCOOH$ (83 %)/H_2O, 7:1:2, at 40°C; under these conditions, the N^{ε}-Boc group of lysine, the tert-butyl groups in the side chains of threonine and serine, and the trityl groups of cysteine residues are not attacked (Fig.8).

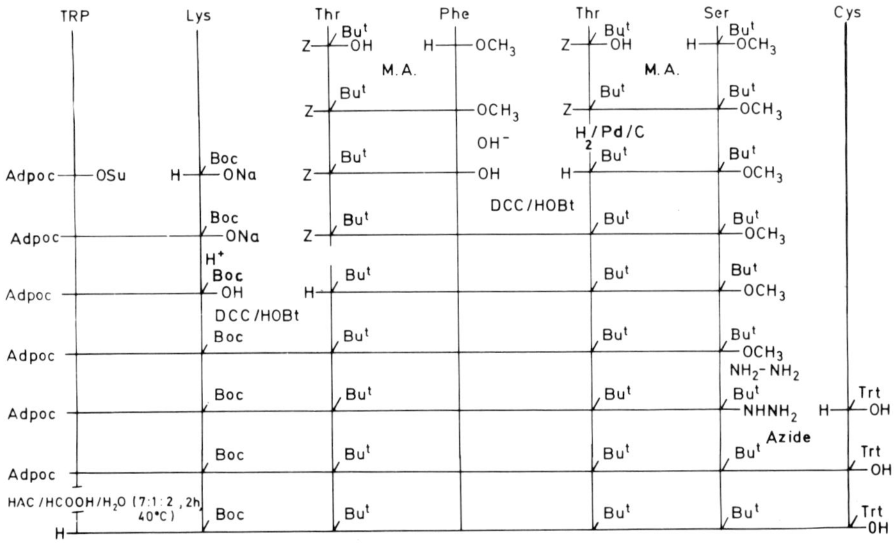

Fig.8. Synthesis of a protected C-terminal heptapeptide with the natural sequence of somatostatin using Adpoc protection for the Trp residue.

Measurements by field desorption mass spectrometry (f.d.m.s.) demonstrate the selecticity of cleavage in the N^{α}-position. Neither in acetic acid/formic acid (83 %)/water (7:1:2, 1h, 40°C) nor in 1 % TFA/CH_2Cl_2(1h, 0°C) deprotection of the protecting groups in the side chains is observed.

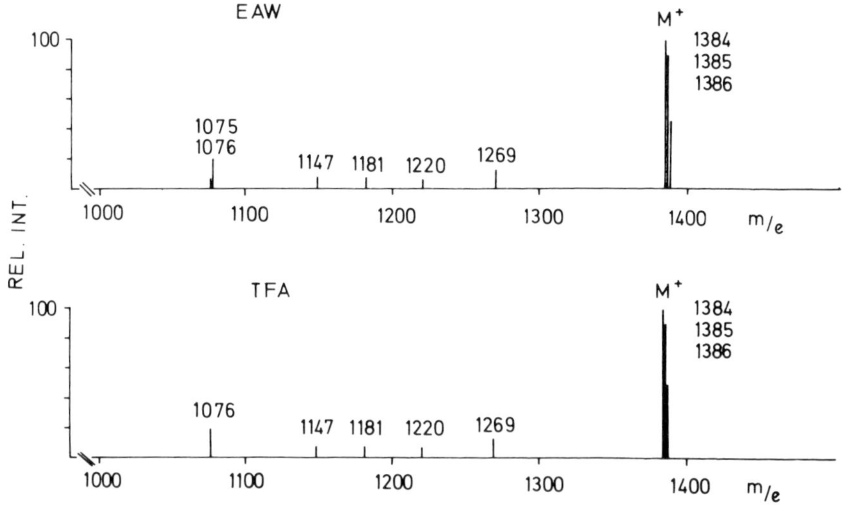

Fig.9. Field desorption mass spectra of H-Trp-Lys(Boc)-Thr-(Bu^t)-Phe-Thr(Bu^t)-Ser(Bu^t)-Cys(Trt)-OH, emitter heating current: 15 mA, EAW: Treated with $CH_3COOH/HCOOH$ (83 % w/w)/H_2O, 7:1:2, at 40°C for 2 h; TFA: Treated with CF_3COOH (1 %)/CH_2-Cl_2 at 0°C for 90 min.

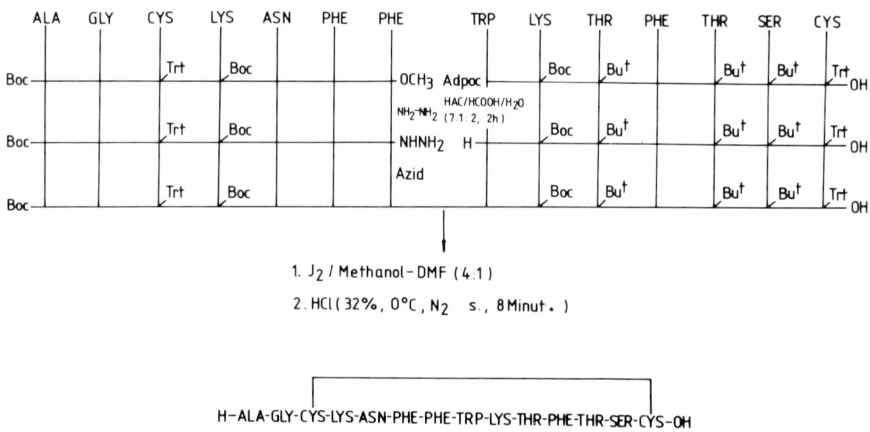

Fig.10. Synthesis of somatostatin by 7+7 fragment condensation.

This heptapeptide was condensed with the azide, derived form Boc-Ala-Gly-Cys(Trt)-Lys(Boc)-Asn-Phe-Phe-$NHNH_2$ according Honzl-Rudinger (13) to give protected somatostatin in more than 85 % yield.

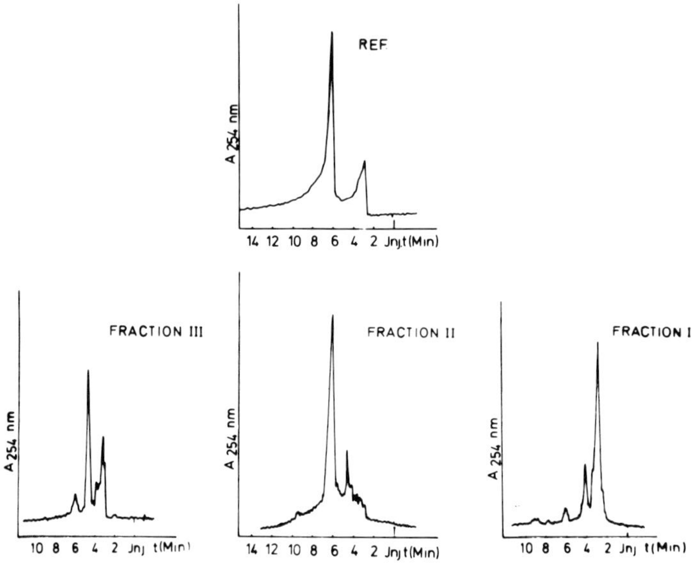

Fig.11. Chromatogram of crude somatostatin received after purification on a Sephadex G-25 column (800 x 6 mm column, solvent 2 N CH_3COOH). Solvent: Isopropanol(35 %)/ 0.01 m NH_4Ac (pH 4.4); column: 250 x 4.6 mm; resin: RP-8; flow rate: 1.5 ml/min; pressure: 210 bar; temperature :25°C; UV detection: 254 nm.

Oxidation of this compound with iodine in methanol/DMF (4:1) (14) followed by a short treatment with hydrochloric acid (0°C, 10 min, N_2 saturated solution) gave crude somatostatin. Purification of this product on Sephadex G-25 (eluent 2 N HAc) yields 3 fractions which are examined by HPLC (Fig.11). The main peak of fraction II correlates with that of a reference sample. Further purification of this fraction by semipreparative HPLC yields a very pure material. The homogeneity of this final product is confirmed by HPLC (Fig.12), TLC in 7 different systems (11), electrophoresis and amino acid analysis.

From the hydrolysis with 4 N methanesulfonic acid/0.2 % 3-(2-aminoethyl)indole at 115°C for 22 h (a) (15) respectively 6 N HCl/2.5 % thioglycolic acid at 115°C for 18 h (b) (11) and the determination of cysteine according to reference 16, the amino acid composition is determined:

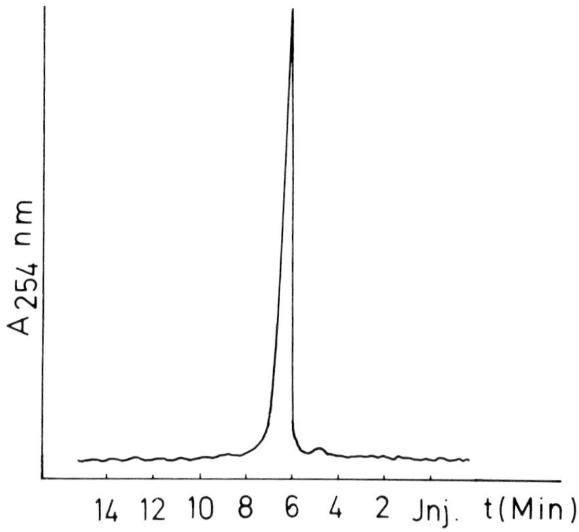

Fig.12. Chromatogram of pure somatostatin. Conditions see Fig. 11.

	Ala	Gly	Lys	Asp	Phe	Thr	Ser	Trp
Found (a)	1.0	1.06	1.93	1.05	2.81	1.77	0.81	0.61
Found (b)	1.0	1.0	2.11	0.91	2.87	2.07	0.71	0.58
Calculated	1	1	2	1	3	2	1	1

For cysteine a value of 1.82 is found (16).

The degree of racemization is detected according to ref. 17 and gives the following values: Ala (2.7 %), Lys (1.6 %), Asp (3.8 %), Phe (2.4 %), Thr (0.39 %), Ser (3.4 %).

These results indicate that the Adpoc group is a useful protecting group with advantageous properties for peptide synthesis.

References

1. Wünsch, E., Synthese von Peptiden, in: Houben-Weyl-Müller (eds.), Methoden der Organischen Chemie, Georg Thieme Verlag Stuttgart 1974

2. Wünsch, E., Jäger, E., Kisfaludy, L., Löw, M.: Angew.Chem. 89, 330 (1977).

3. Sieber, P., Iselin, B.: Helv.Chim.Acta 51, 622 (1968).
4. Birr, C., Lochinger, W., Stahnke, G., Lang, P.: Liebigs Ann.Chem. 763, 162 (1972).
5. Sieber, P., Iselin, B.: Helv.Chim.Acta 51, 614 (1968).
6. Tun-Kyi, A., Schwyzer, R.: Helv.Chim.Acta 59, 1642 (1976).
7. Matsueda, G.R., Stewart, J.M.: in:Walter, R., Meienhofer, J. (eds.): Peptides: Chemistry, Structure and Biology, p.333, Ann Arbor Science Publishers Inc., Ann Arbor, Mich. 1975
8. Sieber, P., Iselin, B.: Helv.Chim.Acta 52, 1525 (1969).
9. Schnabel, E., Schmidt, G., Klauke, E.: Liebigs Ann.Chem. 743, 69 (1971).
10. Kalbacher, H., Voelter, W.: Angew.Chem. 90, 998 (1978).
11. Kalbacher, H.: Dissertation, Universität Tübingen 1980
12. Holt, L.A., Milligan, B., Rivett, D.E., Stewart, F.H.C.: Biochim.Biophys.Acta 499, 131 (1977).
13. Honzl, J., Rudinger, J.: Coll.Czech.Chem.Commun. 26, 2333 (1961).
14. Kamber, B., Rittel, W.: Helv.Chim.Acta 51, 2061 (1968).
15. Simpson, R.J., Neuberger, M.R., Liu, T.-Y.: J.Biol.Chem. 251, 1936 (1976).
16. Needleman, S.B.: Protein Sequence Determination, Springer-Verlag, Berlin-Heidelberg-New York 1975
17. Frank, H., Nicholson, G.J., Bayer, E.: J.Chromat.Sci. 15, 174 (1977).

SYNTHESIS OF THYROLIBERIN ANALOGS WITH MODIFIED PYROGLUTAMYL RESIDUES

Eberhard Schwertner, Siegfried Herrling, Elmar Friderichs, Sung Man Kim, Leopold Flohé

Grünenthal GmbH, Center of Research
D-5100 Aachen, FRG

Introduction

When we started our work, thyroliberin (L-pyroglutamyl-L-histidyl-L-prolinamide; TRH) had been successfully modified in the histidine and proline position only without appreciable loss of biological activity (for review see ref. 1). It had been infered from these studies that the pyroglutamyl group is as essential for the biological activity (2) as is the tripeptide nature of TRH. Our investigations were based on the working hypothesis that in the pyroglutamyl residue the partial structure of formula I

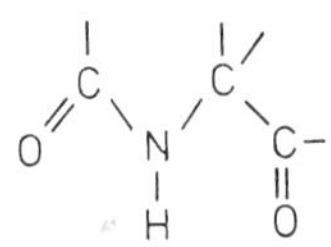

(I)

might be sufficient to maintain biological activity and that the nature of the other members completing the ring might preferentially influence biological stability rather than intrinsic activity. Accordingly, we synthesized analogs of TRH in which the pyroglutamyl group is exchanged by residues of carboxylic acids having one of the two partial structures II or IIa, wherein the ring is a 5- or 6-membered heterocyclic system containing two heteroatoms. Especially when using acids of type IIa the resulting compounds are no longer in the strict sense tripeptides but rather acylated dipeptides.

Structure and Activity of Natural Peptides

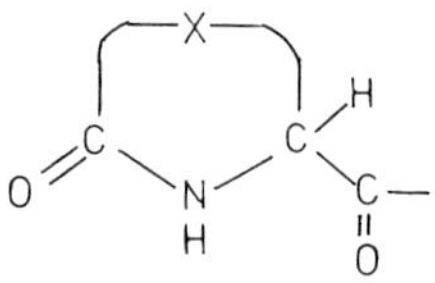

(II)

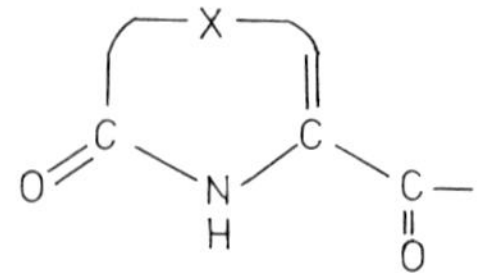

(IIa)

In addition, some analogs with a 5- or 6-membered heterocyclic ring system not containing the partial structure of formula I were synthesized in order to check the above hypothesis.

The present contribution describes the synthesis and some molecular properties of the new analogs. The biological results obtained with these compounds will be presented elsewhere in this issue (3,4).

Materials and methods

Commercially available compounds used in synthetic and analytical experiments were at least synthetic grade or reagent grade products, respectively.

1. Starting materials, if not commercially available, were synthesized according to published methods (ref. in brackets) or in analogy to such procedures:
   5-chloro-orotic acid (5,6);
   5-bromo-orotic acid (7,8);
   5-methyl-orotic acid, 5-ethyl-orotic acid and 5-n-propyl-orotic acid (9,10);
   D-(-)-penicillamine (11);
   3,5-dioxo-tetrahydro-1,2,4-triazine-6-carboxylic acid and 3-thioxo-5-oxo-tetrahydro-1,2,4-triazine-6-carboxylic acid (12);
   3-amino-1,2,4-triazole-5-carboxylic acid (13);

L-2-oxo-imidazolidine-5-carboxylic acid (14);
2-oxo-imidazole-4-carboxylic acid (15);
L-2-oxo-thiazolidine-4-carboxylic acid (16);
(3S)-5-oxo-thiomorpholine-3-carboxylic acid (17);
L-(-)-α-bromo-propionic acid and D-(+)-α-bromo-propionic acid (18) (the corresponding ethyl esters were obtained by esterification catalyzed with 4-toluenesulfonic acid);
(3S, 6RS)-6-phenyl-5-oxo-thiomorpholine-3-carboxylic acid, (3S)-6,6-dimethyl-5-oxo-thiomorpholine-3-carboxylic acid, (3S, 6RS)-6-ethyl-5-oxo-thiomorpholine-3-carboxylic acid, (3S, 6RS)-6-n-propyl-5-oxo-thiomorpholine-3-carboxylic acid and (3S, 6RS)-6-isopropyl-5-oxo-thiomorpholine-3-carboxylic acid (19).

(3S, 6RS)-6-n-butyl-5-oxo-thiomorpholine-3-carboxylic acid (19): $[\alpha]_D^{23} = -99.1^0$ (c = 1 in MeOH), m.p. = 168-169^0 C; crystallization solvent: ethyl acetate.

(3R)-2,2-dimethyl-5-oxo-thiomorpholine-3-carboxylic acid (17): $[\alpha]_D^{23} = -14.2^0$ (c = 1 in MeOH), m.p. = 189-192^{0}C; crystallization solvent: H_2O.

(3S)-6-methyl-5-oxo-thiomorpholine-3-carboxylic acid (A; starting material for compound 17):
$[\alpha]_D^{23} = +41.8^0$ (c = 1 in MeOH), m.p. = 186-187^{0}C; crystallization solvent: H_2O.

(3S)-6-methyl-5-oxo-thiomorpholine-3-carboxylic acid (B; starting material for compound 18):
$[\alpha]_D^{23} = -112.8^0$ (c = 1 in MeOH), m.p. = 161-162^{0}C; crystallization solvent: H_2O.

The acids A and B were obtained in analogy to ref. 19 by reaction of L-cysteine disodium salt with either L, D or D,L-ethyl-α-bromopropionate and subsequent separation of the optical isomers by fractioned crystallization.

2. <u>The synthesis of the TRH analogs</u> were performed according to the routes outlined in scheme 1.

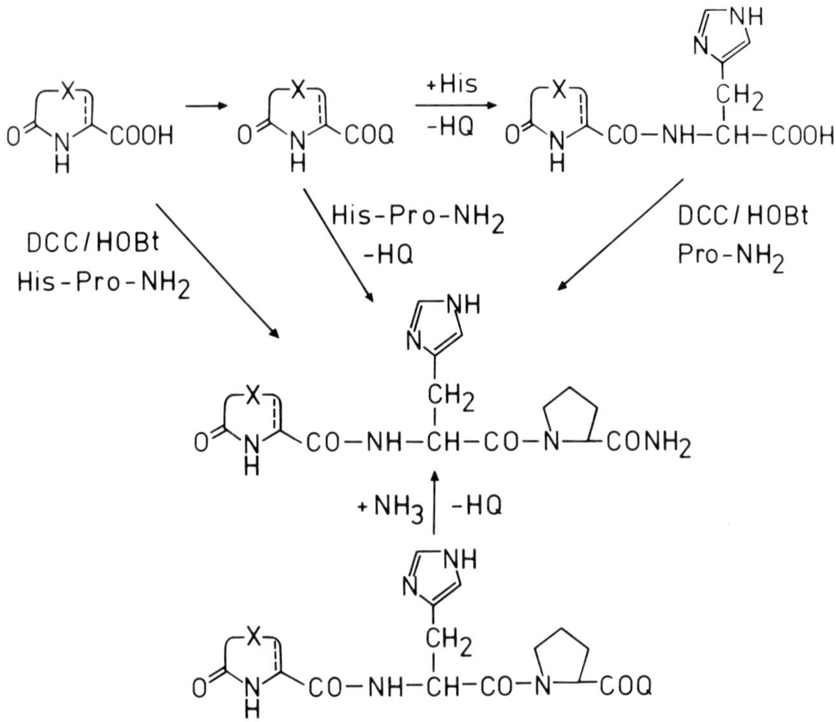

Scheme 1. Synthetic routes to obtain TRH analogs.
-COQ is an activated carboxylic group (e.g. acid chloride, activated ester etc.).

Out of the four possibilities shown in scheme 1 the following procedure proved to be particularly suitable for our purposes: L-His-L-ProNH$_2$. 2 HBr (20) is mixed with equimolar amounts of 1-hydroxy-benzotriazole (HOBt) and the carboxylic acid of formula II or IIa in DMF (e.g. 100 mmoles of each component in 200 ml). If the acid is not soluble under these conditions, it was suspended. At -5^{0}C 2 equivalents of triethylamine and thereafter 1 equivalent of N,N'-dicyclohexylcarbodiimide (DCC) in DMF are added while stirring. The reaction is allowed to proceed for 12 hours during which time the reaction mixture approaches room temperature. The reaction product is obtained following the principle of the process described in ref. 21.

For special purposes an alternative procedure was preferred as outlined in scheme 2.

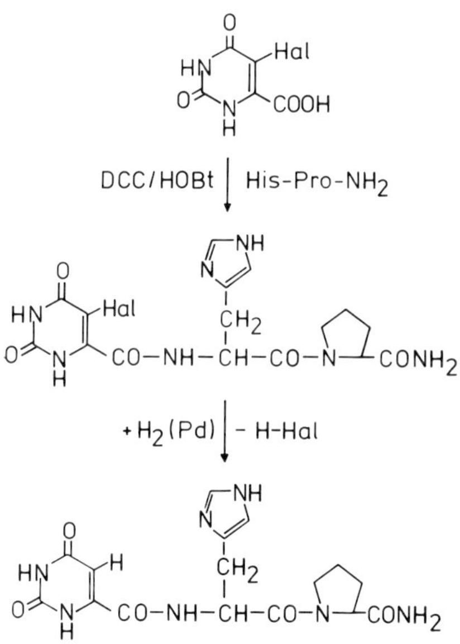

Scheme 2. Synthesis of Orotyl-His-ProNH$_2$.

Orotyl-L-His-L-ProNH$_2$ is obtained from the corresponding 5-halogeno-orotyl derivatives as follows: 2.3 equivalents of Na-acetate and an appropriate amount of 5 % Pd on charcoal in H_2O are treated with an excess of H_2. Then a solution of 2 equivalents of 5-chloro-orotyl-L-His-L-ProNH$_2$ or 5-bromo-orotyl-L-His-L-ProNH$_2$ in H_2O are added and shaking in H_2 atmosphere is continued at normal pressure and room temperature until 2 equivalents of H_2 are consumed. The catalyst is filtered off, the filtrate is extracted with ether and then the aqueous layer is evaporated under reduced pressure to dryness.

3. Purification methods: The following methods were used for the purification of the compounds prepared according to the procedures described above:
Crystallization and recrystallization from water (P 1), water-ethanol (1:1) (P2), n-butanol (saturated with water) (P3), ethanol (P4), methanol-ethanol (1:5) (P5), methanol-water (1:1) (P6); reprecipitation from methanol-chloroform (1:10) (P7); countercurrent distribution in the systems n-butanol-water (P8) and toluene-methanol-chloroform-water (5:8:5:2) (P9); column chromatography on silica gel with methanol as eluant (P10); column chromatography on aluminium oxide with methanol-water (1:1) (P11) or methanol-water (2:1) (P12) as eluting solvents.

For column chromatography silica gel 60 (0.040 - 0.063 mm) and basic aluminium oxide 90 (0.063 - 0.200 mm) of E. Merck AG, Darmstadt were used.

The table shows which particular procedure (P1 - P12) was applied for purification of a given compound.

4. Analytical procedures: Thin layer chromatography was performed on silica gel 60 plates and on silica gel 60 / kieselguhr F_{254} aluminium foil of E. Merck AG and with solvents and detection systems approved in peptide chemistry.

Optical rotations were measured with a Perkin-Elmer 141 polarimeter.

NMR spectra were determined at 60 MHz with a Bruker WP 60 spectrometer.

Melting points were determined with a "Monoskop" of H.Bock, Bad Oeynhausen, and were not corrected.

Results and Discussion

Out of the strategies to synthesize TRH analogs outlined under methods, the first procedure, i.e. simple coupling of His-$ProNH_2$ with the heterocyclic carboxylic acid, proved to be particularly useful to obtain a large number of analogs within a reasonable period of time. All twenty-five compounds listed in the table could be synthesized by this route in satisfying yields and purities.

As could be expected, cyclisation of His-$ProNH_2$ to the respective dioxopiperazine competed with the intended coupling reaction to a variable degree. Other impurities could result from the use of the histidine derivative unprotected in the imidazole moiety, although the corresponding side reactions appeared to be of minor importance according to some orientating experiments. Extensive purification of the raw products, however, was required and resulted in moderate yields in some instances (see table). We deliberately made allowance for these disadvantages of our main procedure and did not systematically try to improve the yields, since the limited amounts of pure compound required for biological testing were always obtained without major difficulties.

Only in special cases, alternative strategies were preferable. For instance, the synthesis of 5-chloro-orotyl-His-$ProNH_2$ followed by hydrogenolysis of the chlorine atom resulted in good yields of Orotyl-His-$ProNH_2$. This way offered a convenient access to radiolabeled compounds simply by substituting tritium for hydrogen in the very last step of the synthesis.

Directed by simultaneous investigations of the biological activity of the TRH analogs (see ref. 3), our synthetic work soon concentrated on derivatives with six membered heterocyclic rings replacing the pyrrolidinone moiety of TRH. Particularly two compounds were further characterized with regard to their oustanding efficacy, Orotyl-His-$ProNH_2$ (CG 3509) and (6-methyl-5-oxo-thiomorpholinyl-3-carbonyl)-His-$ProNH_2$ (CG 3703). The ^{13}C NMR spectra of both compounds show some striking similarities to those of TRH obtained by Voelter et al. (22) (fig. 1-3). Not only are the signals of the His and Pro carbon atoms expectedly close to those of TRH, also the characteristic splitting of the proline signals is observed which according to Zech (23) can be attributed to a cis/trans isomerism of the acylated proline residue. A similar splitting of the histidine signals is most likely caused by the presence of different tautomeric forms (24).

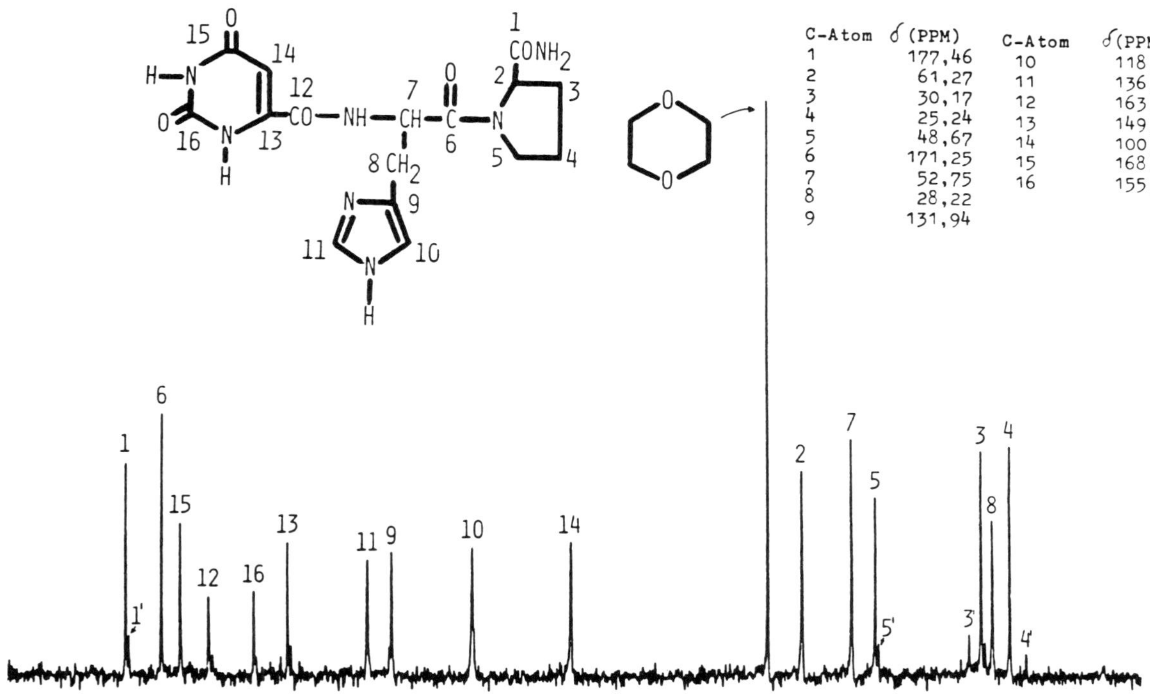

C-Atom	δ (PPM)	C-Atom	δ (PPM)
1	177,46	10	118,24
2	61,27	11	136,13
3	30,17	12	163,28
4	25,24	13	149,71
5	48,67	14	100,89
6	171,25	15	168,09
7	52,75	16	155,55
8	28,22		
9	131,94		

Fig. 1. ^{13}C NMR spectrum of CG 3509. Equipment: Bruker WP 60; solvent: D_2O; int. standard: 1,4-dioxane; sweep width (SW)F_1: 100 Hz/cm; offset (A) F_1: 8100 Hz; times (NS): 96832 scans; $\delta_{TMS} = \delta_{dioxane} - 67.3$ ppm.

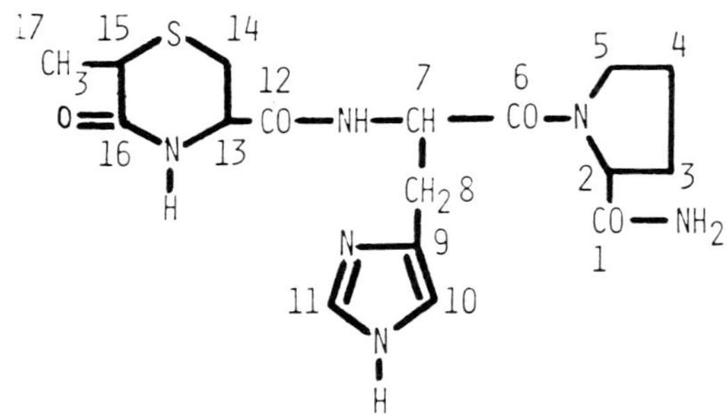

C-Atom	δ (PPM)	C-Atom	δ (PPM)
1	177,12	10	119,24
2	61,72	11	136,64
3	30,68	12	173,41
4	25,75	13	58,13
5	48,63	14	29,40
6	172,18	15	37,50
7	53,26	16	171,46
8	29,89	17	16,86
9	133,72		

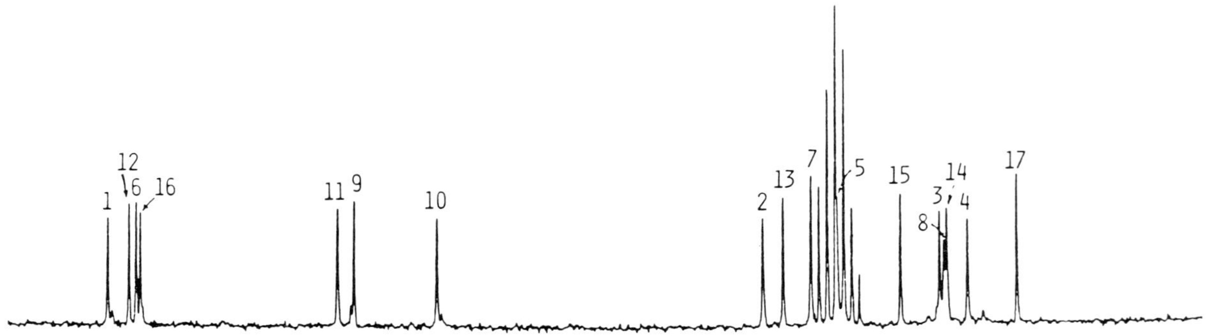

Fig. 2. ^{13}C NMR spectrum of CG 3703. Equipment: Bruker WP 60; solvent: MeOH-d_4; int. standard: MeOH-d_4; sweep width (SW) F_1: 100 Hz/cm; offset (A) F_1: 8050 Hz; times (NS): 72554 scans; $\delta_{TMS} = \delta_{MeOH\text{-}d_4} - 49{,}0$ ppm.

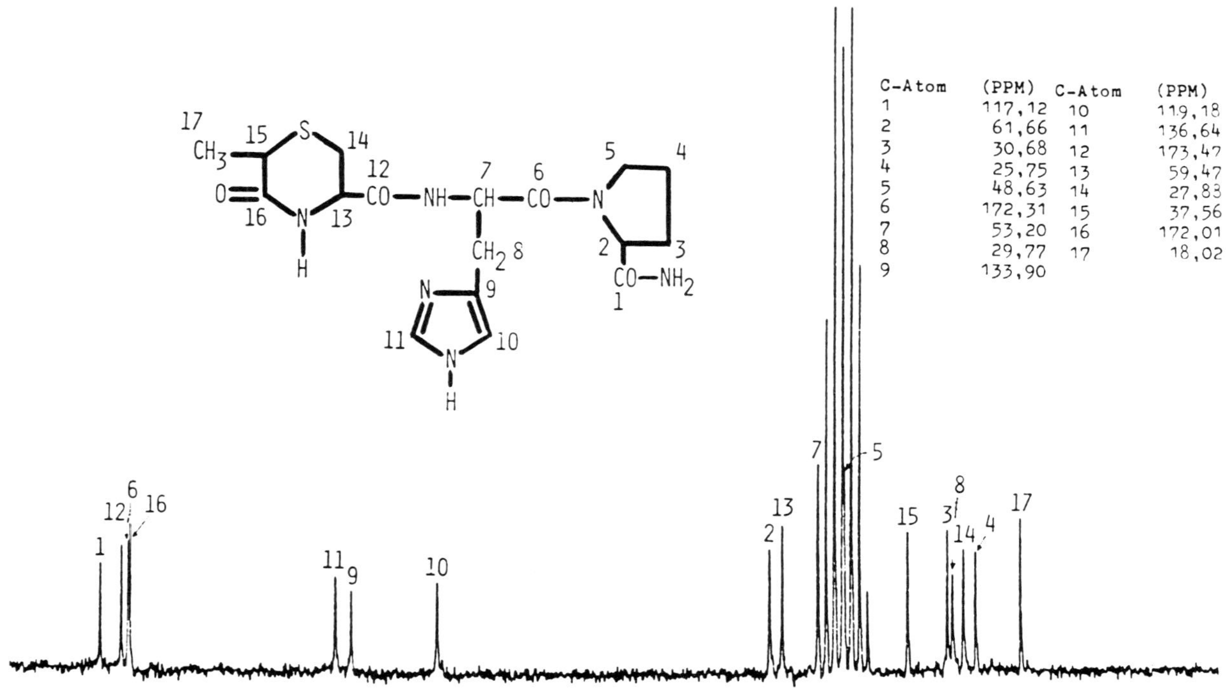

C-Atom	(PPM)	C-Atom	(PPM)
1	117,12	10	119,18
2	61,66	11	136,64
3	30,68	12	173,47
4	25,75	13	59,47
5	48,63	14	27,83
6	172,31	15	37,56
7	53,20	16	172,01
8	29,77	17	18,02
9	133,90		

Fig. 3. ^{13}C NMR spectrum of compound 18. Equipment: Bruker WP 60; solvent: MeOH-d_4; int. standard: MeOH-d_4; sweep width (SW) F_1: 100 Hz/cm; offset (A) F_1: 8050 Hz; times (NS): 9200 scans; $\delta_{TMS} = \delta_{MeOH-d_4} - 49{,}0$ ppm.

In contrast to TRH, CG 3703 possesses an additional chirality center in the 6-methyl-5-oxo-thiomorpholinyl-3-carbonyl moiety (carbon number 15 in fig. 2 and 3, respectively).
Out of the four theoretical isomers, only the two diastereomers with different configuration of the 6-methyl group were synthesized and purified to homogeneity (compounds 17 and 18 in the table). Interestingly, this stereochemical difference caused a considerable shift of biological efficacy (see ref. 3), although the electron densities in the His-ProNH_2 moiety of the molecules appear largely identical and substantial differences of the chemical shift of the NMR spectrum is only obvious in the thiomorpholinone part. Thus, this couple of compounds might provide a suitable tool to investigate the specific influence of minor local changes in steric or electronic configurations on biological parameters of TRH—like molecules.

Anticipating the results to be presented in the forthcoming papers (3,4) we may state that the series of compounds presented here helped to substantiate our working hypothesis outlined initially. In fact, all compounds in which the pyroglutamyl residue of TRH was replaced by a five-or six-membered ring containing the partial structure *CO-NH-C-CO- showed biological activity, while those missing the crucial carbonyl function (asterisk) were inactive or less active. Considerable quantitative or even qualitative differences in specific biological activities (3) or in metabolic stabilities (4) were found depending on the kind of substitution. Therefore this set of analogs may certainly contribute to a more detailed understanding of the metabolic fate and the interaction of TRH with pituitary and CNS receptors.

Table. Synthesized TRH analogs of the general structure R-L-His-L-Pro-NH_2.

Compound No.	R =	Method(s) of Purification	Yield (% of theory)	$[\alpha]_D^{23}$ (c = 1)	C calculated / found	H	N	S	Formula (Molecular weight)
1	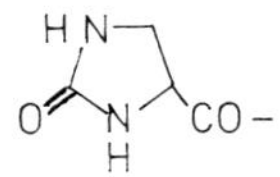	P8	35	-52.4^0	45.10	6.32	24.55		$C_{15}H_{21}O_4N_7 . 2H_2O$
				MeOH	45.73	6.44	24.76		(399.4)
2	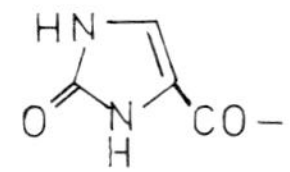	P8,P7	27	-45.3^0	43.34	6.07	23.60		$C_{15}H_{19}O_4N_7 . 3H_2O$
				MeOH	43.66	5.35	23.59		(415.4)
3	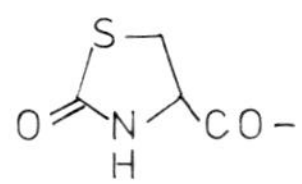	P1	40	-74.9^0	47.35	5.29	22.08	8.43	$C_{15}H_{20}O_4N_6S$
				MeOH	48.30	5.66	21.53	8.39	(380.4)
4	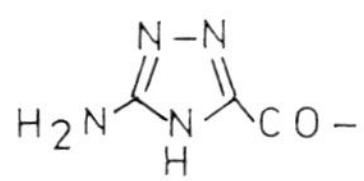	P8	39	-60.5^0	46.60	5.30	34.45		$C_{14}H_{19}O_3N_9$
				H_2O	45.63	5.40	33.56		(361.4)
5	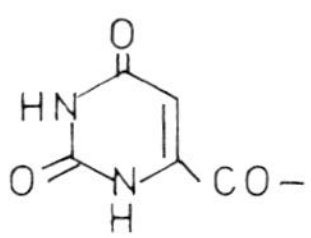 CG 3509	P8 (P12)*	39 (96)*	-46.8^0	47.20	5.19	24.12		$C_{16}H_{19}O_5N_7 . H_2O$
				MeOH	47.95	5.27	24.10		(407.4)

*Obtained from compound 6 by catalytic hydrogenolysis.

Table (ctd.)

Compound No. R =		Method(s) of Purification	Yield (% of theory)	$[\alpha]^{23}_{D}$ (c = 1)	C calculated found	H	N	S	Formula (Molecular weight)
6	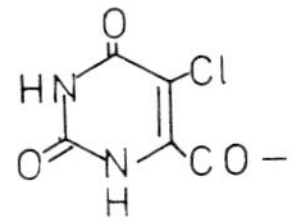	P1,P3 P2	45	-36.2°	45.25	4.27	23.12		$C_{16}H_{18}O_5N_7Cl$
				MeOH	44.99	5.05	22.64 **		(423.8)
7	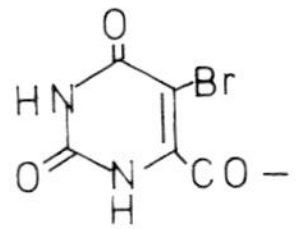	P8,P1	41	-37.5°	41.05	3.87	20.92		$C_{16}H_{18}O_5N_7Br$
				MeOH	40,30	4,58	20,20 **		(468,3)
8	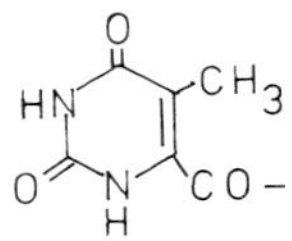	P4,P11	30	-46,8°	44,68	5,95	21,43		$C_{17}H_{21}O_5N_7 \cdot 3H_2O$
				MeOH	44.67	5.98	21.79		(457.5)
9	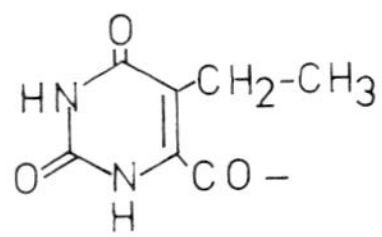	P8,P1	38	-46°	49.64	5.79	22.52		$C_{18}H_{23}O_5N_7 \cdot H_2O$
				MeOH	49.50	5.87	22.59		(435.4)
10	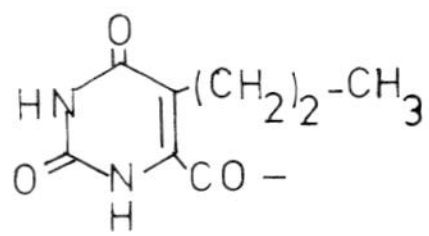	P4,P5, P6	38	-48.6°	50.77	6.06	21.82		$C_{19}H_{25}O_5N_7 \cdot H_2O$
				MeOH	50.86	6.12	22.19		(449.5)

** Found halogen content correspond to theory.

Table (ctd.)

Compound No.	R =	Method(s) of Purification	Yield (% of theory)	$[\alpha]^{23}_{D}$ (c = 1)	C calculated found	H	N	S	Formula (Molecular weight)
11	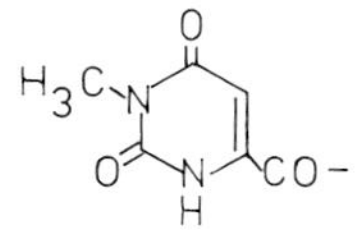	P8	35	-48.7°	50.60	5.24	24.28		$C_{17}H_{21}O_5N_7$
				MeOH	49.90	4.98	23.27		(403.4)
12	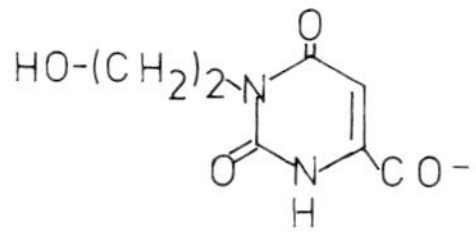	P8	42	-49.5°	48.00	5.58	21.70		$C_{18}H_{23}O_6N_7 \cdot H_2O$
				MeOH	48.63	5.46	21.83		(451.5)
13	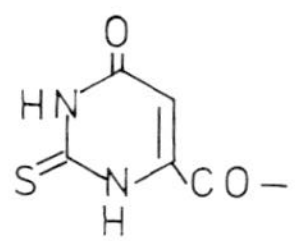	P8	26	-88.5°	47.40	4.72	24.15	7.90	$C_{16}H_{19}O_4N_7S$
				H_2O	46.40	4.90	24.13	7.86	(405.5)
14	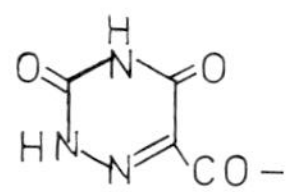	P1	47	-12°	44.10	4.93	27.40		$C_{15}H_{18}O_5N_8 \cdot H_2O$
				DMF	44.93	4.98	26.36		(408.4)
15	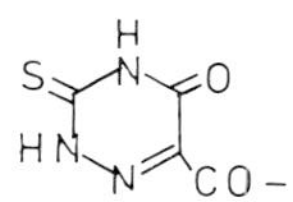	P1	32	-50.7°	40.70	4.32	25.30	7.23	$C_{15}H_{18}O_4N_8S$
				0.1 HCl	42.10	4.80	24.60	7.86	(406.4)

Table (ctd.)

Compound No.	R =	Method(s) of Purification	Yield (% of theory)	$[\alpha]_D^{23}$ (c = 1)	C calculated found	H	N	S	Formula (Molecular weight)
16	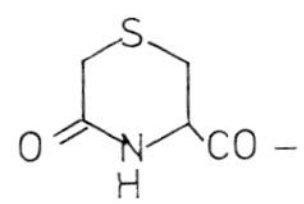	P8	56	-48.5^{0}	46.60	5.86	20.35	7.92	$C_{16}H_{22}O_4N_6S.H_2O$
				MeOH	46.50	5.92	20.92	7.97	(412.5)
17	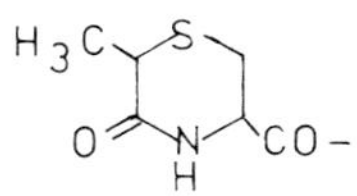 CG 3703	P2,P3 P1	45	-33.5^{0}	44.15	6.55	18.17	6.93	$C_{17}H_{24}O_4N_6S.3H_2O$
				MeOH	44.69	6.47	18.52	6.54	(462.5)
18	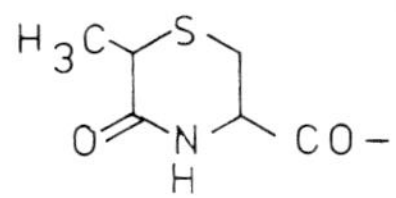	P10,P8	34	-108.5^{0}			n. d.		
				MeOH					
19	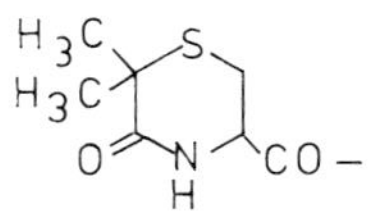	P8	48	-67.8^{0}	51.20	6.21	19.89	7.58	$C_{18}H_{26}O_4N_6S$
				MeOH	50.17	6.32	18.40	8.19	(422.5)
20	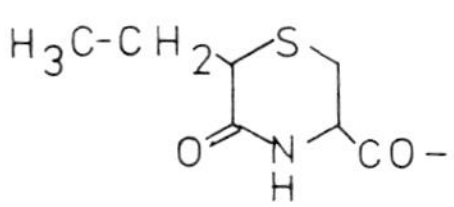	P8	39	-87.5^{0}			n. d.		
				MeOH					

Table (ctd.)

Compound No.	R =	Method(s) of Purification	Yield (% of theory)	$[\alpha]_D^{23}$ (c = 1)	C calculated found	H	N	S	Formula (Molecular weight)
21	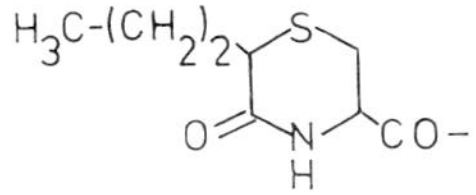	P8	51	-87.1^{o}	52.35	6.48	19.27	7.35	$C_{19}H_{28}O_4N_6S$
				MeOH	51.90	6.79	19.16	7.29	(436.5)
22	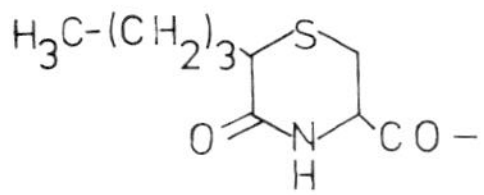	P8	35	-104°	53.30	6.67	18.65	7.12	$C_{20}H_{30}O_4N_6S$
				MeOH	51.60	6.72	18.03	6.90	(450.6)
23	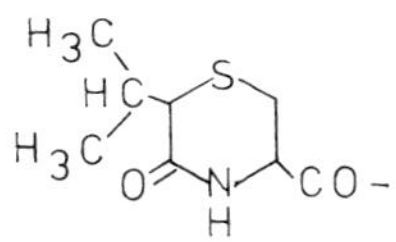	P8	34	-102.5^{o}	52.35	6.48	19.27	7.35	$C_{19}H_{28}O_4N_6S$
				MeOH	51.98	6-86	19.10	7.21	(436.5)
24	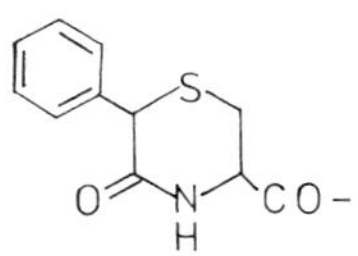	P8	49	-63.1^{o}	56.20	5.57	17.86	6.81	$C_{22}H_{26}O_4N_6S$
				MeOH	55.26	5.70	16.68	6.60	(470.6)
25	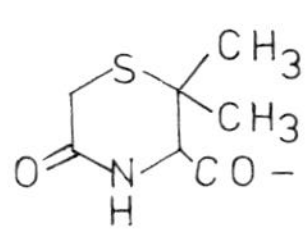	P8,P9	25	-83°	49.07	6.41	19.08	7.28	$C_{18}H_{26}O_4N_6S.H_2O$
				MeOH	48.80	6.43	19.08	7.24	(440.5)

References

1. Voelter, W.: Chem. Ztg. 98, 554 - 561 (1974).

2. Steiner, H., Motta, M., Piva, F., Gillessen, D., Studer, R.O.: Acta Endocrinol. (Copenhagen) Suppl. 155, 5 (1971).

3. Friderichs, E., Schwertner, E., Herrling, S., Günzler, W.A., Ötting, F., Flohé, L.: this issue.

4. Bauer, K.: this issue.

5. Johnson, T.B.: J. Am. Chem. Soc. 65, 1218 - 1219 (1943).

6. Gershon, H.: J. Org. Chem. 27, 3507 - 3510 (1962).

7. Morâvek, J., Leseticky, L.: Collect. Czech. Chem. Commun. 33, 1352 - 1355 (1968).

8. Crosby, D.G., Berthold, R.V.: J. Org. Chem. 23, 1377 - 1378 (1958).

9. Laursen, P.H., Thews, W.A., Christensen, B.E.: J. Org. Chem. 22, 274 - 277 (1957).

10. Clerq-Bory, C., Mentzer, C.: Bull. Soc. Chim. France 1958, 436 - 438.

11. The Distillers Co. Ltd. (Inventors: Ridgway, F., Green, J.N., Cross, N.M.). Brit. Pat. No. 854 339.

12. Barlow, R.B., Welch, A.D.: J. Am. Chem. Soc. 78, 1258 - 1259 (1956).

13. Cipens, G.E., Grinsteins, V.: Latvijas PSR Zinatnu Akad. Vestis, Kim. Ser. 1965, 204 - 208. (C.A. 63, 13244a (1965)).

14. Shiba, T., Koda, A., Kusumoto, S., Kaneko, T.: Bull. Chem. Soc. Japan 41, 2748 - 2753 (1968).

15. Hofmann, K.: Imidazole and its Derivatives, part I, in "The Chemistry of Heterocyclic Compounds", Interscience Publishers, Inc., New York 1953

16. Kaneko, T., Shimokobe, T., Ota, Y., Toyokawa, E., Inui, T., Shiba, T.: Bull. Chem. Soc. Japan 37, 242 - 244 (1964).

17. Nishimura, H., Tahara, S., Okuyama, H., Mitzutani, J.: Tetrahedron 28, 4503 - 4513 (1972).

18. Claeson, G., Pedersen, J.: Acta Chem. Scand. 22, 3155 - 3159 (1958).

19. Nakanishi, M., Muro, T.: Yakugaku Zasshi 90, 570 - 575 (1970) (C.A. 73, 25383x (1971)).

20. Inouye, K., Namba, K., Otsuka, H.: Bull. Chem. Soc. Japan 44, 1689 - 1691 (1971).

21. Abbott Labs. (Inventors: Cole, J.W., Kick, F.E.). German Pat. Appl. No. 23 36 550.

22. Voelter, W., Oster, O., Zech, K., Angew. Chem. 86, 46 - 47 (1974) (Internat. Edit. 13, 131 - 132 (1974)).

23. Zech, K., Dissertation, Tübingen 1973.

24. Reynolds, W.F., Peat, I.R., Freedman, M.H., Lyerla, J.R. jr.: J. Am. Chem. Soc. 95, 328 - 331 (1973).

PLANNING OF PEPTIDE SYNTHESES BY THE AID OF ELECTRONIC DATA-PROCESSING

Wolfgang Voelter and Erich Schmid-Siegmann
Institut für Organische Chemie der Universität Tübingen,
D-7400 Tübingen, Auf der Morgenstelle 18, Germany

Introduction

The development and application of new protecting groups, the large number of possible different amino acid sequences in peptides, the increasing methods of peptide coupling and the continuous discoveries and syntheses of new biologically active peptides and their analogues makes the collection of this information, suitable for electronic data-processing, necessary. Contrary to data collections of spectroscopic methods (absorption, infrared, proton nuclear magnetic resonance, carbon nuclear magnetic resonance, electron spin resonance, mass spectroscopy(1-17))a lot more parameters are necessary for a data compilation used by synthetic peptide chemists such as: Starting materials, end products, sequences of the peptides, molecular formulae, molecular weights, boiling points, melting points, specific rotation values, refractive indices, yields, solvents, reaction conditions used for the synthesis or references including authors and journals. Furthermore, all procedures have to be presented suitable for electronic data-processing.

In the data collection a peptide synthesis is composed of different single steps. All information, stored in the compilation, allows the peptide chemist to perform the synthesis in the laboratory. On the basis of computer file stored melting points, specific rotation values, molecular weights

Structure and Activity of Natural Peptides

or references amino acids or peptides respectively their derivatives are listed by means of computer programs on the computer output. Of even greater importance for the peptide chemist are computer programs which allow on the basis of the stored data suggestions for optimal strategies for the syntheses of peptides. The parameters and their corresponding figures used for the data collection are listed in table 1.

Table 1. Figures and corresponding parameters used for the peptide data collection.

0	-	Number of peptide synthesis
1	-	Type of reaction
2	-	Trivial name
3	-	**Peptide sequence of end product**
3a	-	Peptide sequence of end product plus structural formula
4	-	Peptide sequence of starting material A
4a	-	Pretreatment of starting material A
5	-	Peptide sequence of starting material B
5a	-	Pretreatment of starting material B
6	-	Description of main reaction
6a	-	Comments covering several lines
7a	-	Reactants
7b	-	Reference
8	-	Molecular formula
9	-	Molecular weight
10	-	Yield
11	-	Starting material
11a	-	Solvent
12a	-	Melting point ⟶ decomposition
12a	-	Solvent ⟶ alternative solvent
13	-	Specific rotation value
13a	-	Concentration for which specific rotation value is determined
13b	-	Solvent in which optically active material is dissolved for determination of the specific rotation value

Table 1.(continued)

13c	-	Wavelength at which specific rotation is determined
13d	-	Temperature at which specific rotation value is determined
14	-	Boiling point
14a	-	Pressure [torr]
15	-	Refractive index
15a	-	Wavelength at which refractive index is determined
15b	-	Temperature at which refractive index is determined
16	-	Specific gravity
16a	-	Temperature at which specific gravity is determined
17	-	Author(s)
18	-	Journal (volume, year, page)

General Information about the Peptide Data Compilation

Numbering of the Compounds. The compounds in the data collection are ordered on the basis of the synthetic end products. Amino acids and their derivatives are listed first, followed by di-, tri- and tetrapeptides (etc.). The order parameters for amino acids and their derivatives are listed in table 2.

Table 2. Order parameters for amino acids and their derivatives of the peptide data compilation.

1. Amino acids	-	unsubstituted amino acids
	-	C-H substituted compounds
	-	amino acids for which no generally accepted abbreviation is known
2. Prefix	-	L
		D
	-	D, L

Table 2.(continued)

	-	all other letters in alphabetical order
3. Groups for protection of amino residues	-	common amino protecting groups abbreviated as generally accepted
	-	bifunctional amino protecting groups connecting within the chain
	-	bifunctional amino protecting groups connecting different chains
	-	amino protecting groups for which no generally accepted abbreviation is known
4. Groups for protection of carboxyl residues	-	common carboxyl protecting groups abbreviated as generally accepted
	-	bifunctional carboxyl protecting groups connecting different chains
	-	carboxyl protecting group for which no generally accepted abbreviation is known
5. Protecting groups bound to amino acid side chains		
6. Special derivatives	-	anhydrides
	-	N-alkylamino acids
	-	derivatives with peptide bonds in the side chain.

General description of the syntheses. In order to give detailed information about the syntheses a great number of different parameters have to be collected in the peptide compilation. As computer programs are applied to use the collected information the different parameters are always listed in the same order. The following table (table 3) shows the order of the listed parameters for derivatization and coupling reactions.

Table 3. Order of collected parameters for derivatization and coupling reactions.

Derivatization	Coupling
1 - Type of reaction	....
	2 - Trivial name
3 - End product	....
4 - Starting material A	4 - N-terminal product (A)
	4a- Pretreatment of A
	5 - C-terminal product(B)
	5a- Pretreatment of B
6 - Description of main reaction	....
8 - Molecular formula	....
9 - Molecular weight	....
10 - Yield	....
11 - Starting material A	11 Starting materials A and B
11a- Solvent	....
11b- Quantity of solvent	....
12 - Melting point	....
12a- Solvent	....
13 - Specific rotation value	....
13a- Concentration	....
13b- Solvent	....
13c- Wavelength	....
13d- Temperature	....
14 - Boiling point	....
14a- Pressure	....
15 - Refractive index	....
15a- Wavelength	....
15b -Temperature	....
16 - Specific gravity	....
16a- Temperature	....

Table 3. (continued)

Derivatization	Coupling
17 - Authors	....
18 - Journals	....

1. Type of reaction. The different types of reaction are characterized by the following names: DERIVATIVE, CLEAVAGE, COUPLING, CYCLIZATION, RING OPENING, SALT FORMATION, LIBERATION, POLYMERIZATION, SYNTHESIS and SUBSTITUTION.

Some reaction types are characterized by additional information such as:

C = DERIVATIVE (ACT ESTER)
↓
C = Carboxyl- activated ester
N = Amino -
O = Hydroxy -
S = Thiol -
C = CLEAVAGE (HYDROGEN)
↓
hydrogenation

If the functional group is in the side chain the prefix is put in parentheses. Substitution reactions at a CH group are characterized by C-SUBSTITUTION.

2. Trivial name. Concerning the nomenclature the rules of the IUPAC Commission on Biochemical Nomenclature are applied, however, the application of computer programs made some deviations from these rules necessary.

3. Peptide sequences of end product. The order of the amino acids is given as usual: The N-terminal part is written on the left and the C-terminal part on the right side of the page. The position of the amino acid is given by numbers. The abbreviations used are those suggested by IUPAC.

References

1. A.P.I. Research Project 44: Ultraviolet Spectra Data, Carnegie Institute and U.S. Bureau of Standards

2. Sadtler Standard Spectra (Ultraviolet), Heyden and Son Ltd., London-Philadelphia-Rheine

3. Hershenson, H.M.: Infrared Absorption Spectra, Index for 1945-1957 (1959), Index for 1958-1962 (1964), Academic Press, London-New York

4. Sadtler Standard Spectra (Infrared), Heyden and Son Ltd., London-Philadelphia-Rheine

5. The Aldrich Library of Infrared Spectra, Aldrich Chemical Company, Milwaukee

6. Bhacca, N.S., Johnson, L.F., Shoolery, J.N.: High Resolution NMR Spectra Catalogue I,II, Varian Associates, Palo Alto 1962/1963

7. Bovey, F.A.: NMR Data Tables for Organic Compounds, Wiley-Interscience, New York-London-Sydeny-Toronto, 1967

8. The Aldrich Library of NMR Spectra, Aldrich Chemical Company, Milwaukee

9. Sadtler Standard NMR Spectra, Sadtler Research Laboratories, Philadelphia

10. Johnson, L.F., Jankowski, W.C.: Carbon-13 NMR Spectra, A Collection of Assigned, Coded and Indexed Spectra, John Wiley and Sons, New York-London-Sydney-Toronto 1972

11. API Research Projetct 44: Selected Nuclear Resonance Spectral Data; Selected ^{13}C Nuclear Magnetic Resonance Spectral Data, Carnegie Institute and U.S.Bureau of Standards

12. Breitmaier, E., Haas, G., Voelter, W.: Atlas of Carbon-13 NMR Data, Heyden and Son Ltd., London-Philadelphia-Rheine 1975

13. Bremser, W., Ernst, L., Franke, B.: Carbon-13 NMR Spectral Data, Verlag Chemie, Weinheim-New York 1979

14. Hershenson, H.M.: NMR and ESR Spectra Index, Academic Press, London-New York 1965

15. Stenhagen, E., Abrahamsson, S., McLafferty, F.W.: Atlas of Mass Spectral Data, Interscience Publishers, New York-London-Sydney 1969

16. Eight Peak Index of Mass Spectra, Mass Spectrometry Data Center, Aldermaston 1975

17. Cornu, A., Massot, R.: Compilation of Mass Spectral Data, Heyden and Son Ltd., London-Philadelphia-Rheine 1976

SYNTHESIS AND ANTIGENIC PROPERTIES OF A NEW STEROID HAPTEN: 7-α-CARBOXYETHYLMERCAPTO-5-ANDROSTENE-3β,17β-DIOL-BOVINE SERUM ALBUMIN

Manfred Zwirner, Reinhold Göser, Adolf E. Schindler, Wolfgang Voelter

Universitäts-Frauenklinik Tübingen, D-74oo Tübingen, Germany

Introduction

The observations of Midgley and Niswender (1) concerning the formation of specific antibodies against progesterone lead to the conclusion that a hapten containing a steroid with free functional groups must generate specific antibodies when injected into animals (2-6). We synthesized a steroid hapten which can be attached covalently to the macromolecule from the α-side of the steroid nucleus. We reasoned that this derivate might lead to a specific recognition by the antibody of conformationally important β-hydroxy groups of the androstenediol. In this communication we present details of the synthesis of the steroid hapten, the 7-α-carboxyethylthioether (Fig. 1), the production of antiserum and its application to the radioimmunoassay (RIA) of androstenediol.

Experimental

Solvents and reagents: All solvents were reagent grade and were used after drying on a basic alumina column. The following chemicals were used as furnished by the supplier: 3-mercaptopropionic acid methyl ester, n-butyllithium (2o% in hexane), N-bromosuccinimide, potassium hydroxide (Merck), Freund's adjuvant (Calbiochem), bovine serum albumin (Behring), Unisolve I (Zinsser), Norit A (Serva), Dextran T 7o (Pharmacia).

Structure and Activity of Natural Peptides

The unlabelled steroids were purchased from Steraloids. Proof of chemical purity was established by infrared spectra, melting point determinations, microanalyses, mass spectra and thin-layer chroamtography. 5-Androstene-3ß,17ß-diol-[1.2-^{3}H-(N)] (o.25mCi) was purchased from New England Nuclear Corp.. Melting point determinations were made on a Büchi capillary melting point apparatus and are uncorrected. Mass spectra were made on a LKB 9ooo instument using a direct inlet system. Infrared spectra were made using a potassium bromide disc and a Perkin-Elmer model 221 grating spectrometer. UV spectra were recorded in methanol with a Cary model 15. NMR spectra were obtained with a Varian WP 6o spectrometer in deuterochloroform and are given in ppm from the internal TMS standard. The microanalyses were performed by the microanalytical laboratory of the Institute of Organic Chemistry of the University of Tübingen.

Material and Methods

The preparation of the antigen is shown in Fig. 1. 5-Androstene-3ß,17ß-diol-3-acetate-17-benzoate 2 has served as starting material. Only this derivate of the androstenediol gave a stable 7-bromo product after bromination with N-bromo-succinimide. On the basis of the downfield shift of, both, the C-19 and C-18 methyl singlets, from δ = 1.o to 1.2 and δ = o.9 to 1.o5 respectively and of the 7-α-proton multiplet from δ = 3.7 to 4.9, the ß-configuration was assigned to the C-7-bromo substituent. Metalation of the 3-mercaptopropionic acid methyl ester 1 with n-butyllithium in THF followed by nucleophilic addition of the steroid derivate gave the 7-α-thioether 4 in 2o% yield. Hydrolysis with o.5% potassium methylate resulted in 5. The carboxyethylthioether 5 was coupled to bovine serum albumin (BSA) using the "mixed anhydride" method described by Erlanger et al. (6).

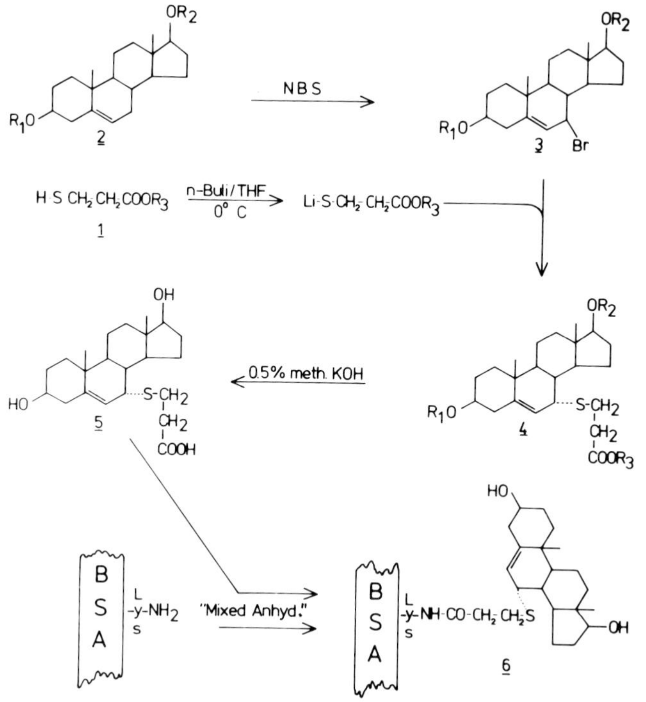

Fig. 1. Synthesis of the steroid hapten,
$R_1 = -COCH_3$, $R_2 = -COC_6H_5$, $R_3 = -CH_3$.

The following procedures describe the syntheses of the products in detail:

7-β-Bromo-androst-5-ene-3β,17β-diol-3-acetate-17-benzoate 3

A mixture of 5g androstenediol-3-acetate-17-benzoate, 2.5g N-bromosuccinimide and 1omg α,α'-azo-bis-isobutyronitrile as catalyst in 1oo ml freshly dried carbon tetrachloride was refluxed for 2o minutes. After cooling the precipitated succinimide was removed by filtration. The filtrate was extracted successively with water, sodium bisulfite solution and water and dried over sodium sulfate. The solvent was removed in vacuo at 5o^0 C. The residual brown oil was dried in the dark.

Crystallization from diethyl ether gave 3 in 7o% yield with a m.p. 147-148^0 C.

IR bands (KBr): 295o, 175o, 173o, 128o cm^{-1}.

^{1}H NMR signals ($CDCl_3$, δ [ppm], TMS = o): 1.2 (s,-CH_3), 1.o5 (s,-CH_3), 4.9 (m,-CH_3).

Anal: Calcd for $C_{28}H_{37}O_4Br$ C: 65.57, H: 7.2o, Br: 15.45, found C: 65.8o, H: 7.o5, Br: 15.76.

5-Androstene-3ß,17ß-diol-3-acetate-17-benzoate-7-yl-2'methoxycarbonylethyl sulfide 4

Compound 1 (o.2 ml) was dissolved in 1o ml of absolute THF and cooled to 0^0 C. 1.3 ml n-butyllithium (2o% in hexane) was added under nitrogen during 15 minutes to the stirred solution. Then 5oo mg of (3) in 5o ml absolute THF was added dropwise to the stirred solution. Stirring was continued for another hour. After washing with water the mixture was extracted with ethyl acetate and crystallized with acetone to give 5o mg of 4 with a m.p. 268-27o^0 C.

IR bands (KBr): 29oo, 175o,171o, 1275, 1245 cm^{-1}.

^{1}H NMR signals ($CDCl_3$, δ [ppm], TMS = o): o.95 (s,-CH_3), 1.o5 (s,-CH_3), 4.2 (m,-CH).

Anal: Calcd for $C_{32}H_{44}O_6S$ C: 69.o4, H: 7.97, S: 5.74, found C: 69.98, H: 7.5o, S: 6.o8 .

3ß,17ß-Dihydroxyandrost-5-en-7-yl-2'-carboxyethyl sulfide 5

The ester 4 (5o mg) was dissolved in methanolic potassium hydroxide (o.5%) and stirred under nitrogen for 24 hours. The methanol was evaporated, ethyl acetate and water were added and, thus, the neutral fractions separated from those soluble in alkali. The aqueous phase was cooled and acidified to pH2.5 with hydrochloric acid, extracted with ethyl acetate and crystallized from ethanol/ether to give 28 mg of 5 with a m.p. 228-23o^0 C.

IR bands (KBr): 338o, 171o, 165o, 125o cm^{-1}.

Anal: Calcd for $C_{22}H_{33}O_4S$ C: 67.4o, H: 8.25, S: 8.19, found C: 66.95, H: 8.51, S: 8.32 .

Preparation of the Steroid-Bovine Serum Albumin Conjugate

7-α-Carboxyethylmercaptoandrostenediol 5 was coupled to bovine serum albumin (BSA) using the "mixed anhydride" method described by Erlanger et al. with the modification described by Rao et al. (7). By the ninhydrin method of Moore and Stein (8), approximately 18 moles of the steroid were incorporated per mole of conjugate.

Immunization Procedure

Six female New Zealand white rabbits, four months of age were used for immunization. The primary injection dose for each rabbit was o.5 mg of the conjugate, dissolved in 1 ml of o.9% sodium chloride solution and mixed with 1 ml of Freund's complete adjuvant. 1.2 ml of this mixture was injected at multiple sites along the back of the rabbit, the remaining amount was applied intramuscularly in one thigh. Four booster injections were carried out in three week distances with the same concentrations of the conjugate in sodium chloride solution but now mixed with Freund's incomplete adjuvant.

Assay Procedure

Standards were prepared from stock solutions of unlabelled steroid in absolute ethanol (1oo ng/ml). The working standard solution containing 1 ng/ml was prepared in o.1% gelatine buffer solution containing o.1 m Na_2HPO_4, o.5 m NaH_2PO_4, 1 m NaCl and 3 mg of thiomersal in 1ooo ml of water. The labelled androstenediol-[1.2-^{3}H-(N)] was prepared in assay buffer with an activity of 1o.ooo cpm per o.1 ml. 5oo µl of plasma were added to 12o x 12 mm assay tubes followed by addition of 1o µl of the tracer solution to correct the recovery. This mixture was incubated after mixing on a Vortex at 4°C for 2o hours. Diethyl ether (5.5 ml) was added for extraction and after mixing 15 minutes on a Vortex the samples were centrifugated and cooled to -3o^{o}C. Five ml of the organic layer were transferred to a second tube and evaporated to dryness. The sidewalls of the dried tubes were rinsed with 1 ml ethanol and the tubes evaporated to dryness again. Assay buffer (1 ml) was added, 7oo µl of the solution were removed for recovery and 1oo µl in duplicate for assay. The standard curve was established by setting up triplicate assay tubes containing o, 15, 3o, 6o 12o, 25o, 5oo, 1ooo pg of the steroid in a total volume of 1oo µl. To all tubes antibody solution (1oo µl) at a dilution of 1:3ooo in the assay incubation volume of 3oo µl was added followed by addition of 1oo µl of tracer solution and mixing on a Vortex. After incubation for about 2o hours at 4°C to all tubes o.5 ml of a cold dextran-coated charcoal suspension (containing 2.5g Norit A and o.25g Dextran T 7o in 5oo ml assay buffer) was added. After mixing the tubes were put to a cold place for 15 minutes and then centrifuged for 2o minutes at 4ooo g. The supernatant liquid was decanted into a counting vial containing 1o ml Unisolve I scintillating medium. A typical standard curve for androstenediol produced under conditions previously described is shown in figure 2.

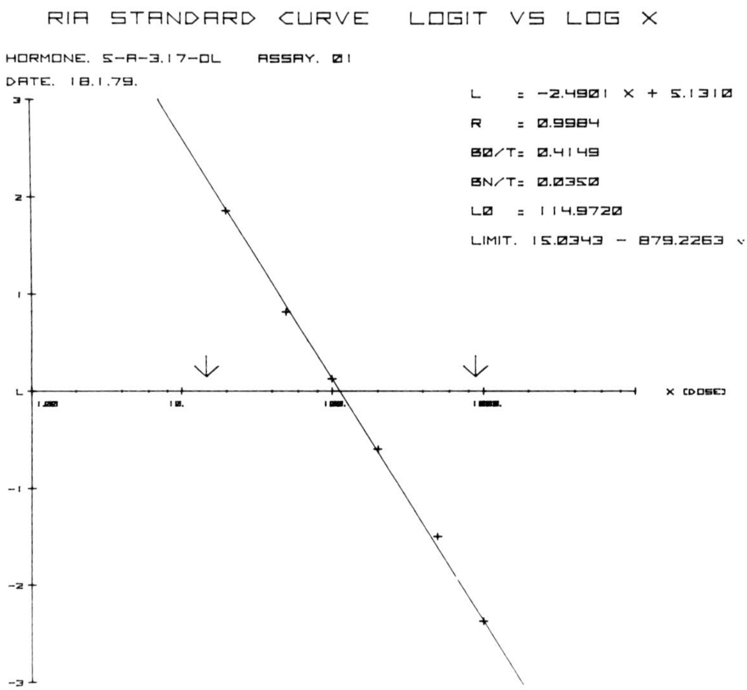

Fig.2. RIA standard curve for androstenediol .

Results and Discussion

Four out of the six rabbits immunized with the androstenediol-7-BSA thioether produced antisera. The titer was determined from the ability of the antibody to bind a constant amount of labelled steroid. The antiserum was tested at dilutions from 1:1oo upwards as necessary to determine the 5o% level. The percent cross reaction of the antiserum with structurally similar steroids in a concentration of 1 ng/ml is shown in table 1.

Tab. 1. Cross Reactivity Data.

		%
5-Androstene-3β,17β-diol		1oo.o
5-Androstene-3β,17α-diol		3.4
5-Androstene-3α,17β-diol		4.6
5-Androstene-3β,17β-diol-3-acetate		1.1
5-Androstene-3β,17β-diol-17-acetate		2.8
5-Androstene-3β-ol-17-one		6.7
5-α-Androstane-3β,17β-diol		73.3
5-α-Androstane-3β,17α-diol		1.9
5-α-Androstane-3α,17β-diol		1.3
5-β-Androstane-3β,17β-diol		o.8
5-α-Androstane-17β-ol-3-one	(Dihydrotestosterone)	2.9
4-Androstene-17β-ol-3-one	(Testosterone)	o.9
4-Androstene-3,17-dione	(Androstenedione)	8.5
Estrone		o.o
Estradiol		o.o
Estriol		o.o
Equilin		o.o
Progesterone		o.o
Cortisol		o.o

As indicated in table 1 the α-configuration of the hydroxyl groups at both, 3 and 17 position, and other functional groups at these positions is required. The strictly major cross reactant was 5-α-androstane-3β,17β-diol which might be due to a strong screening of the double bond in the conjugate by the big sulfur atom of the thioether.

Validation of the Androstenediol Radioimmunoassay

The accuracy of the androstenediol assay was assessed by the determination of the recovery of increasing amounts of unlabelled steroid added to plasma samples with previously determined low values. The results obtained are shown in figure 3 which demonstrate a recovery of 100%, calculated in a family regression test.

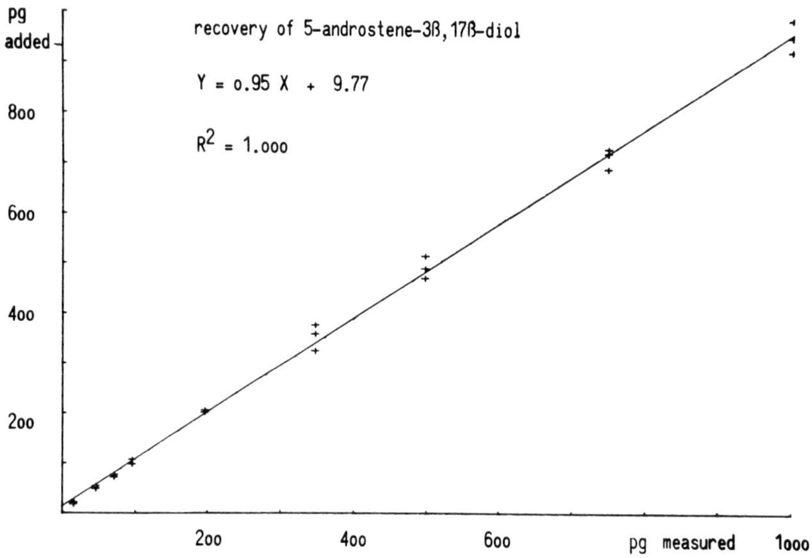

Fig. 3. Recovery of androstenediol in plasma samples.

The high specifity of the anti-androstenediol serum is clearly demonstrated. Due to the high cross reactivity with androstanediol a previous chromatographic separation of this steroid is absolutely necessary to get an accurate determination of plasma androstenediol.

Nomenclature

The following trivial names and abbreviations were used: Androstenediol = 5-androstene-3ß,17ß-diol, androstanediol = 5-α-androstane-3ß,17ß-diol, cortisol = 11ß,17α,21-trihydrxy-4-pregnene-3,2o-dione, estrone = 3-hydroxyestra-1,3,5 (1o)-trien-17-one, estradiol = 1,3,5 (1o)-estratriene-3,17-diol, estriol = 1,3,5 (1o),7-estratriene-3,16,17-triol, equilin = 1,3,5 (1o),7--estratetraen-3-ol-17-one, progesterone = pregn-4-ene-3,2o-dione.

References

1. Midgley, A.R., Niswender, G.D.: Acta Endocrin. 147, 32o - 331 (197o).
2. Lindner, H.R., Perel, E., Friedlander, A.: Research on Steroids (Finkelstein, M., Cont, C., Klopper, A., Cassano, C., eds.) Vol. 4, 197-252, Pergamon Press, Oxford 197o
3. Exley, D., Johnson, M.G., Dean, P.D.: Steroids 18, 6o6-61o (1971).
4. Jeffcoate, S., Searle, J.E.,: Steroids 19, 181-188 (1972).
5. Kuss, E., Goebel, R.: Steroids 19, 5o9-518 (1972).
6. Erlanger, B.F., Borek, F.M., Beiser, S.M., Liebermann, S.: J. Biol. Chem. 228, 713-727 (1957).
7. Rao, P.N., Moore, P.H.: Steroids 24, 793-811 (1974).
8. Moore, S., Stein, W.H.J.: J. Biol. Chem. 211, 9o7 (1954).

V. MISCELLANEOUS AND BIOLOGICAL ACTIVITY OF PEPTIDES

DEGRADATION OF TRF AND TRF ANALOGUES BY BRAIN AND SERUM ENZYMES

Karl Bauer, Horst Kleinkauf
Max-Volmer-Institut, Abt. Biochemie, Technische Universität Berlin, D-1000 Berlin-10, Germany

Leopold Flohé
Grünenthal GmbH., Center of Research, D-5100 Aachen, Germany

Introduction

After its isolation and identification as a hypothalamic hypophysiotropic hormone, TRF (<Glu-His-Pro-NH_2) has also been found in extrahypothalamic brain and other tissues (for review see 1). The localization of TRF in nerve terminals from both, hypothalamic and extrahypothalamic sites, suggests that TRF may also act as a neuromodulator in the central nervous system. The observation that TRF excerts various effects on behaviour and mood state further supports this suggestion (see 2 and references cited there). Since TRF is known to be rapidly inactivated by blood and tissue enzymes (for review see 3-7) further studies about the action of TRF as a centrally active drug might be hampered by limited availability of systematically applied TRF at particular target sites. The characterization of the TRF-inactivating enzymes is therefore of considerable interest and might provide a basis for the design and synthesis of enzyme resistant analogues suitable for further biological studies.

<u>Degradation of TRF by brain enzymes.</u> The pathway of TRF catabolism (Figure 1) has been elucidated by the work of several laboratories (for review see 3-6). The inactivation of TRF

Structure and Activity of Natural Peptides

is catalyzed by 2 enzymes, namely by a pyroglutamate aminopeptidase and by a TRF-deamidating enzyme.

The pyroglutamate aminopeptidase (MW≃ 28 000) hydrolyzes not only TRF (7-9) but also other pyroglutamyl-containing peptides such as LH-RH, neurotensin, gastrin etc. as well as synthetic peptides specifically at the < Glu-X bond. This general proteolytic enzyme is found ubiquitously distributed throughout various tissues (10-16).

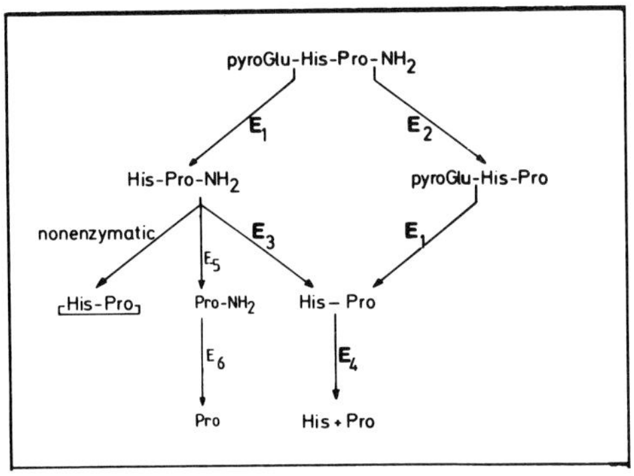

Fig.1. Pathway of TRF catabolism by brain enzymes. E_1, pyroglutamate aminopeptidase; E_2, post proline cleaving enzyme; E_3, post proline dipeptidyl aminopeptidase; E_4, imidodipeptidase; E_5, histidyl-proline iminopeptidase (according to Matsui et al. (28)); E_6, iminopeptidase.

The deamidation of TRF (17,18) is catalyzed by an enzyme (MW≃ 76 000) which has been descriptively termed "TRF-deamidating enzyme" (19,20). Recently this enzyme has been characterized as a post proline cleaving enzyme (21-23). An enzyme with the same substrate specificity and similar chemical characteristics has been originally discovered, isolated and characterized from lamb kidneys (24-26). The post proline cleaving enzyme(s) exhibit a high degree of specificity for the cleavage of proline-containing peptides at the carboxyl side of a proline residue. Although this enzyme catalyzes the deamidation of TRF its substrate specificity is that of an endopeptidase hydrolyzing preferentially internal peptide bonds. Various biologically active peptides such as LH-RH, angiotensin, oxytocin, vasopressin etc. as well as synthetic peptides are effectively hydrolyzed by this enzyme (22,24). As shown in Fig.1, the resulting primary cleavage products are subsequently further metabolized. The degradation of deamido-TRF by the pyroglutamate aminopeptidase and the deamidation of His-Pro-NH_2 by the post proline dipeptidyl aminopeptidase results in the formation of His-Pro as an intermediary product which is subsequently hydrolyzed to the free amino acids. According to Matsui et al. (27), Pro-NH_2 is also formed as an intermediary product of TRF catabolism by brain enzymes. Histidyl-proline diketopiperazine (⌐His-Pro┐) (9, 28), however, is formed nonenzymatically by intramolecular cyclization. This condensation product is not further metabolized but excreted rapidly into the urine (Bauer, unpublished results) from where it has been isolated already in 1965 (29).

Degradation of TRF by serum enzymes. The pathway of TRF catabolism by serum is shown in Fig.2. Only minute activities of the TRF-inactivating tissue enzymes can be detected in serum. The rapid inactivation of TRF by serum (30) is almost exclusively due to a hitherto unknown enzyme which catalyzes the hydrolysis of the pyroglutamyl-histidine peptide bond (31,32). This serum enzyme (MW≃ 260 000) exhibits chemical characteristics distinctly different from the known pyroglutamate aminopeptidases catalyzing the same reaction. For example, the TRF-degrading serum enzyme is inhibited by EDTA and dithiothreitol while the pyroglutamate aminopeptidases are activated by these agents. With regard to the substrate specificity striking differences are also obvious. <Glu-Ala and <Glu-ß-naphthylamide, the substrates of choice for the pyroglutamate aminopeptidases, are not effectively hydrolyzed by the TRF-degrading serum enzyme. Likewise, while it is known that the pyroglutamyl-histidine bond of LH-RH is rea-

dily cleaved by the pyroglutamate aminopeptidase (33), LH-RH is not degraded by the TRF-degrading serum enzyme which, therefore, might be adequately termed "thyroliberinase" rather than a pyroglutamate aminopeptidase. The extraordinary degree of substrate specificity together with the observation that in rat the activity of this enzyme is controlled by thyroid hormones (34-36) and drastically alters with developmental changes (37,38) supports the hypothesis that this enzyme might represent a functional control element within the mechanisms regulating the availability of the neuropeptide at the trophic cells of the pituitary.

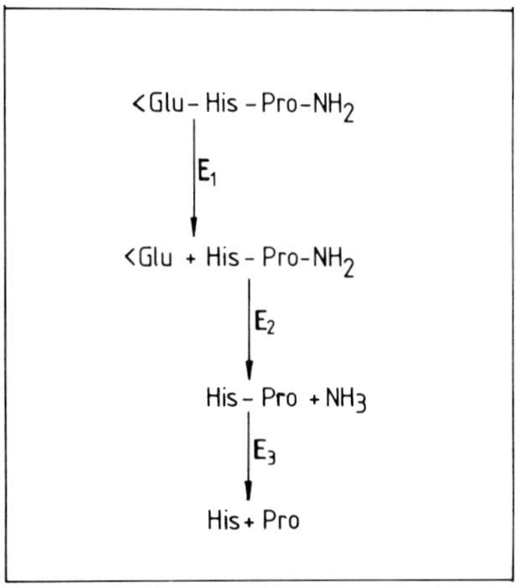

Fig.2. Pathway of TRF catabolism by serum enzymes. E_1, "TRF-degrading serum enzyme" ("Thyroliberinase"); E_2, post proline-cleaving enzyme; E_3, imidodipeptidase.

Materials and methods

TRF and TRF analogues used were synthesized by Schwertner et al. (39) and tested for biological activity by Friderichs et al. (2). The pyroglutamate aminopeptidase and the post proline-cleaving enzyme were purified from brain extracts by ion-exchange chromatography on DEAE cellulose and by gel filtration on Sephadex G-150 as described for the adenohypophyseal enzyme (22). The TRF-degrading serum enzyme was purified by ion-exchange chromatography on DEAE, adsorption chromatography on hydroxyapatite and by gel filtration on Sephadex G-200. The procedure is essentially that described by Taylor and Dixon (31). For monitoring the enzymatic activity the test described by Bauer and Nowak (32) was used.

Assay conditions. The buffers used had the following composition: Buffer A: 100 mM potassium phosphate buffer pH 7.4, 1 mM dithiothreitol, 1 mM EDTA; buffer B: 50 mM potassium phosphate buffer pH 7.4, 2 mM dithiothreitol, 2 mM EDTA; buffer C: 50 mM tris.HCl, pH 7.4.

Degradation by the post proline-cleaving enzyme. The enzyme preparation (200 µl in buffer A) was mixed with 500 µl buffer A containing the substances to be tested (10 mM) and incubated at $37^{o}C$. At given time intervals 100 µl aliquots of the reaction mixture were withdrawn. The liberation of ammonia was determined by the ninhydrin method according to Moore and Stein (40) using 1 mM Na acetate instead of Li acetate buffer. The standard curves were established with ammonium sulfate and methylammonium hydrochloride.

Degradation by the pyroglutamate aminipeptidase. The enzyme preparation (300 µl in buffer B) was mixed with 200 µl buffer B containing the substances to be tested (20 mM) and incubated at $37^{o}C$. At given time interals aliquots of the reaction mixture (100 µl) were withdrawn and mixed with 1.4 ml potassium phosphate buffer (200 mM, pH 8.35). Under vigorous stirring 0.5 ml of a solution of Fluram (30 mg/100 ml acetone) was added. Fluorescence was determined using a Perkin-Elmer MPF 4 spectrofluorometer (excitation at 390 nm, emmission at 470 nm). A standard curve was established with His-Pro-NH_2·H_2SO_4.

<u>Degradation by the TRF-degrading serum enzyme.</u> The enzyme preparation (400 µl) was incubated at 37°C with 100 µl of the substrate solution (50 mM in buffer C). The liberation of His-Pro-NH_2 was determined as described for the pyroglutamate aminopeptidase.

Results

All TRF-analogues in which the pyroglutamyl is replaced by a 6-membered ring are neither degraded by the pyroglutamate aminopeptidase nor by the TRF-degrading serum enzyme ("thyroliberinase"). As expected, the "thyroliberinase" is more specific than the pyroglutamate aminopeptidase. The "thyroliberinase" catalyzes only the degradation of the thiazolidinone derivative but not the respective imidazolidinone analogue of TRF .In contrast, the pyroglutamate aminopeptidase hydrolyzes the imidazolidinone derivative even more effectively than TRF itself. As expected, all TRF analogues containing the strcuture X-His-Pro-NH_2 are deamidated by the post proline-cleaving enzyme (Figure 5). TRF, the 5-membered ring analogues of TRF and the orotyl-containing TRF analogues are degraded at comparable rates while the thiomorpholinone-containing analogues of TRF are deamidated more slowly. In agreement with our earlier observations (22) these results indicate that the affinity of the substrate for the enzyme is apparently also influenced by structural elements remote from the scissile peptide bond. Such structural requirements might be expected to represent necessary requirements for an enzyme which acts as an endopeptidase. While orotyl-His-Pro-NH_2 is deamidated faster than TRF itself, orotyl-His-Pro-$NHCH_3$ is degraded only very slowly (Figure 5) and orotyl-His-Pro-cyclohexylamide (not shown) is totally resistent to enzymatic hydrolysis by the post proline-cleaving enzyme.

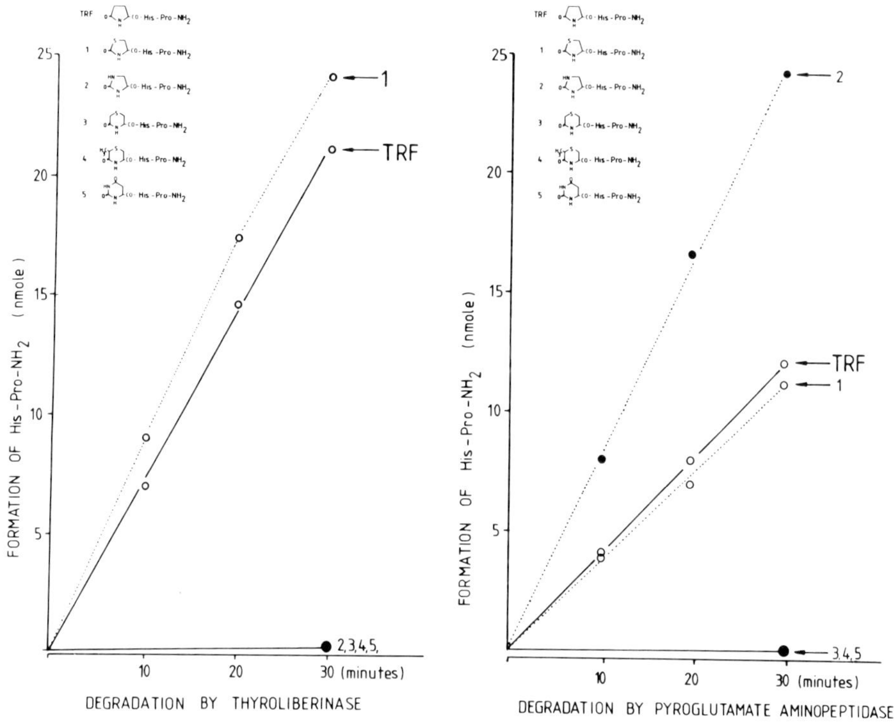

Fig. 3 and 4. Degradation of TRF and TRF analogues by the TRF-degrading serum enzyme ("thyroliberinase") (Figure 2) and by the pyroglutamate aminopeptidase (Figure 3). The degradation was determined as desdribed in methods.

Since TRF is hydrolyzed at the pyroglutamyl-histidine bond by the known pyroglutamate aminopeptidases from various tissues as well as by the highly specific TRF-degrading serum enzyme, it was of special interest to study the degradation of TRF analogues modified at the pyroglutamyl residue. With the purified enzyme preparations used, the degradation of the TRF analogues could be directly and specifically determined. It could be shown that replacement of the pyroglutamyl residue by six-membered ring structures results in analogues which are not degraded by these enzymes. It is interesting to note

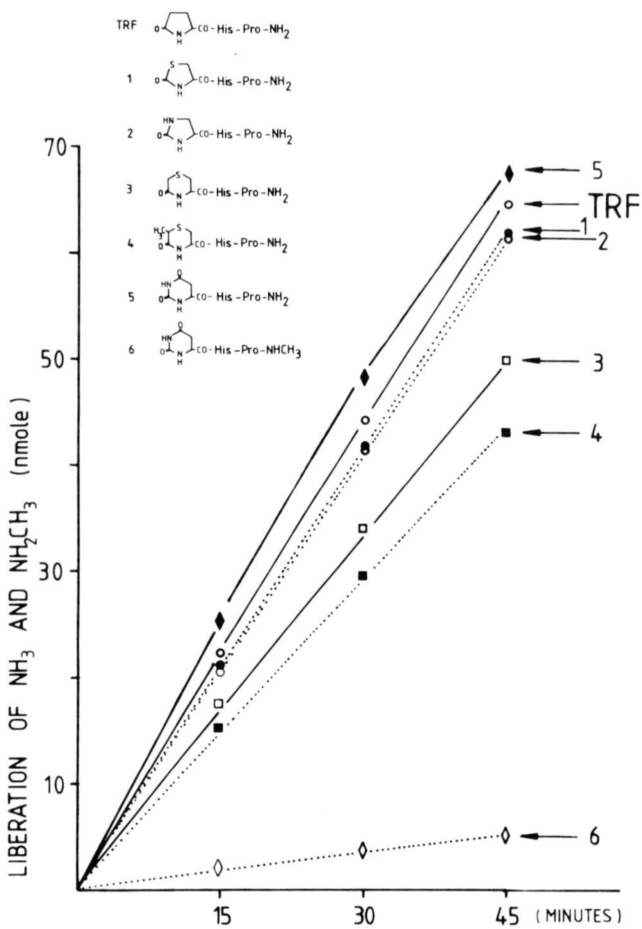

Fig. 5. Degradation of TRF and TRF analogues by the post proline-cleaving enzyme. The deamidation was determined by following the liberation of ammonia and methylamine as described in methods.

that for some of these analogues long lasting efficacy as centrally active drugs could be observed by Friderichs et al. (2). These findings might indicate that the availability of TRF at particular target sites is influenced by the pyroglutamate aminopeptidases and/or by the TRF-degrading serum enzyme. By further studies it should be possible to differentiate whether this effect can be ascribed to either one of these enzymes.

At present time it is not possible to evaluate whether the bioavailability of TRF might be influenced by the TRF-deamidating post proline-cleaving enzyme. The observation, however, that alkylation of the amide residue yields TRF analogues which are not or only slowly degraded by the post proline-cleaving enzyme is of special interest. As known, the post proline-cleaving enzyme also catalyzes very effectively degradation of LH-RH at the Pro-Gly peptide bond (22). The high biological activity observed for the (desglycineamide)-LH-RH alkylamide analogues (41) might again not only be due to changes favorable for receptor interaction but might also be correlated with the resistance of these analogues against enzymatic attack by the post proline-cleaving enzyme. If this is true, our results suggest that the efficacy of TRF might also be strongly influenced by the post proline-cleaving enzyme. Further studies along these lines should enable us to characterize the neuropeptide-degrading enzymes most extensively and to synthesize enzyme-resistant analogues. The determination of the biological activity of these analogues and, most importantly, complementary receptor studies should then help in determining the yet unknown function of neuropeptide-degrading enzymes within the mechanisms of neurotropic transmission.

References

1. Reichlin, S., Saperstein, R., Jackson, I.M.D., Boyd, A.E., Patel, Y.: Annu.Rev.Physiol. 38, 389-424 (1976).
2. Friderichs, E., Schwertner, E., Herrling, S., Günzler, W.A., Ötting, F., Flohé, L.: this issue.
3. Marks, M.: Frontiers in Neuroendocrinology, Ganong, W.F., Martini, L., eds., Vol.5, pp. 329-377 Raven Press, New York 1978

4. Griffiths, E.C., Kelly, J.A., Forbes, R., Jeffcoate, S.L., White, N., Milner, R.D.G., Visser, T.J.: Brain Peptides, A New Endocrinology, Gotto, A.M., Peck, E.J., Boyd, III, A.E.,eds., Elsevier/North Holland Biomedical Press, 197-209 (1979).

5. Bauer, K.: Colloques Internationaux du Centre de la Recherche Scientifique, Vincent, J.D., Kordon, C., eds., pp. 477-486 (1978).

6. Bauer, K.: Brain and Pituitary Peptides, Ferring Symposium, Munich, S. Karger AG , Basel, in press.

7. Mudge, A.W., Fellows, R.E.: Endocrinology 93, 1428-1434 (1973).

8. Prasad, C., Peterskofsky, A.: J.Biol.Chem. 251, 3229-3234 (1976).

9. Bauer, K., Gräf, K.J., Faivre-Baumann, A., Beier, S., Tixier-Vidal, A., Kleinkauf, H.: Nature 274, 174-175 (1978).

10. Doolittle, R.F., Armentrout, R.W.: Biochemistry 7, 516-521 (1969).

11. Szewczuk, A., Mulczyk, M.: Eur.J.Biochem. 8, 63-67 (1969).

12. Armentrout, R.W., Doolittle, R.F.: Biochem.Biophys. 132, 80-90 (1968).

13. Armentrout, R.W.: Biochem.Biophys.Acta 191, 756-758 (1969).

14. Szewczuk, A., Kwiatkowsky, J.: Eur.J.Biochem. 15, 92-96 (1970).

15. Mulczyk, M., Szewczuk, A.: J.Gen.Microbiol. 61, 9-13 (1970).

16. Kwiatkowsky, J., Torain, B., Glenner, G.G.: J.Biol.Chem. 249, 7729-7736 (1974).

17. Bauer, K., Sy, J., Lipmann, F.: Fed. Proc.32, 489 (1973).

18. Bauer, K., Lipmann, F.: Endocrinology 99, 230-242 (1976).

19. Taylor, W.L., Dixon, J.E.: Biochim.Biophys.Acta 444, 428-434 (1976).

20. Prasad, C., Peterkofsky, A.: J.Biol.Chem. 251, 3229-3234 (1976).

21. Knisatschek, H., Bauer, K.: 12th FEBS Meeting Dresden, GDR, Abstract 2819 (1977).

22. Knisatschek, H., Bauer, K.: J.Biol.Chem. 254, 10936-10943 (1979).

23. Hersh, L.B., McKelvy, J.F.: Brain Research 168, 553-564 (1978).

24. Koida, M., Walter, R.: J.Biol.Chem. 251, 7593-7599 (1976).

25. Walter, R., Shlank, H., Glass, J.D., Schwartz, I.L., Kerenyi, T.D.: Science 172, 827-829 (1971).

26. Yoshimoto, T., Orlowski, R.C., Walter, R.: Biochemistry 16, 2942-2948 (1977).

27. Matsui, T., Prasad, C., Peterkofsky, A.: J.Biol.Chem. 254, 2439-2445 (1979).

28. Prasad, C., Matsui, T., Peterkofsky, A.: Nature 268, 142-144 (1977).

29. Perry, T.L., Richardson, S.C., Hanson, S., Friesen, A.J. D.: J.Biol.Chem. 240, 4540-4544 (1965).

30. Redding, T.W., Schally, A.V.: Proc.Soc.Exptl.Biol.Med. 131, 415-420 (1969).

31. Taylor, W.L., Dixon, J.E.: J.Biol.Chem. 253, 6934-6940 (1978).

32. Bauer, K., Nowak, P.: Eur.J.Biochem. 99, 239-246 (1979).

33. Amos, M., Burgus, R., Blackwell, R., Vale, W., Fellows, R., Guillemin, R.: Biochem.Biophys.Res.Commun. 44, 205-210 (1971).

34. Bauer, K.: Nature 259, 591-593 (1976).

35. White, N., Jeffcoate, S.L., Griffiths, E.C., Hooper, K. C.: J.Endocrinol. 71, 1-7 (1976).

36. Dupont, A., Labrie, F., Levasseur, L., Dessault, J.H., Schally, A.V.: Clin. Endocrinol. 5, 323-328 (1976).

37. Oliver, C., Taurog, A., Porter, J.C.: Nouv.Presse Med.3, 1941-1944 (1974).

38. Neary, J.T., Kieffer, J.D., Federico, P., Mover, H., Maloof, F., Soodak, M.: Science 193, 403-405 (1976).

39. Schwertner, S., Herrling, S., Friderichs,E., Kim, S.M., Flohé, L.: this issue.

40. Moore, S., Stein, W.H.: J.Biol.Chem. 211, 907-913 (1954).

41. Kitada, C., Nakayama, R., Yamazaki, I.: Biochem.Biophys. Res. Commun. 49, 863-869 (1972).

STERIC AND ELECTRONIC STRUCTURE OF THE THYROTROPIN-RELEASING HORMONE

Mariann Farkas

Computer Centrum of Medical University Budapest, 1089 Budapest, Kulich Gy.tér 5, Hungary

Wolfgang Voelter

Institut für Organische Chemie der Universität Tübingen, Auf der Morgenstelle 18, 7400 Tübingen, FRG

Introduction

The thyrotropin-releasing hormone is a hypothalamic regulatory hormone, which, because of its very small natural concentration, was difficult to isolate and elucidate its structure (1-3, Fig. 1). In the last 9 years a large number of TRH analogues have been synthesized and tested for their biological activity, but one managed to find few relationships between their biological activity and chemical structure only. In our work only such a series of derivatives is treated in which a minimal change in the atomic but a significant change in the electronic structure results in a significant change in the physiological activity, e.g. TRH (L-Pyr-L-His-L-Pro-NH_2, 100 % activity), His $(\tau Me)^2$-TRH (L-Pyr-3-Me-L-His-L-Pro-NH_2, 800 % activity), and His $(\pi Me)^2$-TRH (L-Pyr-1-Me-L-His-L-Pro-NH_2, 0,04 % activity (4,5)).

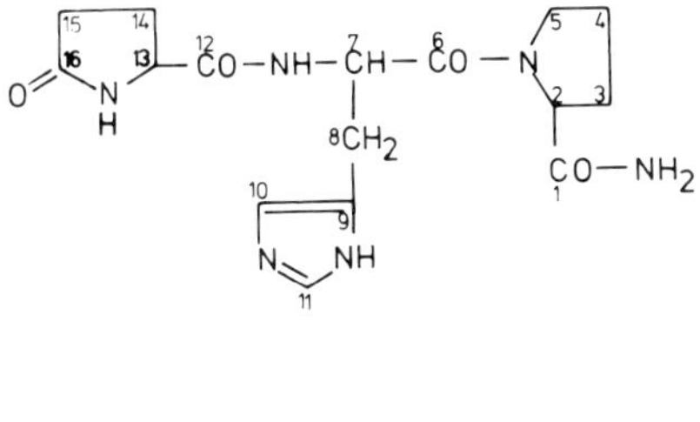

Fig.1. Structure of TRH.

Structure and Activity of Natural Peptides

Calculation of ^{13}C chemical shifts of model compounds. The measured ^{13}C NMR spectrum shows that only one of the two possible tautomeric structures of the imidazole ring is stable. In order to decide which one is the stable tautomer (6,7) we calculated the molecular wave functions of the two tautomeric models by the CNDO/S method according to Bene and Jaffe (8,9) using ^{13}C chemical shifts and the 128 ppm value of the benzene carbon as a reference shielding:

$$1. \quad \delta_{A_{Calc}} = \sigma_{A_{Calc}} - (\sigma_{^{13}C(Benzene)_{Calc}} - 128).$$

According to Saika and Slichter the shielding has different terms:

$$2. \quad \sigma_A = \sigma_A^{dia} + \sigma_A^{para} + \sum_{B \neq A} \sigma_{AB}.$$

The diamagnetic part (δ_A^{dia}) was calculated by the Lamb formula:

$$3. \quad \sigma_A^{dia} = \frac{e^2}{3\,mc^2} \cdot \sum_{\nu_A} \left\langle r_{\nu_A}^{-1} \right\rangle_{2s,2p},$$

$\langle r_{\nu_A}{}^{-1} \rangle$ is the mean inverse distance of the electron ν from the nucleus of atom A and the summation runs over all the electrons on the atom under consideration. The second term in the equation 2. is the paramagnetic contribution which results from mixing of ground and excited states by the external field. The remaining contribution comes from currents on neighboring atoms. The paramagnetic term dominates in ^{13}C chemical shifts - it follows most sensitively the changes in the chemical structure.

The calculation of paramagnetic shielding according to Ramsey (10) is too complicated for larger molecules. Therefore, on the basis of the theory of Ramsey, Pople derived a simplified approximation, which is commonly used in LCAO - MO theory (11) by linear combination of gauge invariant atomic orbitals. In this approximation

$$4. \quad \sigma_A^{para} = -2N^{-1} \cdot \chi_A^{para} \cdot \left\langle r^{-3} \right\rangle_{2p}$$

N is the Avogadro number, χ_A^{para} the paramagnetic susceptibility of atom A in the molecule, and the approximation of $\langle r^{-3} \rangle_{2p}$ is:

$$5. \quad \langle r^{-3} \rangle_{v_A} = \left(\frac{Z'_A}{a_0} \right)^3 ,$$

where a_0 is the Bohr radius and Z'_A is the core charge of atom A in the molecule,

$$6. \quad Z'_A = Z_A - 0.35\, q_A ,$$

and where q_A is the electron charge of atom A in the molecule. The paramagnetic susceptibility in the "average excitation energy" approximation (12,13) is:

$$7. \quad \chi_A^{para} = N \cdot e^2 \cdot h^2 \cdot \frac{1}{m^2 c^2} \cdot \frac{1}{\Delta \varepsilon_{av}} \cdot (Q_{AA} + \sum_{A \neq B} Q_{AB}),$$

where h is the Planck constant, c the velocity of light and Q_A:

$$8. \quad Q_A = \frac{1}{3} \cdot (2 \sum_{\mu,\nu}^{x,y,z} P_{\mu_A \mu_A} \cdot P_{\nu_A \nu_A} + \sum_{\mu}^{x,y,z} P^2_{\mu_A \mu_A})$$

and Q_{AB}:

$$9. \quad Q_{AB} = -\frac{1}{3} (P_{y_A y_B} \cdot P_{z_A z_B} + P_{x_A x_B} \cdot P_{z_A z_B} + P_{x_A x_B} \cdot P_{y_A y_B}) + \frac{1}{3} (P_{y_A z_B} \cdot P_{z_A y_B} + P_{x_A z_B} \cdot P_{z_A x_B} \cdot P_{x_A y_B} \cdot P_{y_A x_B}).$$

$P_{\mu\nu}$ are the orbital populations:

$$10. \quad P_{\mu\nu} = 2 \cdot \sum_i^{occ} C_{i\mu} \cdot C_{i\nu}$$

and $C_{i\mu}$ are the μ orbital coefficients on the energetic level i and the summation runs over all occupied orbitals. Calculation of the average excitation energy (ΔE_{av}) involves many difficulties. If, however, one calculates not with the real value of ΔE_{av}, it results in wrong values for shieldings, but this mistake changes the calculated value of ^{13}C chemical shifts only by a constant factor. By this way the correlation of calculated and measured ^{13}C shifts for the atoms of a larger molecule is:

$$11. \quad \delta(\text{measured}) = A \cdot \delta(\text{calculated}) + B .$$

If ΔE_{av} is not the real one then $A \neq 1$ and $B \neq 0$, but if the calculated model is the same as the real one the correlation must be linear.

Conformational analysis by the method of Scheraga. Our aim was, to make clear, whether the methyl substitution on different positions of the imidazole ring changes the stereochemical circumstances significantly or not. Scheraga (14) made conformational analyses of TRH. In these calculations it is assumed that bond lenghts and bond angles are constant and the total energy of the molecule depends on the dihedral angles. Therefore, in the method of Scheraga (15,16,17) only those parts of the total energy are calculated which are dependent on the dihedral angles. This conformational energy is treated as a sum of electrostatic, nonbonded, hydrogen bond and torsional contributions. The electrostatic energy is calculated as a sum of Coulomb potentials between atom - centered monopoly partial charges q_A which are calculated by the CNDO/S (8,9) method,

$$12. \quad U_{el} = 332 \cdot \frac{q_A \cdot q_B}{D \cdot r_{AB}} ,$$

where D is the dielectric constant, taken as 2.0 (17), r_{AB} is the internuclear distance in Angström units, and 332 is the factor to convert the energy to units of kcal/mole. The nonbonded interaction energy of an atom pair A and B (type K and L) is calculated using a modified Lennard - Jones 6 - 12 potential:

$$13. \quad U_{nonbonding} = F \cdot \varepsilon_{KL} \cdot \left(\frac{r_{o_{KL}}}{r_{AB}}\right)^{12} - 2 \cdot \varepsilon_{KL} \cdot \left(\frac{r_{o_{KL}}}{r_{AB}}\right)^{6} ,$$

where F = 0.5 for 1-4 interactions and F = 1.0 for 1-5 interactions. ε_{KL} is the energy minimum and r_o the corresponding internuclear distance. Hydrogen bond energy (U_{HB}) between the electron donor hydrogen (amine, amide, hydroxyl, carboxyl group) and the acceptor atom (uncharged ring and amide nitrogen or hydroxyl, ester and carbonyl oxygen) is calculated by

$$14. \quad U_{HX} = \varepsilon \cdot \left(\frac{r_o}{r}\right)^{12}_{HX} - 2.0 \cdot \varepsilon \left(\frac{r_o}{r}\right)^{10}_{HX}$$

and the torsional energy (U_{TOR}) by

$$15. \quad U_{TOR} = \frac{U_o}{2} \cdot (1 \pm \cos n \cdot \Theta),$$

where U_0 is the difference between the experimental barrier height and that calculated from the electrostatic, nonbonded and hydrogen bond potential functions, Θ is the dihedral angle and n gives the symmetry of the barrier.

The minimal energy as a function of sterical angles was calculated by the minimalization procedure of Fletcher and Powell. Earlier NMR results showed (18) that the ring of the proline part is stabilized by a NH ··· O=C hydrogen bridge. Therefore we treated only all other sterical angles in the minimalization procedure. Stezowsky et al.(19) made an X-ray structure analysis of Phe^2-TRH. For the CNDO calculations the input geometry was built up using his data set for bond distances and bond angles. For sterical angles we used the results of Scheraga (see Fig.2).

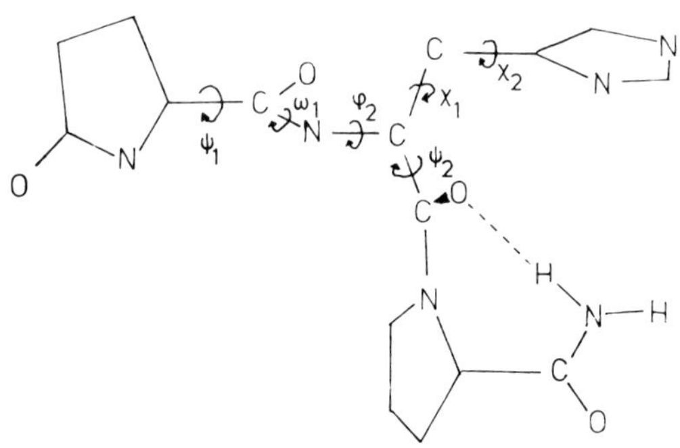

Fig. 2. Sterical angles of TRH.

Results

1. <u>Results of calculated ^{13}C chemical shifts.</u> Comparison of measured ^{13}C chemical shifts of TRH with calculated values of the two possible tautomeric model compounds (see Table 1) shows that the correlation is good enough in both cases (the regression is 0.99 for the 1-NH tautomer and 0.98 for the 3-NH tautomer) with the exception of the imidazole carbon atoms of the 3-NH tautomer (see Fig.3). The results make the existence of the 1-NH tautomer of TRH probable.

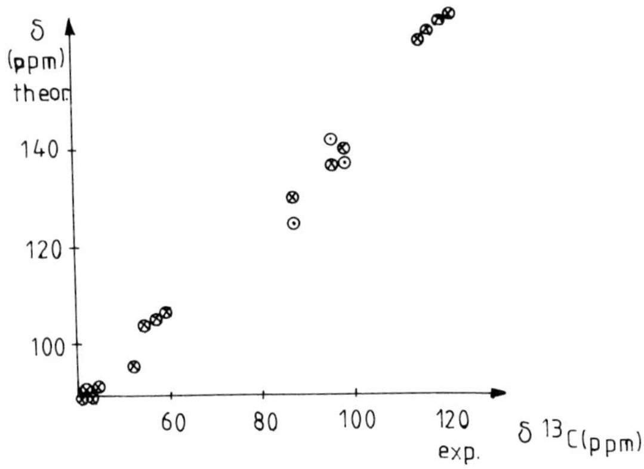

Fig.3. Comparison of experimental ^{13}C chemical shifts of TRH with calculated values of the two tautomeric model compounds, x His-1-NH tautomer, o His-3-NH tautomer.

2. <u>Results of the conformational analyses.</u> If one compares the lowest energy conformations of TRH, $(1\text{-Me-His})^2$-TRH and $(3\text{-Me-His})^2$-TRH (see Tab.2) there is no significant difference in the sterical angles. As expected, maximal difference, however, is also not too large (6°), in the sterical angle

Table 1. Comparison of measured ^{13}C chemical shifts of TRH with calculated values of the two possible tautomeric model compounds(His-1-NH and 3-NH); for numbering see Fig.1.

C-Atom	^{13}C chemical shifts, δ ppm, relative to TMS = 0		
	experimental	calculated	
		His-1-NH	His-3-NH
1	176.75	167.1	167.1
2	60.5	106.5	106.4
3	29.45	90.7	90.6
4	24.5	90.7	90.7
5	47.9	95.0	95.0
6	174.5	166.5	166.8
7	51.8	105.8	105.8
8	29.1	92.8	92.9
9	132.8	136.7	142.6
10	117.4	131.1	124.8
11	136.5	139.7	138.9
12	171.5	166.3	166.4
13	56.6	105.7	105.8
14	25.1	90.3	90.3
15	28.1	91.5	91.5
16	182.1	169.4	169.4

r (regression)	0.99	0.98
A	1.9968	1.9916
B	-152.55	-151.82

of the imidazole ring.

3. Electron structure of TRH and derivatives. For comparison of the electron structure of the compounds studied we used our CNDO/S results. The electron structures of the pyroglutamyl-

Table 2. Lowest energy conformers of TRH and imidazole ring methyl substituted derivatives.

Substituent	-	1 - Me	3 - Me
Φ_2	121.5	121.6	121.6
Ψ_2	-34.8	-33.9	-34.4
ω_2	-129.4	-130.9	-131.1
χ_1	159.5	159.7	158.0
χ_2	85.7	86.4	91.7

and prolineamide part do not change significantly on methyl substitution at the imidazole ring. Some characteristic data of the electron structures of the aromatic part are seen in table 3 (see also Fig.4).

In all 3 cases the two ring nitrogens have significant negative charges and C_{11} has a positive one, but neither of the charges of the ring atoms changes parallel to the physiological activity. The two negative nitrogen atoms may help the ring to approach the electron acceptor centrum of the receptor (Fig.4).

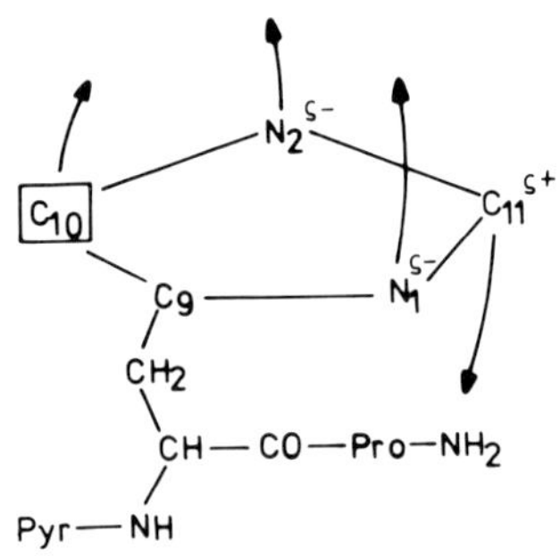

Fig. 4. Possible approach of TRH to the positive centrum of its receptor.

Table 3. Electron charge distribution and frontal electron density of the aromatic part of TRH and derivatives.

	$(1\text{-Me-His})^2$-TRH	TRH	$(3\text{-Me-His})^2$-TRH
Biological activity	0.04	100	800
Atom		Q (net)	
C_9	+0.1015	+0.0190	+0.0676
C_{10}	+0.0028	+0.0158	-0.0327
C_{11}	+0.1744	+0.1633	+0.2169
N_1	-0.0826	-0.0958	-0.0975
N_2	-0.2676	-0.2705	-0.3063
	Frontal electron density		
C_9	0.0014	0.7756	0.4997
C_{10}	0.0010	0.3449	0.6676
C_{11}	0.0004	0.5087	0.4629
N_1	0.0001	0.0018	0.0986
N_2	0.0002	0.1604	0.0484

For further control if this suggested biochemical mechanism is correct the CNDO/S calculations for Phe^2-TRH were repeated. In this case the imidazole ring is replaced by a phenyl ring (Fig.5.).

In the case of Phe^2-TRH the ortho carbon atom has the largest negative charge, so it can attack the electron acceptor centrum. The frontal electron density of this atom is 0.216. The physiological activity of Phe^2-TRH is 10 % of that of TRH. If one compares finally with the results of the other substituted derivatives (Tab.4), then there is quite a good correlation between biological activities and frontal electron densities of the ortho carbon atoms of the aromatic part of TRH and TRH analogues.

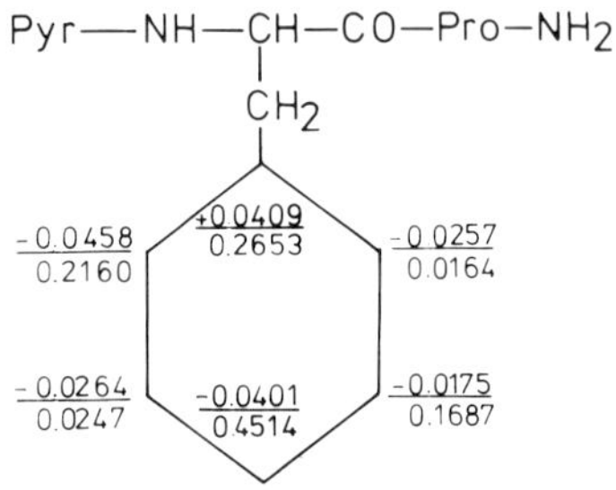

Fig. 5. Electron charge distribution of the aromatic part (underlined numbers) and frontal electron density of Phe^2-TRH.

Table 4. Correlation of biological activity and the frontal electron density of ortho carbon atoms of the aromatic part (f.e.d. C_{ortho}) for TRH and different analogues.

	$(1\text{-Me-His})^2$-TRH	Phe^2-TRH	TRH	(3-Me-His^2)-TRH
Biological activity(%)	0.04	10	100	800
f.e.d.C_{ortho}	0.0010	0.2160	0.3440	0.6676

References

1. Schally, A.V., Bowers, C.Y., Redding, T.W., Barrett, J.F.: Biochem.Biophys.Res.Commun.25, 165 (1966).
2. Burgus, R., Dunn, T.F., Desiderio, D.,Ward,D.N., Vale, W., Guillemin, R.: Nature 226, 331 (1870).
3. Folkers, K., Chang, J.K., Currie, B.L., Bowers, C.Y., Weil, A., Schally, A.V.: Biochem.Biophys.Res.Commun. 39, 110 (1970).
4. Vale, W., Rivier, J., Burgus, R.: Endocrinol. 89, 1485 (1971).

5. Rivier, J., Vale, W., Monahan, M., Ling. N., Burgus, R.: J.Med.Chem. 15, 479 (1972).

6. Tokutiro, T., Fraenkel, G.: J.Am.Chem.Soc. 91, 5005 (1969).

7. Farkas, M., Voelter, W., Vajda, M.:Acta.Chim.Hung.Acad. Sci. 96, 373 (1978).

8. Pople, J.A., Segal, G.A.: J.Chem.Phys. 43, 136 (1965), 44, 3289 (1966).

9. Bene, J.D., Jaffe, H.H.: J.Chem.Phys. 48, 1807 (1968).

10. Ramsey, N.F.: Phys.Rev. 78, 699 (1950); 86, 243 (1952).

11. Pople, J.A.: J.Chem.Phys. 37, 53 (1962).

12. Karplus, M., Pople, J.A.: J.Chem.Phys. 38, 2803 (1963).

13. Pauling, L., Wilson, E.B.:Introduction to Quantum Mechanics, McGraw-Hill Book Company, New York 1935

14. Burgess, A.W., Momany, F.A., Scheraga,H.A.:Proc.Natl.Acad. Sci. 70, 1456 (1973).

15. Scheraga, H.A.: Advan.Phys.Org.Chem. 6, 103 (1968).

16. Scheraga, H.A.: Chemical Reviews 71, 195 (1971).

17. McGuire , R.F., Momany, F.A., Scheraga, H.A.: J.Phys. Chem. 76, 375 (1972).

18. Gupta, D., Voelter,W.: Hypothalamic Hormones - Structure, Synthesis and Biological Activity, Verlag Chemie, Weinheim 1975

19. Stezowsky, J.J.: this issue.

ACTIVITY OF THYROLIBERIN ANALOGS WITH A MODIFIED PYROGLUTAMYL RESIDUE ON THE CENTRAL NERVOUS SYSTEM

Elmar Friderichs, Eberhard Schwertner, Siegfried Herrling, Wolfgang-Amandus Günzler, Fritz Ötting, Leopold Flohé

Grünenthal GmbH, Center of Research
D-5100 Aachen, FRG

Introduction

Thyroliberin (pGlu-His-ProNH_2; TRH) induces stimulation of the central nervous system (CNS) in addition to its well-documented endocrine effects (1). This activity appears to be independent from its action on the pituitary as releasing hormone, since it is retained in hypophysectomized animals (2). In pharmacological studies in human volunteers the CNS activity of TRH could be classified as "psychostimulant" by means of EEG analysis (3). Clinical trials with TRH in a variety of affective disorders, however, yielded inconsistent results (1) which could be due to erratic bioavailability and/or the short half-life of the natural releasing factor (4, 5). For these reasons it appeared promising to investigate a series of analogs of TRH which were synthesized with the intention to increase the biological stability and efficacy and to achieve a more favourable relation between endocrine and CNS effects (6). The biological activity of these analogs is of particular interest, since they are modified in the pyroglutamyl moiety of the molecule which has been inferred from previous studies to be indispensable for efficacy (7, 8) and is known as a major site of metabolic degradation (9, 10). We thus expected a more pronounced change in the pattern of thyroliberin-like effects and in addition a clearer insight into the structural requirements for receptor binding of TRH.

Structure and Activity of Natural Peptides

Materials and Methods

Compounds used in pharmacological and biochemical experiments were all reagent grade products.
TRH analogs were newly synthesized compounds of Grünenthal GmbH, Aachen, GFR (6).

Pharmacological Investigations

Group motility of mice was investigated in infrared beam activity cages (Lehigh Valley). 10 male mice (18-22 g body weight) per dose were placed into the cage immediately after dosing and the spontaneous locomotor activity of the groups was counted for every 10 min until activity approached base line levels.

Antagonism against reserpine-induced hypothermia was investigated in male NMRI mice (16-18 g body weight). 18 hours before testing the animals received 8 mg/kg reserpine dissolved in 5 % ascorbic acid solution intraperitoneally. Rectal temperature was measured before reserpine treatment, immediately before application of test compounds and every 30 min thereafter. 8 animals were used per dosage. ED_{50} values were calculated from the number of animals in which the reserpine hypothermia was reversed to at least 50 % by the test compounds. For documentation of efficacy ED_{50} values at peak activity are given (30-60 min or 90-120 min after i.p. or oral application, respectively).

Antagonism of barbiturate narcosis was investigated in each 10 male NMRI mice (20-25 g) per dosage. The animals received 50 mg/kg secobarbital-Na i.p. 15 min after application of the test compounds and were observed at constant temperature (31.5^{0}C) for reappearance of the righting reflex, spontaneous movements, twitches and tremor of extremities. ED_{50} values were obtained from the number of animals exhibiting a flattening of narcosis as judged upon the above parameters.

Potentiation of amphetamine-induced group toxicity was investigated in groups of 10 male NMRI mice (18-22 g). 15 min after i.p. application of the test drugs the animals received 5 mg/kg d,l-amphetamine subcutaneously and were placed into small glas vessels (Ø = 20 cm). The number of animals which died during 4 hours were counted. In order to keep the size of the group constant during the experiments the dead animals were replaced by untreated controls.

The "wet dog shaking" (shaking movements of head and body, as shown by wet dogs; WDS) was investigated in female Wistar rats of 150-200 g body weight. After dosing each animal was observed at 5 min intervals for a period of 1 min and the number of WDS reactions was counted. For comparison of efficacy of test compounds an ED_{10}-value was defined as the effective dose provoking 10 WDS reactions within an overall observation time of 60 min according to the above evaluation scheme (i.e. within 12 min).

General behaviour of mice and rats was evaluated according to Irwin (11).

Induction of TSH release by TRH and the test compounds was investigated as follows: Groups of 5-6 male Sprague Dawley rats were injected i.p. with 6 ng - 3.8 μg/kg TRH or TRH analogs and decapitated 15 min later. Serum samples were stored frozen and subsequently analyzed for TSH by means of a homologous radioimmunoassay (kindly performed by Prof. Dr. D. Gupta, Universitätskinderklinik, Tübingen).

Preliminary studies on the biological stability of CG 3509 (Orotyl-His-ProNH$_2$) [^{3}H]-labeled in position 5 of the orotyl moiety were performed in rats, mice and human serum, respectively. [^{3}H] CG 3509 (240 mCi/mMol) was obtained by reductive dehalogenation of the 5-chloro analog as outlined in ref. 6.

The profile of metabolites was investigated in the urine of 3 female NMRI mice (20 g) sacrificed 90 min after i.v. application of 4.7 µg (2.9 µCi) [^{3}H] CG 3509. Aliquots of the urine collected in the bladders by clamping the urethras during the experiment were pooled, acidified with trichloroacetic acid (TCA) and analysed as described below.

In vitro metabolic degradation of CG 3509 was studied by incubation of rat or human serum or rat tissue homogenate with 1.7 µg (0.9 µCi) [^{3}H] CG 3509/ml at 37^0 C. The incubation was stopped by TCA at the times given.

For separation of products the TCA supernatants were subjected to cation-exchange chromatography (Technicon Amino Acid Analyzer TSM; 405 x 5 mm C 3 resin, physiologic fluid program); separation of [^{3}H]orotic acid and [^{3}H] water eluted in the void volume was achieved by subsequent anion-exchange chromatography (100 x 10 mm Dowex 1 x 8). The peaks observed in the eluate at 280 nm were assigned to CG 3509 and its metabolites by comparison with the retention times of reference compounds and were quantitated by ß-counting.

Results and Discussion

In comparative studies with TRH not to be presented here in detail, the analogs, if active, exhibited a pharmacological profile qualitatively identical to that of TRH. At least with the most active test compounds investigated (CG 3509 and CG 3703; see below) the complete scope of pharmacological effects of TRH described in the literature (1, 12) could be reproduced at lower dosages: The general behaviour of the animals (mice and rats) was characterized by signs of central stimulation such as increase of spontaneous locomotor activity associated with agitated running, frequent rearing and grooming, intensified reactions to pain and other stimuli, "wet dog shaking" (rats), jumping and finally in some animals tonic clonic convulsions at high dosages. Correspondingly, pharmacological symptoms of central depressant drugs such as barbiturates, benzodiazepines, neuroleptics, reserpine and tetrabenazine were antagonized, while effects of amphetamine, monoamine oxidase inhibitors and tricyclic antidepressants were enhanced, as observed in mice.

Muscle tone was considerably increased even at low dosages, and a characteristic muscle vibration or trembling of the entire body and in particular of the upper limbs, accompanied by brief tremors was observed predominantly in mice. Occurrence of Straub tail phenomenon indicated stimulation of the dorsal muscle tone (mice and rats).

With regard to the autonomous nervous system salivation, lacrimation, mydriasis, an increase of heart and respiration rate, and particularly in rabbits elevated body temperature were seen.

In agreement with published data the LD_{50} was favourably high in most species (> 1000 mg/kg i.p. in mice and rats) for TRH and also for our analogs. Surprisingly, however, fatalities at moderate dosages were sometimes observed, when TRH or analogs were applied either alone or more frequently in combination with other CNS-active compounds. This obvious increase of toxicity appeared not to depend on the pharmacological profile of the combinant drug, but on the mere fact of its CNS activity, whether sedating or stimulatory, and in addition on exogenous factors such as handling of the animals, environmental temperature, density of housing and so forth.

This brief overview of findings characterizes TRH and the active analogs under consideration as strong CNS stimulants. Most symptoms suggest an enhanced catecholaminergic, particularly dopaminergic activity in the CNS. Some observations, however, could also be attributed to enhanced tryptaminergic transmission.

Structure-activity relationship of TRH analogs was elaborated by means of convenient and easily quantifiable test systems only (see methods).

Table 1. Biological activity of TRH and derivatives with modified pGlu residue (five membered ring systems).

Modification of pGlu	Stimulation of motility (threshold, mice, mg/kg i.p.)	Enhancement of amphetamine group toxicity (mice, ED_{50}, mg/kg i.p.)	Antagonism of reserpine hypothermia (mice, ED_{50}, mg/kg i.p.)	Antagonism of barbiturate anesthesia (mice, ED_{50}, mg/kg i.p.)	Relative TSH-releasing activity (rats, i.p.; TRH = 1)
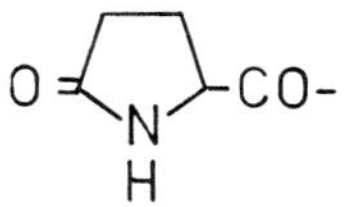	20 - 50	4.1 (1.8 - 9.4)	3.3 (1.9 - 5.6)	~ 10	1.00
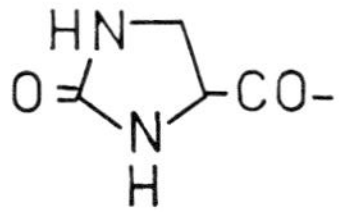	10 - 20	4.6 (2.4 - 8.8)	4.2 (2.1 - 8.3)	5.3 (2.9 - 9.7)	0.08
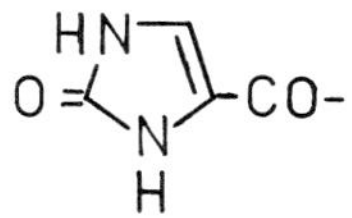	10 - 20	11.1 (4.0 - 31.2)	6.3 (3.4 - 11.5)	7.9 (5.1 - 12.4)	-
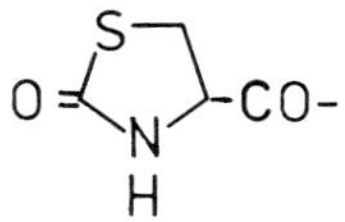	10 - 20	10.3 (3.4 - 31.6)	4.2 (2.5 - 7.0)	15.0 (8.9 - 25.4)	0.05
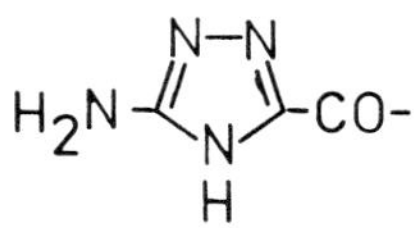	30 - 100	~ 50	34.1 (23.6 - 49.2)	10.7 (7.1 - 16.3)	-

Table 2. Biological activity of TRH analogs with modified pGlu residue (six membered ring systems).

Modification of pGlu	Stimulation of motility (threshold, mice, mg/kg i.p.)	Enhancement of amphetamine group toxicity (mice, ED_{50}, mg/kg i.p.)	Antagonism of reserpine hypothermia (mice, ED_{50}, mg/kg i.p.)	Antagonism of barbiturate anesthesia (mice, ED_{50}, mg/kg i.p.)	Relative TSH-releasing activity (rats, i.p.; TRH = 1)
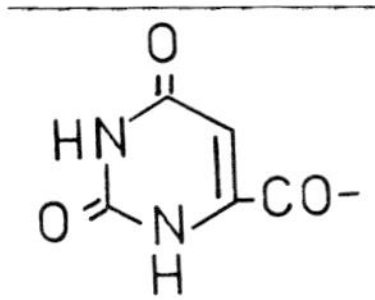	2 - 5	0.7 (0.4 - 1.1)	1.0 (0.5 - 2.1)	0.6 (0.3 - 1.1)	0.08
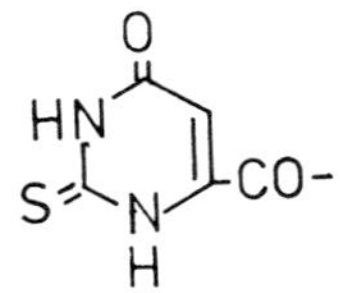	10 - 30	~2	1.3 (0.8 - 2.1)	~ 2	-
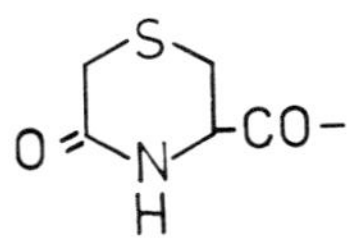	5 - 10	1.6 (0.7 - 3.5)	0.3 (0.2 - 0.4)	0.7 (0.4 - 1.2)	0.07
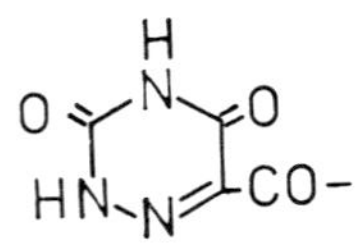	inactive (100)*	> 200*	inactive (200)*	inactive (100)*	-
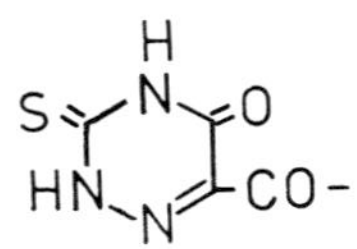	inactive (100)*	78 (37 - 166)	> 100 *	inactive (100)*	-

*maximum dosage tested

The first set of experiments was designed to work out which part of the pyroglutamyl moiety of the natural TRH molecule is responsible for its biological activity. As can be deduced from table 1 and 2, the pyrrolidinone ring can be replaced by imidazolidinone, imidazolinone, thiazolidinone, pyrimidinedione or thiomorpholinone rings without essential loss of biological activity. The pyrimidinedione and thiomorpholinone derivatives showed consistently higher CNS activity. TSH-releasing activity is essentially retained in the same group of compounds, although it usually occurs at higher dosages when compared to TRH (Tab. 1 and 2). This dissociation between CNS and endocrine activity might be explained by differences in the TRH binding sites in the tissues under consideration. Alternatively, it could be visualized that a lower intrinsic activity or affinity of the analogs prevails at both receptors, but is overcompensated by higher biological stability and availability in the CNS, while a sustained supply to pituitary receptors remains futile due to rapidly developing tolerance. Minor discrepancies of the relative activities between the four test systems for CNS stimulation are also most likely based on the different time requirements of testing.

Neither consistent CNS nor pituitary effects were detected with two 1,2,4-triazine analogs (Tab. 2), whereas some activity is retained in the 3-amino-1,2,4-triazolyl-5-carbonyl-derivative (Tab. 1) and appreciable potency is observed with the thioorotyl derivative (Tab. 2). This leads to the tentative conclusion that, in order to enable TRH-like activity, a $\overset{|}{C}O\text{-}NH\text{-}\overset{\backslash /}{C}\text{-}CO\text{-}$ or isosteric group has to be retained in a five or six membered ring replacing the pyroglutamyl moiety.

Since the TRH analogs containing an orotyl or thiomorpholinyl-carbonyl instead of the pyroglutamyl moiety showed the highest specific activities, further substitutions of these derivatives were tried (Tab. 3 and 4).

Table 3. Comparative activity of ring-substituted Orotyl-His-ProNH$_2$ (CG 3509) derivatives.

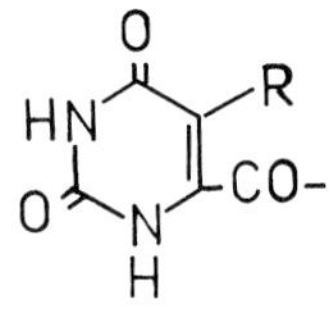

R =	Antagonism against reserpine hypothermia in mice: ED_{50} (mg/kg i.p.)	relative activity (TRH = 1)
-H (=CG 3509)	1.0 (0.5 - 2.1)	3.3
-Br	4.7 (2.5 - 8.9)	0.7
-Cl	8.7 (3.2 - 24)	0.4
$-CH_3$	50 (34 - 73)	0.07
$-C_2H_5$	36 (22 - 58)	0.09
$-n-C_3H_7$	4.2 (2.7 - 6.6)	0.8

Table 4. Comparative activity of ring-substituted 5-oxo-thiomorpholinyl-3-carbonyl-His-ProNH$_2$ derivatives.

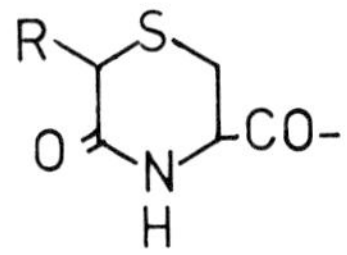

R =	Antagonism against reserpine hypothermia in mice: ED_{50} (mg/kg i.p.)	relative activity (TRH = 1)
-H	0.3 (0.2 - 0.4)	11
$-CH_3$ (CG 3703)	0.04 (0.02 - 0.10)	83
$-C_2H_5$	1.1 (0.5 - 2.4)	3
$-n-C_3H_7$	0.3 (0.1 - 0.8)	11
$-n-C_4H_9$	1.4 (0.8 - 2.4)	2.4
$-C_6H_5$	0.2 (0.1 - 0.4)	16.5

As can be seen from table 3 insertion of halogens or alkyl groups in position 5 of the orotic acid derivatives rather deteriorates than improves efficacy. In contrast, alkyl substitution of the thiomorpholinone series in position 6 does not considerably decrease activity and in the case of the methyl derivative (CG 3703) yields a highly active compound (Tab. 4). Substitution in position 2 of the thiomorpholine ring appeared less favourable, as was demonstrated by comparison of the 6,6'- and 2,2'-dimethyl derivatives (ED_{50} of reserpine antagonism 0.2 and 8.0, respectively). Interestingly, the two diastereomers of the 6-monomethyl derivative, only differing in the configuration of the 6-methyl group (6) showed marked differences in biological activity. The less laevorotatory isomer ($[\alpha]_D^{23} = -33.5^0$; CG 3703; data listed in Tab. 4) is about five times as potent as the strong rotating one ($[\alpha]_D^{23} = -108,5$).

In the reversal of reserpine-induced hypothermia CG 3703 is 83 times as active as TRH, while it is only 2.5 times as active as TRH in TSH release. Thus, the ratio of CNS to pituitary effects, as compared to TRH, is increased by a factor of about 33. But still the CNS effects cannot be induced without previously triggering a pituitary response.

Duration of action of the most active analogs of the pyrimidinedione (CG 3509) and the thiomorpholinone (CG 3703) series was investigated in comparison with TRH. As no direct receptor assay for TRH-like CNS effects is available, consideration of maximum effect in relation to duration of action should provide a preliminary insight into the problem, whether increased efficacy of the analogs is due to higher intrinsic activity, affinity or bioavailability.

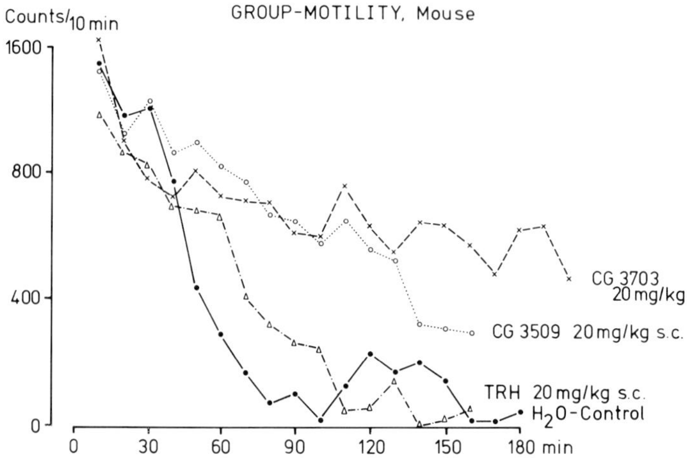

Fig. 1. Locomotor activity of male mice (N = 10/dose) counted by means of infrared beam activity cages.

As can be seen from figure 1, group motility studies in mice with TRH, CG 3509 and CG 3703 at identical dosage do not reveal any essential difference within the early observation period. However, while the TRH effect levels off rapidly, the motility remains high up to three hours after application of the analogs. With CG 3703 at a dosage of 20 mg/kg the motility does not yet approach control levels after 200 minutes.

The onset and time course of CNS activity is more clearly demonstrated, if a parameter is observed which is not contained in the behavioural pattern of untreated animals. For this purpose we chose the "wet dog shaking" reaction of rats repeatedly described as a TRH-induced phenomenon(13-15).

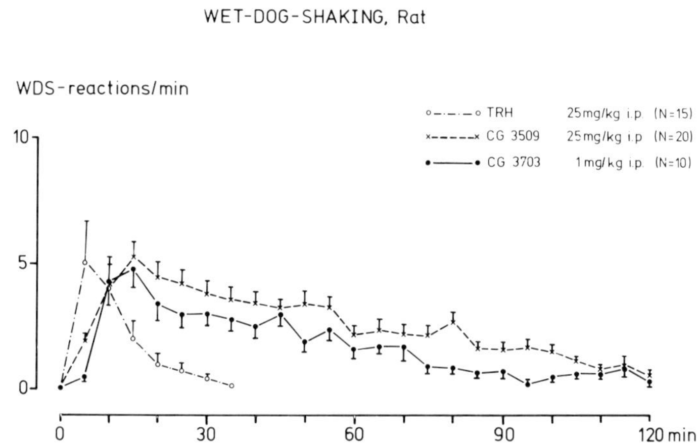

Fig. 2. Wet dog reactions ($x \pm s\bar{x}$) of male rats.

Figure 2 shows that i.p. injection of TRH almost instantaneously triggers this behaviour, but even at high dosage the effect does not last longer than half an hour. CG 3509 at the same dosage induces the same behaviour with some delay, the peak activity is not significantly different, but the effect lasts for about two hours. CG 3703 at a dosage of only 1 mg/kg i.p. gives a peak response almost identical to that of CG 3509 at 25 mg/kg i.p. and also lasting for 2 hours. These results clearly demonstrate that the comparably "high" activity of CG 3509 in our experiments, which necessarily requires an observation time beyond some minutes, is predominantly due to its longer biological survival (see below). It remains to be elucidated whether the additional improvement of efficacy seen with CG 3703 can also be explained by this simple assumption alone.

The metabolic fate of CG 3509 could be investigated since an orotyl-radiolabeled derivative was easily prepared. At least three enzymes are discussed to be involved in the initial breakdown of TRH in tissues and serum. In tissues, a "post proline cleaving" enzyme of low substrate specificity acts as a TRH desamidase. In addition, a pyroglutamyl aminopeptidase cleaves the pyroglutamyl moiety of TRH (9). An isoenzyme of the latter enzyme, tentatively called "Thyroliberinase", is supposed to inactivate TRH in serum (10,16).

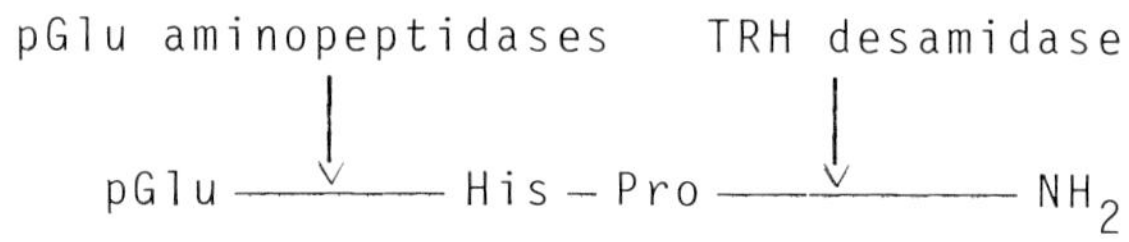

By detection of the presumed radioactive fragments of [^{3}H] Orotyl-His-Pro-NH_2 ([^{3}H]Orotyl-His-Pro, [^{3}H]Orotyl-His and [^{3}H]orotic acid) it seemed possible to evaluate the contribution of these enzymes to the metabolism of CG 3509.

We found that [^{3}H] CG 3509 in vitro was stable for two hours in rat and human serum. If present at all, the amount of the desamidation product [^{3}H] Orotyl-His-Pro and of free [^{3}H] orotic acid was less than two per cent each. Since Bauer (17) demonstrated that CG 3509 is cleaved by desamidase purified from brain tissue as fast as TRH, our results support the conclusion of Visser et al. (18) and Emerson et al. (19) that desamidation of TRH in contrast to earlier assumptions (2o) is not responsible for the short half-life of the natural releasing factor in blood.

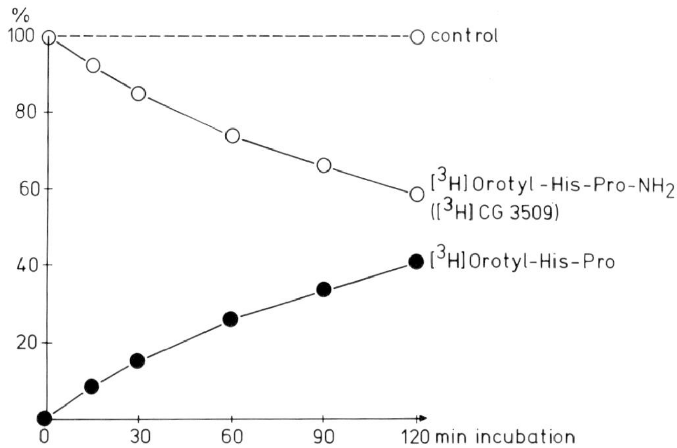

Fig. 3. Degradation of [^{3}H] Orotyl-His-ProNH$_2$ ([^{3}H]CG 3509) by rat hypothalamus homogenate. Conditions: 1 vol of a 10% homogenate (w/v) in 0.5 M phosphate buffer of pH 7 was incubated with 0.1 vol of [^{3}H] CG 3509 (final concentration 1.7 μg ≙ 0.9 μCi/ml) at 37^{0}C. control measurements were performed with homogenate denatured for 10 min at 100^{0}C. Results are given in % of initial (t = 0) [^{3}H]CG 3509 content.

Desamidation of CG 3509, however, takes place in homogenate of rat hypothalamus at a considerable rate. As can be seen from figure 3, the disappearance of [^{3}H] CG 3509 is accounted for almost completely by the formation of [^{3}H] Orotyl-His-Pro, while again free [^{3}H] orotic acid could not be detected. A trace (<3%) of the [^{3}H]Orotyl-His could not be excluded with certainty at the end of the incubation period. This implies that the long lasting activity of CG 3509 is attributable to its resistance to the pyroglutamate aminopeptidases. This conclusion is in agreement with the pronounced substrate specificity of these enzymes (10,16) and could be substantiated by Bauer (17), who found no degradation by pyroglutamate aminopeptidases of any of the TRH analogs in which the pyrrolidinone moiety is replaced by a six-membered ring. The long-lasting activity of the analogs with six membered rings thus gives additional support to the hypothesis that pyroglutamate aminopeptidase is the dominant TRH-degrading enzyme.

Table 5. Metabolites of [^{3}H] Orotyl-His-ProNH_2; [^{3}H]-CG 3509) in urine of mice 90 min after i.v. application of 235 µg [^{3}H] CG 3509. Results are given in % of total radioactivity of the sample.

CG 3509	Orotyl-His-Pro	Orotyl-His	Orotic acid + water
85	7	4	2

The minor importance of the TRH desamidase also reveals from the observation that CG 3509, which is resistant to pyroglutamate aminopeptidase but susceptible to desamidation, is largely excreted unchanged in the urine (Tab. 5). At 90 min after application of [^{3}H] CG 3509, when about 50 % of the administered radioactivity is excreted (not shown),85 % of the radioactivity present in urine represents the unchanged compound, while only 7 % of the desamidation product and only traces of other cleavage products are detectable. Although the relevance of local metabolic variations cannot be predicted at present, the overall half-life of pyroglutamate aminopeptidase-resistant TRH analogs appears to be determined predominantly by renal excretion and only to a minor degree by metabolic breakdown.

Oral availability of the most active analogs was investigated in mice and rats in respect to its eventual therapeutic relevance (Tab. 6). Consistent activity of TRH, CG 3509 and CG 3703 could be detected in mice after oral application. The i.p./p.o. ratio of equally effective dosages was similar for these three compounds and ranged between 7 and 20 (Tab. 6). In contrast, TRH and CG 3509 via oral route was inactive or only marginally active in rats, while CG 3703 proved orally available by inducing typical WDS reactions (Tab. 6). It remains to be investigated whether the oral availability of CG 3703 in rats is due to its high activity or to a particular resistance to proteolytic degradation.

Table 6. Comparison of oral and intraperitoneal activity of TRH, CG 3509 and CG 3703 in mice (antagonism of reserpine hypothermia; ED_{50}) and rats (wet dog shaking; ED_{10}).

Species	Compound	Dose i.p.	Dose p.o.	Dose Ratio i.p./p.o.
Mouse ED_{50}	TRH	3.3 (1.9 - 5.6)	21.6 (10.9 - 42.9)	1 : 6.5
	CG 3509	1.0 (0.5 - 2.1)	14.1 (9.9 - 20.2)	1 : 4.1
	CG 3703	0.04 (0.02 - 0.10)	~1.0	1 : 25
Rat ED_{10}	TRH	12.7 (4.6 - 25)	(1000)*	-
	CG 3509	1.8 (1.1 - 2.4)	> 100	> 1 : 50
	CG 3703	0.25 (0.20 - 0.32)	22.8 (15.0 - 30.9)	1 : 91

*inactive

Conclusions

1. The claim that the pyroglutamyl moiety of TRH is essential for biological activity has to be specified in so far as only the partial structure

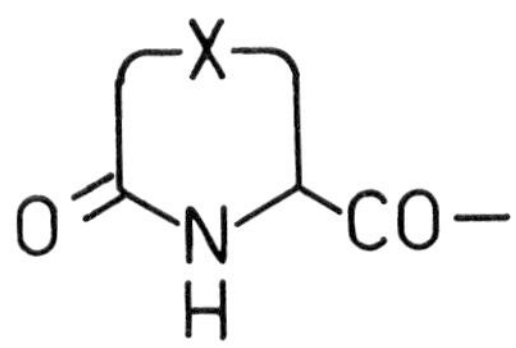

or

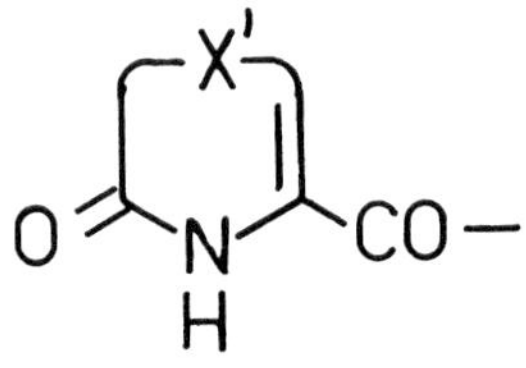

is required.

2. 6-Membered ring derivatives (e.g. orotyl- or 5-oxo-thiomorpholinyl-3-carbonyl-derivatives) appear to be more active than 5-membered ring derivatives and are definitely longer acting than TRH in models of CNS stimulation, while the endocrinological activity is less affected. 6-Methyl-5-oxo-thiomorpholinyl-3-carbonyl-histidyl-prolineamide (CG 3703) was the most active compound of the series.

3. The comparatively long lasting CNS effects of some analogs may be attributed to their resistance to cleavage by pyroglutamate aminopeptidases.

4. As could be exemplified with CG 3509, the biological half life of analogs resistant to pyroglutamate amino-peptidases appears to be determined primarily by renal clearance of the undegraded compounds and only to a minor degree by desamidation. This lends further support to the hypothesis that TRH inactivation is

largely determined by cleavage of the pyroglutamyl residue.

5. In principle, a dissociation between CNS stimulant and endocrine activity can be achieved, but may partially be due to increased bioavailability of the analogs in the CNS, since the CNS effects appear to reflect the sustained tissue levels, while the pituitary rapidly develops tolerance to continuous stimulation. However, differences in receptor specificity in the CNS and the pituitary may also be considered.

6. As TRH itself, all active compounds investigated proved to be orally available in mice to some degree. In contrast to TRH, the most active compound CG 3703 showed also oral activity in rats.

7. In general, the CNS effects of TRH and its more stable analogs in particular may be characterized as "psychostimulant". Most of the symptoms observed are generally attributed to increased catecholaminergic, particularly dopaminergic transmission, while the "wet dog shaking" in rats reminds of tryptaminergic activity. The described stable TRH analogs thus represent interesting psychopharmacological tools and eventually pharmacotherapeutic agents in affective disorders.

References

1. Prange, A. J. jr., Nemeroff, Ch.B., Loosen, P.T., Bisette, C., Osbahr III, A.J., Wilson, I.C., Lipton, M.A.: Central nervous system effects of hypothalamic hormones and other peptides (Collu et al. eds), pp. 75 - 96, Raven Press, New York 1979.

2. Plotnikoff, N.P., Prange, A.J., Breese, G.R., Anderson, M.S., Wilson, I.C.: Science 178, 417 - 418 (1972).

3. Itil, T.M., Patterson, C.D., Polvan, N., Metha, D. Bergey, B.: Psychopharmac. Bull. 1, 29 - 37 (1975).

4. Bassiri, R.M., Utiger, R.D.: J. Clin. Invest. 52, 1616 - 1619 (1973).

5. Morley, J.E., Garvin, T.J., Pekary, A.E., Utiger, R.D., Nair, M.G., Baugh, C.M., Hershman, J.M.: J. Clin. Endocrinol. Metab. 48, 377 - 380 (1979).

6. Schwertner, E., Herrling, S., Friderichs, E., Kim, S.M., Flohê, L.: This issue.

7. Steiner, H., Motta, M., Piva, F., Gillessen, D., Studer, R.O.: Acta Endocrinol. (Copenhagen) Suppl. 155, 5 (1971).

8. Voelter, W.: Chem. Ztg. 98, 554 - 561 (1974).

9. Prasad, C., Peterkofsky, A.: J. Biol. Chem. 251, 3229 - 3234 (1976).

10. Taylor, W.L., Dixon, J.E.: J. Biol. Chem. 253, 6934 - 6940 (1978).

11. Irwin, S.: Gordon Res. Conf. on Med. Chem. p. 133 (1959).

12. Kruse, H.: Naunyn-Schmiedebergs Arch. Pharm. 297, 61 - 62 (1977).

13. Wei, E., Sigel, S., Loh, H., Way, E.L.: Nature 253, 739 - 740 (1975).

14. Yarbrough, G.G., Lotti, V.J.: Pharmacologist 20,207 (1978).

15. Martin, B.R., Dewey, W.L., Chau Pham, T., Prange A.J. jr.: Life Sci. 20, 715 - 722 (1977).

16. Bauer, K., Nowak, P.: Eur. J. Biochem. 99,239-246 (1979).

17. Bauer, K.: This issue.

18. Visser, T.J., Klootwijk, W., Docter, R., Hennemann, G.: Acta Endocrinol. (Copenhagen) 86, 449 - 456 (1977).

19. Emerson, C.H., Mishal, A., Mahabeer, H.L., Currie, B. L.: J. Clin. Endocrinol. Metab. 49, 138 - 140 (1979).

20. Nair, R.M.G., Redding, T.W., Schally, A.V.: Biochemistry 10, 3621 - 3624 (1971).

INTERMITTENT PERNASAL GONADOTROPIN-RELEASING HORMONE THERAPY IN MALE PATIENTS WITH DELAYED PUBERTY

Joachim Happ, Volker Deigentasch, Volker Schmitz, Ulrich Krause, Jürgen Beyer
Abteilung für Klinische Endokrinologie, II. Medizinische Klinik, Universität Mainz, D-6500 Mainz, Germany

Marianne von der Ohe
Medizinische Abteilung, Hoechst AG, D-6000 Frankfurt, Germany

Introduction

We previously reported effectiveness of gonadotropin-releasing hormone (GnRH) therapy in male patients with delayed puberty; similar results have been reported by other authors: Significant stimulation of Leydig cell function could be achieved by subcutaneous (1) or pernasal (PN) (2) treatment with GnRH as well as by PN treatment with potent GnRH analogs (3). However, great differences were observed in the individual patients' responsiveness to the treatment. In some patients, lacking effectiveness of the treatment was observed, probably due to primary testicular deficiency which accompanied their deficiency of gonadotropin secretion. On the other hand, in some patients, lacking effectiveness proved to be the result of decreasing responsiveness of the gonadotropins as assessed by standard GnRH tests carried out every 1 or 2 weeks during treatment. It could be shown by animal experiments (4, 5) that supraphysiologic stimulation by GnRH or potent analogs has an inhibitory effect on gonadotropin secretion. However, decreasing responsiveness of gonadotropins was not only found with the use of potent GnRH analogs (3, 6, 7) but was also observed with low dose GnRH therapy in male patients with delayed puberty receiving 200 µg PN, 6 to 8 times daily (2). In the patients concerned, the phenomenon did not start earlier than 2 to 4 weeks of continuous long term treatment and finally lead to unresponsiveness of the gonadotropins within the following weeks of treatment. In order to avoid this event, we investigated intermittent PN GnRH therapy consisting of treatment periods of 4 weeks followed by 2 week intervals.

Methods

Nine male patients, 15 to 20 years of age, with delayed puberty, 2 of them anosmic, received an intermittent GnRH therapy consisting of treatment periods of 4 weeks which were separated from each other by treatment intervals of 2 weeks. Before the commencement of these studies the patients or their parents had given their informed consent. The treatment was carried out for 2 1/2 months. The daily dosage was 1200 µg PN (200 µg GnRH 6 times daily, given at 8.00, 9.00 and 12.00 a.m. and at 1.00, 6.00 and 7.00 p.m.). 3 of the patients (K.R.II, E.O., G.K.) had previously undergone continuous GnRH therapy over a period of 3 months which was terminated 2 1/2 to 3 1/2 months prior to the intermittent therapy. The results of the continuous treatment have been published elsewhere (2). In the present study the single doses were administered in form of a nasal spray as used in our previous study (2). The spray was provided by Hoechst AG, Frankfurt. The clinical data of the patients are given in table I. Sellar tumors were excluded by X-rays of the skull. Birth complications were reported to have existed in patient H.H. At an age of 6 years, the patient had been involved in a car accident; abdominal organs had been injured and splenectomy had become necessary. Patient K.R.II was conspicuous by an exaggerated increase of serum prolactin in the thyrotropin-releasing hormone test (200 µg intravenously (IV)). In the other patients, there were no important anamnestic, clinical or humoral abnormalities, in particular no signs of endocrine disturbances other than the ones reported above. 3 patients had previously been treated for unilateral cryptorchidism (orchidopexy in H.A. and B.A.; treatment with human chorionic gonadotropin (HCG) in B.H.). Control groups of other publications were used for height (8) and for sexual development (9, 10). Bone age was determined by X-rays of the left hand and wrist. For measurement of testicular volume an orchidometer (11) was used; the average of the right and the left testicle is recorded.

Table I

Clinical data of the patients with retarded puberty.

name	age (yrs;mths)	bone age (yrs)	(% normal)	height (cm)	(% normal)	testes (ml)	pubic hair (Tanner´s stage)	penis length (cm)
K.R.I	15; 0	13	87	151	90 [c]	3	1-2	5.5
K.R.II	15; 9	12.5	79	148	86	9	3	8
B.H. [a]	16; 1	13.5	84	166	95 [c]	6	4-5	6.5
B.A. [a]	16; 4	13.5	81	161	93 [c]	4	1-2	3.5
H.A. [a]	16; 6	14	85	190	108 [c]	4.5	4	6
E.O.	17; 1	12.5	73	152	87	6	2	7
M.R. [b]	18; 5	14	76	167	95 [c]	4	1	3
H.H.	19;11	12.5	63	170	96 [c]	5.5	2	5
G.K. [b]	20; 0	14	70	172	97 [c]	3	2	5.5

[a] cryptorchidism [b] anosmia [c] eunuchoid proportions

Table II

Normal range[+] of basal serum gonadotropins, of serum gonadotropin increases (△) after 25 ug GnRH i.v. and of basal serum testosterone.

LH (mU[++]/ml)	△LH (mU/ml)	FSH (mU/ml)	△FSH (mU/ml)	T (ng/100ml)
6 - 35	14 - 111	1,3 - 12	0 - 10	272 - 1560

[+] 95% range, n=100

[++] IRP_2-HMG (2nd International Reference Preparation of Human Menopausal Gonadotropin)

Before treatment, every 2 weeks during treatment and at longer intervals after treatment, GnRH tests were performed in the morning using IV bolus injections of 25 µg as in our previous study, mentioned above (2). For measurement of luteinizing hormone (LH) and follicle-stimulating hormone (FSH), blood samples were taken once before, 30 minutes and 60 minutes after the injection of the test dose. The values before and after the stimulation represent basal secretion and responsiveness of the hormones. For measurement of serum testosterone (T) blood was drawn once before and 120 minutes after the injection of the test dose. LH and FSH measurements in serum were done by radioimmunoassay methods described elsewhere (12). The kits were purchased from Union Carbide, Düsseldorf. Serum T was measured radioimmunologically using our own double antibody method. Rabbit-anti-testosterone-3-BSA antibody was kindly provided by E. Nieschlag, Münster and used in the assay in a final dilution of 1/50000. Anti-rabbit gamma globulin RD 17 was purchased from Wellcome and used in a dilution of 1/20. 125-I-testosterone was purchased from Immuno Nuclear Corporation (distributer in Germany: Biermann GmbH, Frankfurt). T was extracted from 0.5 ml serum with 5 ml peroxide-free ether by shaking the mixture for 30 minutes. Under these conditions and in the absence of cold hormone, tracer is bound by the antibody to an extent of 50% of the total activity. In order to exclude interassay variance, sera of the GnRH tests performed during treatment, shortly before or after the treatment were tested in the same assay for each patient.

Normal values (95% range) of serum LH and FSH obtained in GnRH tests carried out in 100 healthy adult males between 20 and 40 years of age are reported in table II. Basal serum T was measured in the same patients (Table II). Serum T increase in the GnRH tests, however, was recorded only in 6 control persons; the results are presented in the figures 2 - 4.

The immediate effect of the double stumulation (single doses applied at 8.00 and 9.00 a.m.) with 200 µg GnRH PN per stimulation, being part of the therapeutic dose regimen was checked in each patient to exclude disturbances of PN absorption of GnRH. In 3 patients who had previously received continuous GnRH therapy (G.K., E.O., K.R.II) the immediate ef-

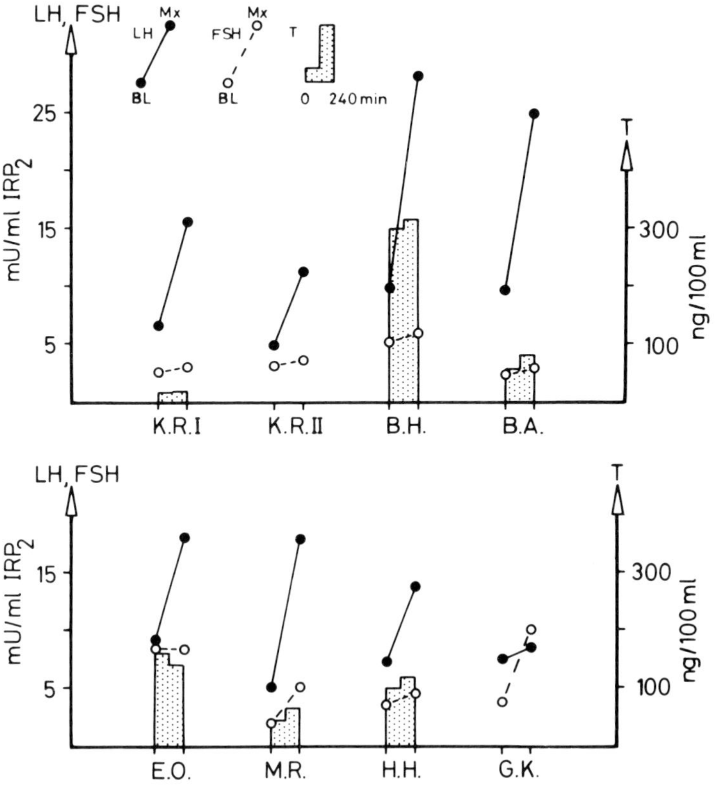

Figure 1. Serum gonadotropins (LH and FSH) and serum testosterone (T) after repeated pernasal stimulation applying 200 µg GnRH at an interval of 1 hour; basal level (BL) and maximum level (Mx) of LH and FSH and concentration of T before (0 min.) and 240 minutes after (240 min.) the application of the first dose.

fects of the PN double doses applied at an interval of 1 hour had been checked before enacting the previous continuous therapy. In the present work, only basal serum levels (BL) and maximum concentrations (Mx) of the gonadotropins after PN stimulation with GnRH are recorded; for this, blood samples were taken before and 30, 60, 90, 120, 180 and 240 minutes after the first dose.

Results

In the standard GnRH tests conducted before the GnRH therapy, basal serum LH was found to be prepubertally low in most of the patients (fig. 2-4); only 2 patients (B.H., E.O.) had serum levels in the lower adult range. LH responsiveness to the test dose of GnRH, however, was found within the lower adult range in most of the patients; only 3 patients (G.K., K.R.I, H.H.) presented LH responsiveness definitely below the adult range. Low basal serum FSH was found in 3 patients (G.K., H.A., K.R.II) but all patients had a FSH response in the GnRH tests. Serum T was far below the normal range for adults in all patients, except patient B.H. who showed a serum concentration at the lower limit of the adult range; in most of the patients, serum T was found in the infantile range, i.e. markedly below 100 ng/100ml and in 3 other patients (E.O., K.R.II, H.H.) serum T was found to be around 100 ng/100ml or slightly above.

The smallest testicular volumes were found in the patients with the lowest FSH serum levels. The patients with testicular volumes above 5 ml showed T concentrations in serum of 100 ng/100ml or above. The expected correlation between serum T and bone age was missing.

The immediate effect of the first 2 of the 6 daily PN doses (fig. 1) checked in all patients, except H.A., approximately reflected the gonadotropin responsiveness of the individual patient known from the standard GnRH test and showed normal PN absorption of GnRH in the individual patient. In the patients who had received a continuous treatment prior to the intermittent treatment and in whom PN absorption had been checked before the first treatment , the effect of the PN stimulation appeared

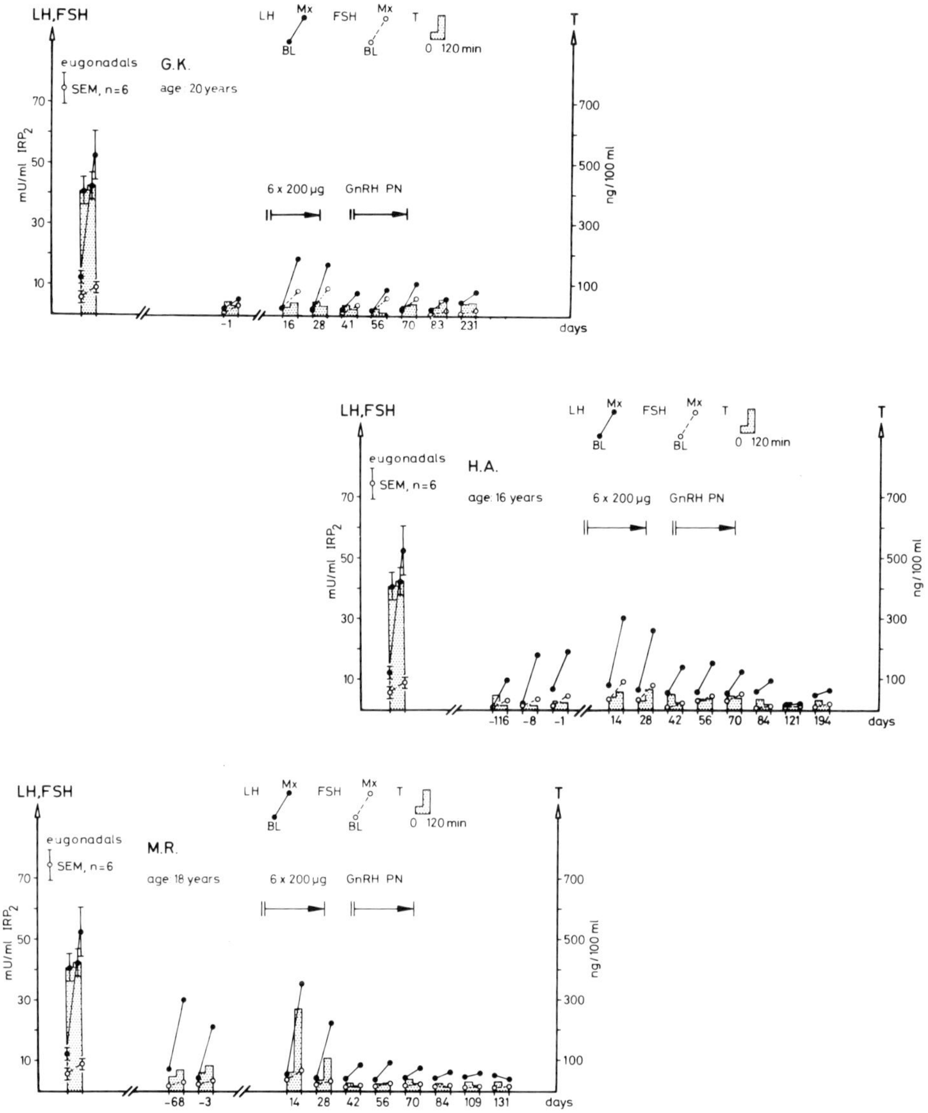

Figure 2. Serum gonadotropins (LH and FSH) and serum testosterone (T) in GnRH tests performed before intermittent pernasal (PN) GnRH therapy, during the treatment period and after therapy in male patients with delayed puberty (G.K. and M.R. anosmic); basal levels (BL) and maximum levels (Mx) of LH and FSH and concentration of T before (0 min.) and 120 minutes after (120 min.) the test doses of GnRH. For comparison GnRH tests in 6 eugonadal adult males.

definitely smaller than the effect of the IV standard test dose (cf. our previous investigation (2)). PN absorption was also tested in the other patients receiving the actual intermittent treatment and showed about the same gonadotropin response after the PN therapeutic dose as after the IV test dose.

The individual patients responded very differently to the intermittent PN GnRH therapy. In 2 patients with a small LH response in the GnRH test before treatment (G.K., K.R.I), LH responsiveness increased within the first 2 weeks of treatment; 1 patient (H.H.) kept a small responsiveness of LH throughout the treatment (fig. 2-4). In 3 patients, gonadotropin responsiveness decreased after the second or fourth week of the first 4 week period of treatment (fig. 2).

In none of the 3 patients with decreasing responsiveness of the gonadotropins after several weeks of treatment (fig. 2), an increase of basal serum T was observed; patient M.R., however, showed an increased T response in the GnRH tests conducted after the second and fourth week of treatment.

In 2 patients, despite good LH responsiveness throughout the treatment, no increase of serum T was found (fig. 3). In patient B.H., LH responsiveness was decreased after the second 4 week period of treatment. 2 weeks after termination of the treatment, basal serum LH had increased; this was no longer found 5 weeks after treatment. In the same patient, serum T, being found at the lower limit of the adult range before treatment, did not show a significant change during treatment. However, basal serum concentration was slightly lowered half way through the first (and the second ?) 4 week period of treatment. Serum concentration had slightly increased during the 2 week interval after the first treatment period. A second transitory increase occured 5 weeks after the second treatment period.

In the last 3 patients (fig. 4), basal serum T at least doubled during intermittent treatment but did not enter the adult range in any of the patients. In the same patients, a common pattern of basal serum T was

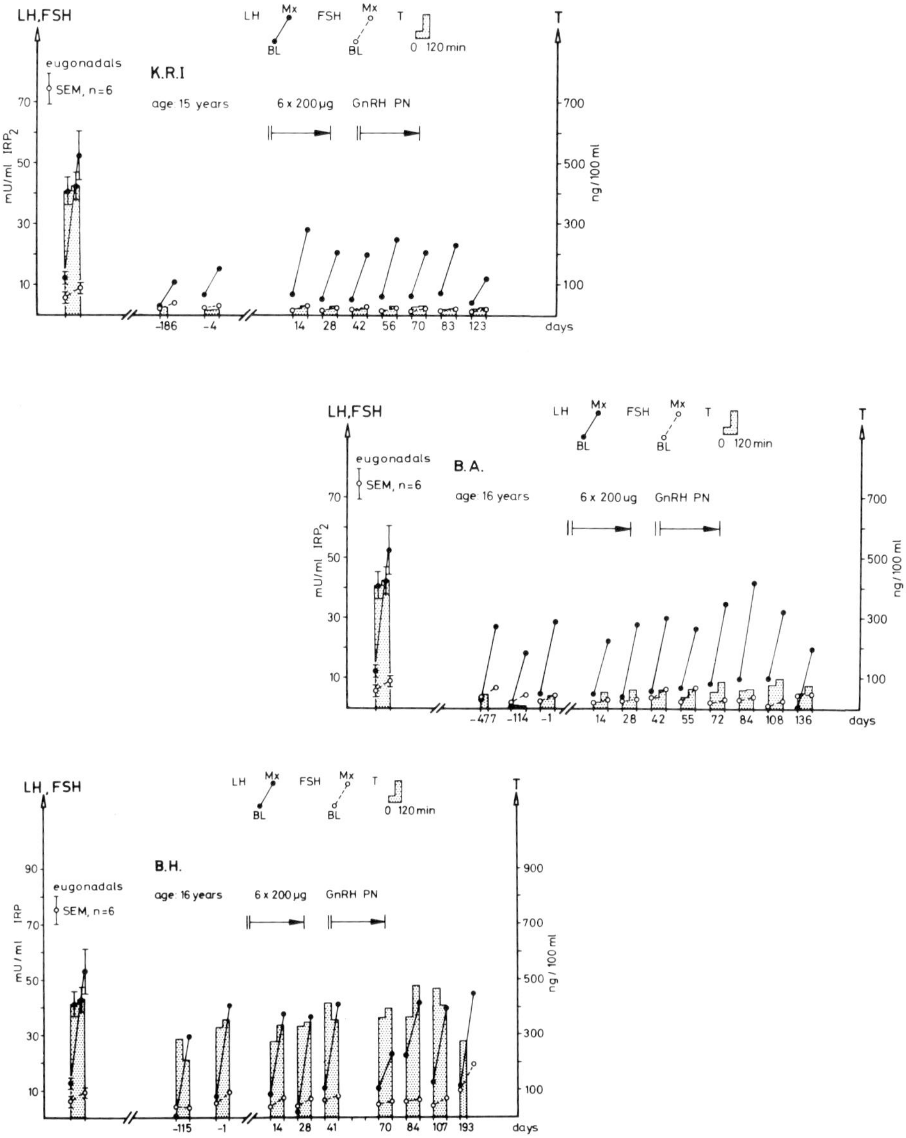

Figure 3. Serum gonadotropins (LH and FSH) and serum testosterone (T) in GnRH tests performed before intermittent pernasal (PN) GnRH therapy, during the treatment period and after therapy in male patients with delayed puberty; basal levels (BL) and maximum levels (Mx) of LH and FSH and concentration of T before (0 min.) and 120 minutes after (120 min.) the test doses of GnRH. For comparison GnRH tests in 6 eugonadal adult males.

found under therapy: The serum level increased during the first 2 weeks of treatment and decreased again during the following 2 weeks of treatment. A new increase was seen after the following 2 weeks without treatment. After the first 2 weeks of the next treatment period, serum T decreased markedly in 2 patients and slightly in 1 patient. At the end of this treatment period, serum T had increased again. After termination of the treatment, serum T decreased again in 2 patients; a further increase, however, was observed in 1 patient (K.R.II). In 1 of the patients in whom serum T decreased after termination of the treatment, an increase was observed 9 months after treatment. Changes in LH responsiveness paralleling the fluctuations in serum T levels were evident only in 1 of the 3 patients (K.R.II).

In 7 patients, a small growth of testicular volume was observed during the 3 month period of intermittent GnRH therapy: The mean volume increased from 5.0 $\pm$ 0.6 ml before treatment to 6.5 $\pm$ 0.8 ml after treatment ($\pm$ standard error of the mean (SEM)). Pubic hair development seemed to have advanced about 1 grade on Tanner´s scale (9) in 4 patients and penile length seemed to have increased between 0 and 2 cm in 5 patients, except patient E.O. showing an increase of 5 cm.

Discussion

Continuous PN GnRH therapy (2) as well as intermittent PN GnRH therapy can induce stimulation of Leydig cell function. This can be expected to occur in patients without Leydig cell deficiency showing good gonadotropin responsiveness throughout the treatment (2). Serum T increase as seen in the present study was in the range observed in our previous investigation of continuous PN GnRH therapy (2). In the present study, serum T decreased again in 3 out of 4 patients after termination of the treatment; in 1 patient (K.R.II), however, a further increase with complete normalization of the serum level occured after treatment. It is of special interest that 3 months prior to the intermittent treatment, this patient had undergone a continuous treatment after which serum T had returned to pre-treatment levels. This example shows that in individual patients, repeated courses

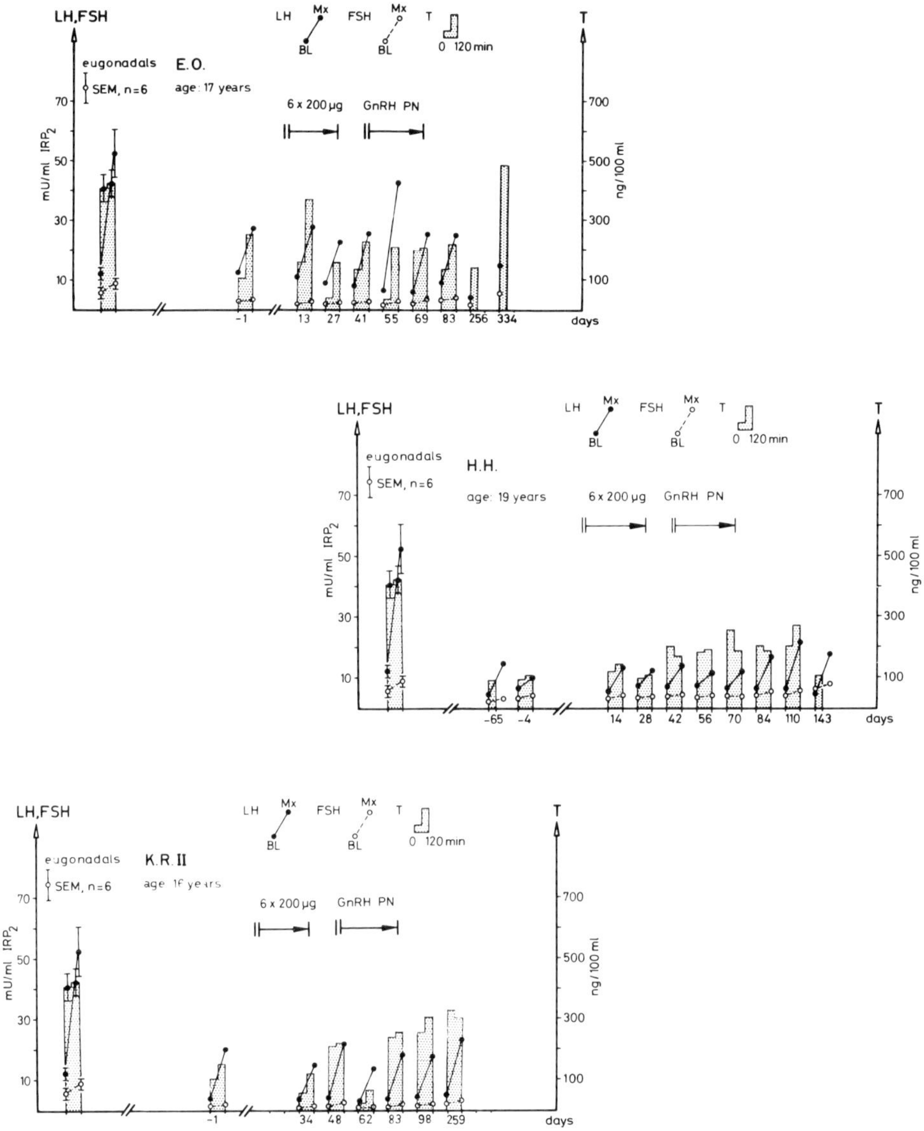

Figure 4. Serum gonadotropins (LH and FSH) and serum testosterone (T) in GnRH tests performed before intermittent pernasal (PN) GnRH therapy, during the treatment period and after therapy in male patients with delayed puberty; basal levels (BL) and maximum levels (Mx) of LH and FSH and concentration of T before (0 min.) and 120 minutes after (120 min.) the test doses of GnRH. For comparison GnRH tests in 6 eugonadal adult males.

of GnRH therapy are necessary to trigger off spontaneous pubertal development. Patient E.O. showed a decrease of serum T 6 1/2 months after termination of the treatment but had a new spontaneous increase 9 months after treatment. The patient was obviously not mature enough for the GnRH therapy to have triggered a further spontaneous pubertal development as it was observed in patient K.R.II immediately after his intermittent treatment.

In 2 patients, no increase of serum T was achieved by intermittent PN GnRH therapy despite good LH responsiveness throughout the treatment. This might be due to a primary Leydig cell deficiency accompanying hypogonadism of hypothalamic origin in these patients. Such a "double defect" seems to occur quite often (2) but still has to be proven in the 2 patients mentioned above, by HCG tests.

Uneffectiveness of releasing hormone therapy in male patients with idiopathic deficiency of gonadotropin secretion may also result from decreasing responsiveness of the gonadotropins during treatment. The mechanisms this phenomenon of tachyphylaxis might be based upon have been extensively discussed by us elsewhere (6): The phenomenon is obviously caused by a dose-dependent direct inhibitory influence of GnRH or potent agonist analogs of GnRH on gonadotropin secretion. The occurrence of the phenomenon seems to be dependent not only on the potency of the substance used, the dosage and the frequency of application (13), but also on individual particularities of the pituitary function. Therefore, it was surprising that during treatment with potent GnRH analogs, decreasing gonadotropin responsiveness was observed, except for rare cases (3, 6, 7, 15). It must be taken into account that D-Ser(TBU)^{6}-des-Gly^{10}-GnRH ethylamide, the analog which we used in most of our patients, was administered in relatively high doses; this is known from previous studies in which GnRH and GnRH analogs were compared using different modes of application (14). During PN GnRH therapy (2), we observed decreasing gonadotropin responsiveness only in patients with Kallmann´s syndrome who preferentially seemed to develop this phenomenon. In the present study, however, not only did the 2 anosmic patients develop unresponsiveness of the gonadotropins to GnRH, the same phenomenon also occured in a patient without anosmia. This either indicates that patients with Kallmann´s syndrome are not the only ones who easily develop tachyphylaxis even during low dose treatment with GnRH, or that the occurrence of the phenomenon in the patient without anosmia might result from a higher effectiveness of the new series of sprayers used in the present study. In the present study, gonadotropin increases after PN single doses of 200 µg GnRH were nearly as high as the ones after the IV doses of 25 µg GnRH used for the standard GnRH test. This indicates a PN absorption of 10% of the applied dose. In previous work (2, 13) we found a smaller effectiveness with PN application which might be due to technical differences between the release mechanisms of the sprayers used in previous studies and the ones used in the present study. The effectiveness of the nasal spray certainly depends on the reliability of the applicator mechanism, on the size of the spray particles and on their homogeneous distribution on the nasal mucosa. In view of an absorption rate of 10%, the daily dose should be reduced to 600 µg GnRH PN. Furthermore, since in the present study, in most of the patients, a repeated increase of the serum gonadotropins did not occur after the second dose of the PN double stimulation, the future dose regimen should consist of either 100 µg GnRH PN, 6 times daily, or 200 µg GnRH PN, 3 times daily. In numerous patients, one can possibly take advantage of the priming effect of GnRH as described in previous studies (14) only if the single doses are applied at intervals of 2 or more hours.

Decreasing responsiveness of gonadotropins, if observed during long term treatment with GnRH or potent GnRH analogs, begins to develop after 2 to 4 weeks of continuous treatment (2, 3, 7). A decrease of gonadotropin responsiveness was also observed with intermittent PN GnRH therapy which was investigated in the present study. The patients in whom the phenomenon was present showed decreasing responses of LH and FSH in the GnRH tests conducted every 2 weeks during the treatment. Regression of gonadotropin responsiveness was not noticed earlier than after 4 weeks of treatment. In spite of an interruption of the treatment for 2 weeks, a further decrease of gonadotropin responsiveness was observed. The treatment-free interval did not restore gonadotropin responsiveness, except in 1 patient (K.R.II). From these observations it can be concluded that treatment periods should not be longer than 1 or 2 weeks. Previous studies, the present study and still unpublished data of recent investigations done by our research group show that after termination of continuous or intermittent treatment with

GnRH or potent analogs, a decreased gonadotropin response in the GnRH test can persist up to 3 to 6 months or, in rare cases, even longer (2, 3).

Regarding the clinical parameters, minor changes were observed during intermittent PN GnRH therapy as with the previously published (2) continuous form of the treatment. Longer periods of treatment are certainly necessary to find a significant correlation between the hormonal changes and the clinical parameters.

References

1. Mortimer, C.H., McNeilly, A.S., Fisher, R.A., Murray, M.A.F., Besser, G.M.: Brit. Med. J. 4, 617-621 (1974).
2. Happ, J., Schmitz, V., Cordes, U., Krause, U., Atzpodien, W., Beyer, J.: Internat. J. Fertility, in press.
3. Happ, J., Schmitz, V., Cordes, U., Krause, U., Beyer, J.: Acta endocr. (Kbh), Suppl. 225, Abstract 134 (1979).
4. Sandow, J., Heptner, W.: Acta endocr. (Kbh), Sppl. 184, 28 (1974).
5. Sandow, J., von Rechenberg, W., Jerzabek, G., Stoll, W.: Fertil. Steril. 30, 205-209 (1978).
6. Happ, J., Schmitz, V., Cordes, U., Krause, U., Beyer, J.: Academic Press, in press.
7. Tharandt, L., Benker, G., Hackenberg, K., Reinwein, D.: V Internat. Congr. Endocrinology, Hamburg, July 18-24, Brühl´sche Universitätsdruckerei, Gießen, 1976, p. 346.
8. Bergmann, R., Bergmann, K., Kollmann, F., Maaser, R., Hövels, O.: Wachstum, Atlas, Publishers: Papillon, Wiesbaden, 1977. p. 26.
9. Tanner, J.M.: Growth and Adolescence, 2nd ed., Blackwell, Oxford 1962.
10. Tanner, J.M., Whitehouse, R.H.: Arch. Dis. Childh. 51, 170-179 (1976).
11. Prader, A.: Triangle 7, 240-243 (1966).
12. Happ, J., Weber, Th., Krause, U., Beyer, J.: Infertility 1, 87-99 (1978).
13. Wiegelmann, W., Solbach, H.G., Kley, H.K. Krüskemper, H.L.: Horm. Metab. Res. 9, 521-522 (1977).
14. Happ, J., Hartmann, U., Weber, Th., Cordes, U., Beyer, J.: Fertil. Steril. 30, 666-673 (1978).
15. Happ, J., Schmitz, V., Cordes, U., Krause, U., von der Ohe, M., Beyer, J.: Hoppe-Seyler´s Zeitschrift für Physiologische Chemie 360, 1154 (1979).

ANTI-INFLAMMATORY PROPERTIES OF DERIVATIVES AND SEQUENCE FRAGMENTS OF THE MCD-PEPTIDE FROM BEE VENOM

Peter Hartter, Wolfgang Martin
Physiologisch-chemisches Institut der Universität Tübingen,
Hoppe-Seyler-Str. 1, D-7400 Tübingen, Germany

The mast cell degranulating peptide (MCD) from bee venom exhibits anti-inflammatory activity in biological inflammation models (1). Its activity depends on special structural requirements. This is demonstrated by the use of derivatives and synthetic fragments of the natural molecule.

1 3 5 15
Ile-Lys-Cys-Asn-Cys-Lys-Arg-His-Val-Ile-Lys-Pro-His-Ile-Cys-Arg-Lys-Ile-

19 22
Cys-Gly-Lys-Asn-NH_2

Figure 1. Primary structure of MCD-peptide.

The primary structure of the MCD-peptide (MW. 2588 D) comprises 22 amino acids. Remarkable molecular characteristics are the high over-all basicity of the peptide chain, caused by five lysine and two arginine residues and the existence of two disulfide bridges.

The original and typical mode of action of the peptide, the degranulation of mast cells, releases vasoactive mediators like histamine and 5-hydroxytryptamine (2). This implies a process of essentially inflammation-generating quality. The reason for the anti-inflammatory action of the peptide is still unknown, however, its strong edema-suppressing effect is evident.

Structure and Activity of Natural Peptides

Biological Test on Inhibition of Inflammation

A Carrageenan-induced edema at the rat's hind paw serves as inflammation model (3).
Carrageenan, a sulfated polygalactan, produces an unspecific, locally limited and well reproducible irritation of paw tissue. After application an increased vessel permeability results, followed by a strong formation of exsudate. The maximum swelling appears 4 hours after injection.
A suitable method to quantitate this pathophysiological situation is the rat serum albumin test (RSA), which uses ^{125}I-labeled RSA as an indicator of inflammation (4).

The test is conducted as follows (4):

1) After intraperitoneal injection of possible anti-inflammatory substances the carrageenan is applied subplantar to the right hind paw of the animal. The contralateral extremity remains untreated, for comparison.
2) Subsequently each animal receives an intravenous injection of a constant dose of 10 μC ^{125}I-RSA, dissolved in 0.5 ml physiological saline (specific activity: 13-17 μC/mg RSA).

Four hours after injection of the RSA an accumulation of radioactivity occurs at the inflamed paw, correlating with the inhibitory potency of the applied antiphlogistics.
This extravascular protein, caused by inflammatory increased vessel permeability, is determined.
After killing the animal the hind paws are amputated at the tibio-tarsal articulation. Then the difference of counts (cpm) from Carrageenan-treated right and untreated left paw is calculated. The obtained value is related to the counts (cpm) of 1 ml blood (cardiac punction).

$$\frac{\text{cpm(r)} - \text{cpm(l)}}{\text{cpm/ml blood}} = \text{P (inflammatory increased permeability)}$$

This estimation accounts for the individual characteristics of each animal. Different blood volumes cause different isotope dilutions and therefore variable concentrations of the labeled protein in the inflamed tissue.

Results

To ascertain the inhibitory effects of different anti-inflammatory drugs, the P values calculated after drug application are compared with the average value established with a group of untreated Carrageenan-edema rats.

Applied Substance	Inhibition
Dexamethasone	86 %
Phenylbutazone	47 %
Indomethacin	59 %
MCD-peptide	87 %

Table 1. Inhibition of Carrageenan edema at a dose of 1 mg/kg body weight.

Table 1 shows that the MCD-peptide competes satisfactory with the therapeutically approved drugs Dexamethasone, Phenylbutazone, and Indomethacin. Its edema-suppressing effect of 87 % inhibition corresponds well with the activity of the corticosteroid derivative Dexamethasone.

Inhibitory Effects of MCD-peptide Derivatives

The following compounds are synthesized to decide whether the peptide's anti-inflammatory action is based upon a specific molecular structure or is likely to be altered by chemical modification of functional groups (5).

1) Acetylation of the N-terminal amino group of isoleucine and the ε-amino groups of the lysine residues by acetic anhy-

dride (6). Except the arginine residues, most of the other basic functions of the molecule are blocked by acetyl groups.
The resulting N-acetyl-MCD-peptide becomes far less basic than the natural molecule.

2) Reduction of the disulfide functions by dithiothreitol and subsequent reaction of the thiol groups with iodoacetamide leads to a S-alkyl product, the S-carboxamidomethyl MCD-peptide (7).

3) An oxidative cleavage of the disulfide bridges by forming S-sulfonate groups yields the MCD-peptide-tetra-S-sulfonate, MCD-$(SSO_3)_4$, a substance commonly called "Bunte salt" (8). This derivative possesses, compared to the natural molecule, a minor over-all basicity, caused by four negatively charged S-sulfonate groups.

Applied substance	Inhibition
MCD-peptide	87 %
N-Acetyl-MCD-peptide	2 %
S-Carboxamidomethyl-MCD-peptide	9 %
MCD-peptide-$(SSO_3)_4$	24 %

Table 2. Inhibitory effects of MCD-peptide and its derivatives in the ^{125}I-RSA test, at doses of 1 mg/kg. Number of experimental animals: 5 - 8.

Table 2 shows the failing anti-inflammatory potency of the MCD-peptide derivatives in the ^{125}I-RSA test. Hence it appears that blocking the amino groups and S-alkylation at the cysteine residues leads to a drastic reduction of the anti-inflammatory activity at doses of 1 mg/kg.
The "Bunte salt" with its diminished basicity exhibits no significant inhibitory effect, too.

```
1   2              5                                             15      17          19          21   22
Ile-Lys-Cys-Asn ¦ Cys-Lys-Arg ¦ His-Val-Ile-Lys ¦ Pro-His-Ile-Cys-Arg ¦ Lys-Ile-Cys-Gly-Lys ¦ Asn-NH2
       V              IV              III                  II                     I
```

Protected fragments:		Deprotected, dimeric disulfides:	
MCD-Pos.(17-21):	Boc-Lys(Z)-Ile-Cys(SiPr)-Gly-Lys(Z)	Lys-Ile-Cys-Gly-Lys Lys-Ile-Cys-Gly-Lys	I
MCD-Pos.(12-16):	Boc-Pro-His(Trt)-Ile-Cys(Trt)-Arg(Tos)	Pro-His-Ile-Cys-Arg Pro-His-Ile-Cys-Arg	II
MCD-Pos.(8-11):	Boc-His(Trt)-Val-Ile-Lys(Z)	His-Val-Ile-Lys	III
MCD-Pos.(5-7):	Boc-Cys(SiPr)-Lys(Z)-Arg(Tos)	Cys-Lys-Arg Cys-Lys-Arg	IV
MCD-Pos.(1-4):	Boc-Ile-Lys(Z)-Cys(Trt)-Asn(Mbh)	Ile-Lys-Cys-Asn Ile-Lys-Cys-Asn	V

Figure 2. Synthetic fragments of the MCD-peptide.

Inhibition of Inflammation by Synthetic MCD-peptide Fragments

The protected peptides of figure 2 are synthesized according to l.c. 9, 10.

The deprotected fragments I - V are prepared by removing the protecting groups via acid treatment. Acidolytic cleavage leads to the corresponding thiol peptides (I, II, IV and V). Considering the possible spontaneous disulfide formation in the ^{125}I-RSA test the peptides are converted into their dimeric disulfide form before studying their anti-inflammatory properties.
These peptides are applied to the Carrageenan inflammation model at doses of 10 mg/kg, a ten-fold higher concentration compared to application of natural MCD-peptide. This dose chosen for these small peptides seems to be metabolized faster than MCD-peptide in the rat organism.

Fragment	Inhibition	Dose	Significance P
$(\text{Lys-Ile-Cys-Gly-Lys})_2$	55 %	10 mg/kg	0.001
$(\text{Pro-His-Ile-Cys-Arg})_2$	24 %	10 mg/kg	not significant
His-Val-Ile-Lys	13 %	10 mg/kg	not significant
$(\text{Cys-Lys-Arg})_2$	34 %	10 mg/kg	$0.02 < P < 0.05$
$(\text{Ile-Lys-Cys-Asn})_2$	18 %	10 mg/kg	not significant

Table 3. Inflammation inhibition by MCD-peptide fragments in the ^{125}I-RSA-test. Number of experimental animals: 6.

Only the C-terminal situated fragment I exhibits significant activity in the RSA test. Intravenous application of fragment I even increases its inhibitory effect to 63 %.
A statistical evaluation of the results obtained by the ^{125}I-RSA test reveals confidence limits of $\pm$ 13 % in terms of in-

flammation inhibition. Therefore inhibitory effects below 30 % are regarded as not relevant.
Furthermore, the biological variability and the relative small number of experimental animals used leads to set 30 %-limit of relevance.
At this point it seems worth while to note that the monomer synthetic precursor of fragment I Lys-Ile-Cys(SSO_3)-Gly-Lys is inactive.

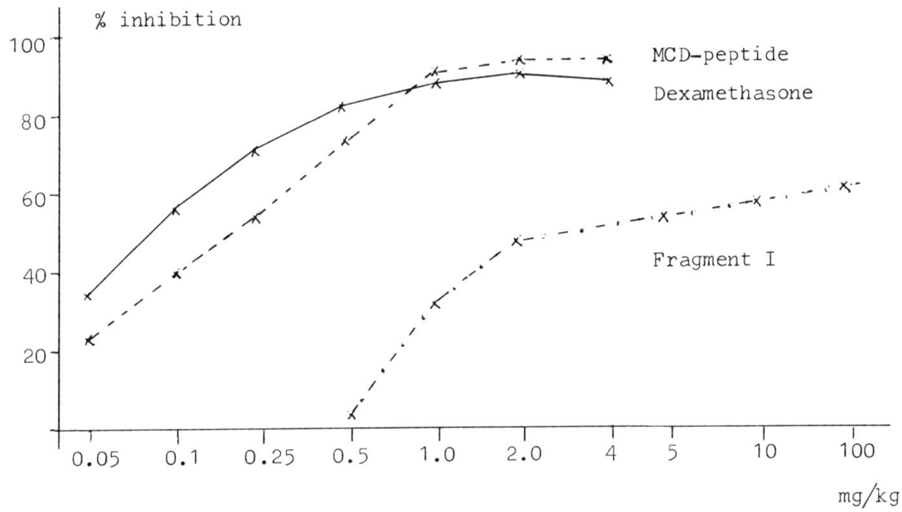

Figure 3. Dose/response relationships of Dexamethasone, MCD-peptide and fragment I.

Figure 3 compares the active fragment I with MCD-peptide and Dexamethasone by means of dose/response relationships.
The curves show that all 3 substances exhibit a saturation of their anti-inflammatory activity at the above dose level. For fragment I this is the case at about 10 mg/kg, for MCD-peptide and Dexamethasone at 2 mg/kg. Within the dose range from 0.05-0.5 mg/kg the curves of Dexamethasone and MCD-peptide have comparable slopes. Peptide I shows a corresponding slope between 1 and 5 mg/kg.

This fact allows an evaluation of relative activities: At an inhibitory effect of 50 % Dexamethasone is the most active substance related to the applied concentrations. The small peptide reaches this 50 % inhibition at a 20-fold higher dosage.

In summary the anti-inflammatory activity of MCD-peptide is determined by its high over-all basicity and its special structural characteristics such as the two disulfide bridges of the molecule. Perhaps the C-terminal region of the peptide which includes the amino acid composition Lys-Ile-Cys-Gly-Lys is essentially involved in the anti-inflammatory mode of action of this basic bee venom peptide.

References

1. Billingham, M.J.E., Morley, J., Hanson, J.M., Shipolini, R.A., Vernon, C.A.: Nature (New Biol.) 245, 163-164 (1973).
2. Breithaupt, H., Habermann, E.: Naunyn-Schmiedeberg's Arch. Pharmakol. Exp. Pathol. 261, 252-270 (1968).
3. Winter, C.A., Risley, E.A., Nuss, G.W.: Proc. Soc. Exp. Biol. Med. 111, 544-547 (1962).
4. Hanson, J.M., Morley, J., Soria-Herrera, C.: Brit. J. Pharmac. 50, 383-392 (1974).
5. Habermann, E.: Naunyn-Schmiedeberg's Arch. Pharmakol. Exp. Pathol. 300, 189-191 (1977).
6. Chauvet, J., Acher, R.: Biochem. Biophys. Res. Commun. 27, 230-235 (1967).
7. Doonan, Sh., Garman, A.J., Hanson, J.M., Loudon, A.G., Vernon, C.A.: Journal Chem. Soc., Perkin Trans. 1, 1157-1160 (1978).
8. Hartter, P., Weber, U.: Hoppe Seyler's Z. Physiol. Chem. 356, 693-699 (1975).
9. Hartter, P.: Hoppe Seyler's Z. Physiol. Chem. 357, 1683-1693 (1976).
10. Hartter, P.: Hoppe Seyler's Z. Physiol. Chem. 358, 331-337 (1977).

THE BEHAVIOR OF URINARY HYDROXYPROLINE CONCENTRATION IN HEALTHY SUBJECTS AND IN PATIENTS WITH PROSTATIC CARCINOMA WITH AND WITHOUT BONE METASTASES

Wolfgang Heller, Rolf Harzmann, Karl-Horst Bichler, Karlheinz Schmidt

Eberhard-Karls-Universität Tübingen, Chirurgische Klinik and Poliklinik and Lehrstuhl and Abteilung für Urologie

Introduction

At present, the so-called staging and classification technique is used to provide a guideline for a specific therapy for cancer of the prostate. These guidelines are based on differential diagnosis and more specifically on the basis of hematogenic and lymphogenic metastases or upon local operability. It is therefore of the utmost importance that the urologist is able to perceive a bone metastasis in order to avoid burdening the cancer patient with a falsely indicated operation. Nevertheless, recognizing local tumor spread or metastatic areas outside the skeletal system still presents problems to the clinician.

An important point in staging and classification is the diagnosis of bone metastases, which we - apart from the well-known routine procedures - undertook with the aid of urine hydroxyproline determination, sternal bone marrow and pelvic crest puncture (Table 1).

Methods and Materials

During a period of 18 months all bed patients with a histologically or cytologically ascertained prostate carcinoma were examined according to the

Structure and Activity of Natural Peptides

Table 1. Diagnosis of prostatic carcinoma. Our own procedure for demonstration of bone metastases, numbered in order of performance.

palpation → biopsy

radiological examination	laboratory data	hematology and histology
urogram	alkaline phosphatase	sternal marrow aspiration
skeleton X-ray	acid phosphatase	pelvic marrow aspiration
skeleton scintigram	prostate phosphatase	pelvic crest biopsy
lymphogram	electrophoresis	
vesiculogram	clotting factors	
	platelet agglutination test	
	urine hydroxyproline	

schema described in Table 1. The ages of the patients were 46 - 90 years with an average of 68 years. The results of lymphography, vesiculography, serum protein electrophoresis and clotting parameters are not considered in the framework of these examinations, since they are unimportant for the diagnosis of osseous metastases.

1. Radiological examination proceded routinely in the areas of the pelvic and lumbar spine and the pelvis.
2. Total body scintography was carried out following injection of 99 ml of TC-methylene diphosphate.
3. Examination of the phosphatases followed in each case in a safe interval after the rectal examinations, cystoscopies and prostatectomies to avoid any falsification of the data.
4. Determination of the hydroxyproline was done according to the method described in Table 2. The values, which were obtained under strict collagene-free diet refer to body surface. Since hydroxyproline is the

major product of collagen destruction, in other words, of the organic matrix of the bone and the connective tissue portions of the skin, analysis of this substance allows conclusions to be drawn about the metabolic activity of the skeletal system provided that dermatological lesions over a large body and the possibility of additional extraneous collagen are excluded (Table 3). In this way pathologically increased values can be obtained for inflammatory, traumatic, degenerative and malignant bone changes.

Interesting, and important in the diagnosis of bone metastases is the fact that steroid hormones and cytostatica with known effect in the skeletal system influence excretion of hydroxyproline. It is therefore necessary to consider endocrinopathies or therapeutic dosis of steroid hormones (e.g. in the treatment of prostatic carcinoma) in the evaluation.

Table 2. Analysis of hydroxyproline in the urine.

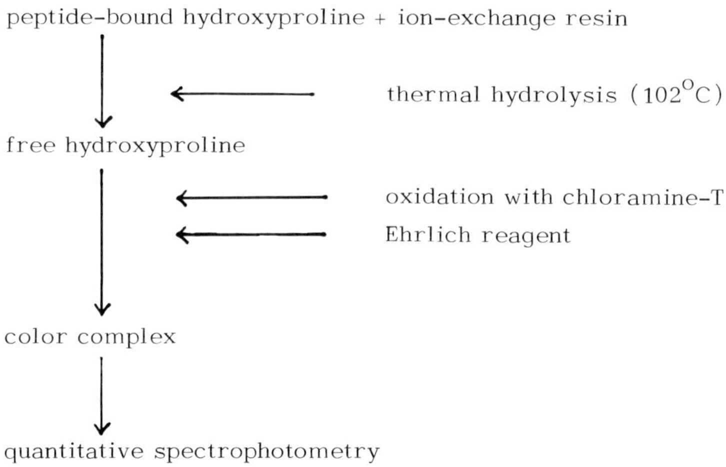

Table 3. Hydroxyproline metabolism.

periosteum	soluble collagen (1-2 %) ⇄ unsoluble collagen (98-99 %)	
	↓	
serum tissue	HYDROXYPROLINE free (25 %) bound (75 %)	
	↙	↘ metabolism
excretion	20 - 25 % urinary hydroxyproline free (5 %) bound (95 %)	75 - 80 % $H_2O + CO_2$ + urease

Since up till now little has been known about normal values for hydroxyproline excretion in the urine, we examined first a standard group of 145 persons of all age groups. The values of Fig. 1 show our results. Growing children of both sexes have excessively high values. Differentiation appears between 14 and 15 years, obviously the result of the beginning of the sex hormone effect in girls. This metabolism is generally more active in men and women until the year 35 of life than in older people. Comparison of the standard values for women before the menopause with those of women above 45 yields further proof of the estrogen dependency of hydroxyproline values.

Studies by Marshall et al. (1) showed that in patients with postmenopausal osteoporosis an increased urine hydroxyproline excretion can be ascertained. The use of the ethinyl estradiol permits the urine hydroxyproline excretion to be reduced to the premenopausal range. This means that the urine hydroxyproline excretion approaches the earlier premenopausal values. It appears that there is a relation between the degree of estrogen deficiency and the rate of bone loss in postmenopausal women and that it may be possible, by measuring urinary hydroxyproline in the fasting state, to identify fast bonelosers and select them for estrogen therapy. In established postmenopausal osteoporosis it is likely that estrogen therapy alone

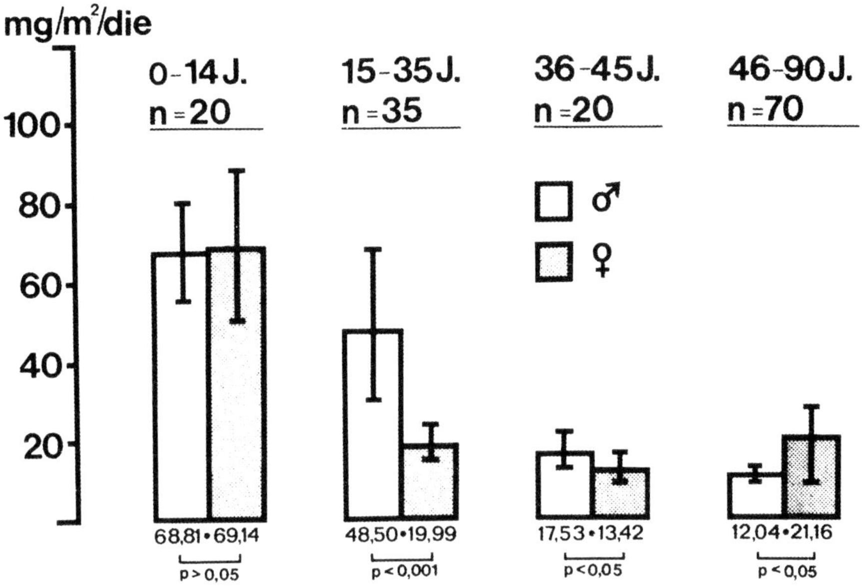

Fig. 1. Urine hydroxyproline in males and females of various age groups (n = 145).

is insufficient and that there is a place for calcium supplementation in cases with normal absorption of calcium (1).

Horsman et al. (2) found similar results. To compare the effects of no treatment, estrogen treatment and treatment with 72 calcium postmenopausal women were controlled for at least two years and examined by means of densitometry and morphometry. Women in the untreated control group continued to lose bone during the two years, whereas the estrogen-treated group lost none. Loss in the calcium-treated group was intermediate. Estrogen appeared to inhibit endosteal bone resorption and may have stimulated subperiosteal bone apposition.

Proceeding from these standard values we determined the hydroxyproline excretion in the urine of prostatic cancer patients, with and without bone metastases. In each of these groups the patients were differentiated as to whether or not they had been treated with estrogens. The values obtained were compaired with those from a group of men of the same age (n = 35), predominantly prostata adenoma patients. Assignment of a patient to the metastasis group proceeded when radiological findings and phosphatase values so indicated, or any of the 3 biopsy methods yielded definite tumor material.

5. Sternal marrow aspiration: 5 Smears were taken for each patient and stained using the Pappenheim method.
6. Pelvic marrow aspiration: This was done in each case in the area of the Spina iliaca posterior superior right in local anesthesia as the first step in the pelvic crest biopsy. Again, five smears per patient were done and stained according to Pappenheim.
7. Pelvic crest biopsy: The bone cylinder obtained is sufficient to certify metastases. Depending on the method of preparation "osteoblastic activity" can also be evaluated.

The order in which the various examinations are undertaken is of practical significance (Table 1) because a biopsy in the area of the sternum or the pelvic crest can simulate an isolated metastasis at the puncture site in a short time.

A similar effect is observed for the hydroxyproline determination: If the biopsies precede the hydroxyproline determination, this simple operation can lead to false pathological values for hydroxyproline. For this reason the biopsies should be performed at the end of the diagnostic program.

Results

Fig. 2 shows the results of hydroxyproline determinations in all patients with prostate carcinoma. In comparing these with those of men of similar age, but without metastases have significantly lower values, while the metastasis group lies significantly above the other two. Further differentation between patients treated respectively untreated with estrogens shows that

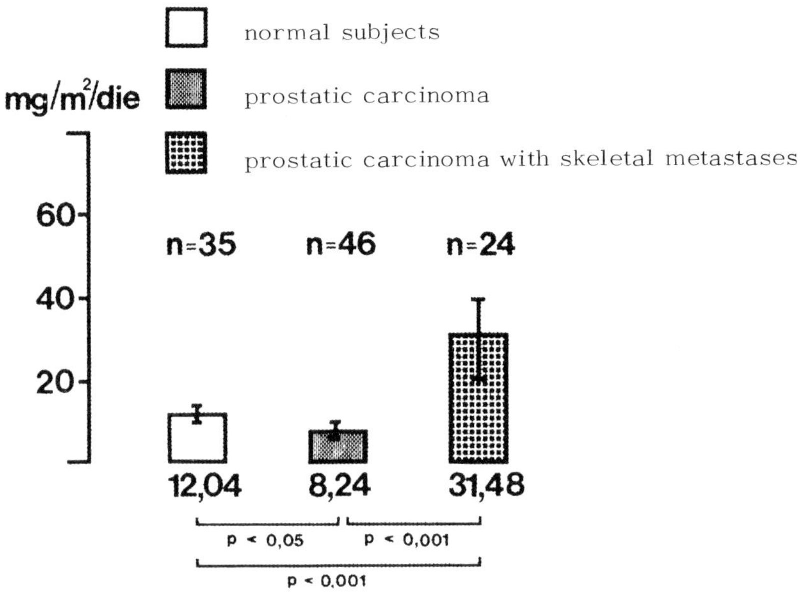

Fig. 2. Urine hydroxyproline in patients with prostate carcinoma (n = 70) and male normal subjects of the same age group (n = 35).

knowledge of estrogen therapy is of essential sificance in the evaluation of the data (Fig. 3). The groups show no difference between untreated metastases-free prostate cancer patients and men in similar age groups without prostate CA, but the group treated with estrogens shows significantly lower values.

Estrogen treatment leads to significant differences even within the metastasis group, although not to a pseudonormalizing of the values: Even estrogen-treated prostate cancer with bone metastases showed significantly higher urine hydroxyproline excretion than the metastasis-free group or the group of normal subjects.

Proceeding from these experiments we contrasted the results of this method of examination with other diagnostic procedures and have found far-reaching agreement with the data from the skeletal scintigram. Biopsy, skeleton X-

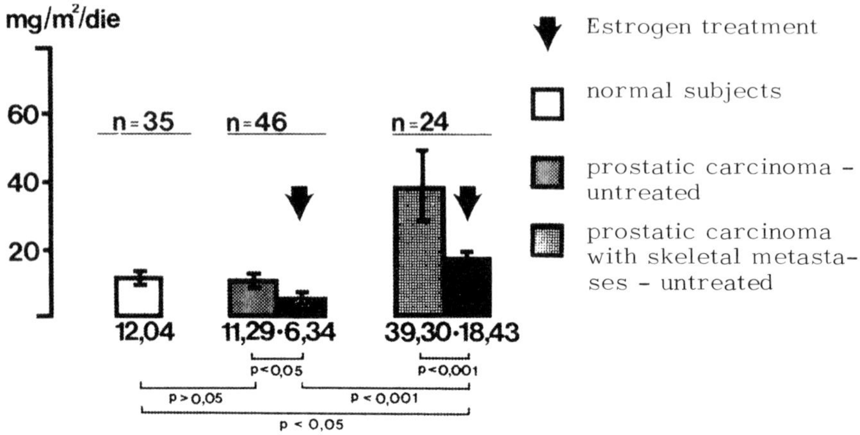

Fig. 3. Urine hydroxyproline in prostate cancer patients without (n = 46) and with (n = 24) skeletal metastases. Difference between patients who were treated or untreated with estrogens is shown.

ray or phosphatase diagnosis proved less effective than that from hydroxyproline concentration and scintigram.

Discussion

With respect to sensitivity and diagnostic success hydroxyproline determination and skeletal scintigram are equally valuable, however, with the limitation, that both provide equal opportunity for false interpretation. Traumatic, inflammatory and degenerative changes can simulate effects found for metastasis patients. This limitation is of clinical importance since degenerative changes of the skeletal system affect the same group as does prostate cancer.

Borderline findings from hydroxyproline determination and skeleton scintigram are therefore needed which can be received with the help of biopsy. Since it can correct a questionable finding on either side and thus contribute to the decision for or against a prostatovesiculectomie, this stepwise procedure seems justified. In our study the biopsy proved to be a valuable addition to the hydroxyproline values and to the scintigram.

Our studies show the lesser efficiency of radiological diagnosis of skeletal metastases. All radiologically-proven metastasis-free patients or patients with questionable findings are in need of the above mentioned closer explanation. In 4 patients tumor cells were found in either sternal or pelvic bone marrow in spite of unsuspicious radiological and scintigraphic findings. Determination of the phosphatases was about as successful as biopsy or radiological diagnosis. Recent studies have shown that measurement of the isoenzymes of alkaline phosphatase and of tartrate-inhibiting phosphatase with the help of radioimmunoassay have a considerable greater value in diagnosis.

These studies should demonstrate the methods bone metastases can be proven quickest and with the greatest accuracy, so that the decision for or against prostatovesiculectomy can be made with the greatest sureness.

Summary

An essential part of the classification and staging of prostate cancer is the diagnosis of bone metastases. This was carried out on 70 patients with the help of X-ray, scintography, determination of the acid and alkaline phosphatases, pelvic crest biopsy, pelvic bone marrow aspiration and sternal marrow aspiration. Additionally the concentration of hydroxyproline in the 24 hour urine was determined.

A previously undertaken study of a standard group (n = 145) showed an age- and sex-dependency for hydroxyproline values which had until then been hardly noticed. Premenopausal women had significantly higher values than men of the same age. Examination of prostate cancer patients (n = 70) showed that patients who had been treated with estrogens, with or without bone metastases, had significantly lower hydroxyproline values than did prostate cancer patients who had not received estrogen therapy. Patients with skeletal metastases (n = 24) had a significantly higher hydroxyproline excretion in the urine than those without (n = 46) or than healthy men of the same age group (n = 35).

Comparison of the results of hydroxyproline determination with other diagnostic procedures for the determination of skeletal metastases showed to be of equal value with the scintigram. Pelvic crest biopsy, pelvic bone marrow and sternal marrow aspiration proved to be valuable subsidiary methods.

References

1. Marshall,D.H., Horsman,A., Nordin,B.E.C.: Acta Obstet.Gynecol.Scand.Suppl. 65, 49 (1977).
2. Horsman,A., Gallagher,J.C., Simpson,M., Nordin,B.E.C.: British Med. J. 24, 789 (1977).
3. Borel,J.P., Caranjot,J.M., Jayle,M.F.: Clin.Chim.Acta 16, 409 (1967).
4. Burkhardt,H., Wepler,R., Rommel,K.: Dtsch. med. Wschr. 101, 1394 (1967).
5. Harzmann,R., Bichler,K.-H., Heller,W., Schmidt,K.: 28. Deutscher Urologenkongreß, Innsbruck 1976.

PHYSIOLOGICAL AND CLINICAL ASPECTS OF HYPOTHALAMIC-PITUITARY-OVARIAN MATURATION

Erich Keller, Reinhold Göser, Wolfgang Zubke, Adolf E. Schindler
Universitäts-Frauenklinik Tübingen, D-7400 Tübingen, Germany

Introduction

Only several years after the onset of menarche, a psychologically most meaningful symbol for female sexual maturation, the hypothalamic-pituitary-ovarian system achieves its definitive functional integrity which is necessary for the monthly release of an ovum. According to Frisch and Revelle (1,2) the onset of menarche is associated with a critical body weight of 48 kg. At this stage of the female sexual maturation process characteristic hormonal changes occur in the hypothalamic-pituitary-ovarian system (for review see:3). Disturbances of this maturation process may provoke amenorrhea. Thus, it was the aim of the present study to investigate the influence of body weight and of psychosomatic diseases (e.g. anorexia nervosa, anorectic reaction) on the functional state of the hypothalamic-pituitary-ovarian system in amenorrheic women.

Material and methods

A single dose of 25 µg luliberin (LH-RH) - a gift from Hoechst AG, Frankfurt - was administered intravenously to n = 490 women with functional amenorrhea (hypergonadotropic or hyperprolactinemic status excluded) and to n = 26 women with psychogenic amenorrhea (anorexia nervosa: n = 11; anorectic reaction: n = 15) by using a standardized test procedure (4). The patients with psychogenic amenorrhea received a second luliberin (LH-RH) test after a nine months' interval during which they were on psychotherapy. Prior to therapy these patients had an

Structure and Activity of Natural Peptides

average weight loss of 16.4 kg. Plasma lutropin (LH), follitropin (FSH), prolactin and estradiol were measured by specific radioimmunoassays as previously described (5,6). Luliberin (LH-RH) test results are expressed in terms of cumulative response (CUM) (7, 8), i.e. area under the curve of luliberin (LH-RH)-induced gonadotropin release over a period of 60 min. All values are mean values.

Results

Figure 1 shows the endogenous (basal) lutropin (LH) and follitropin (FSH) levels in correlation to the body weight in n = 490 women with functional amenorrhea: Lutropin (LH) levels

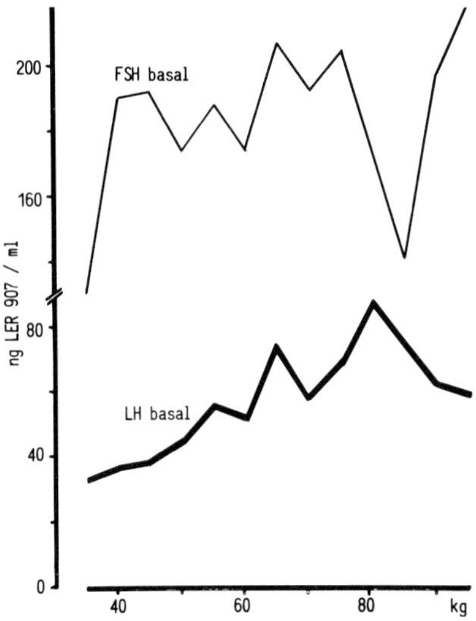

Fig.1. Mean endogenous (basal) lutropin (LH) and follitropin (FSH) levels in correlation to body weight in n = 490 women with functional amenorrhea.

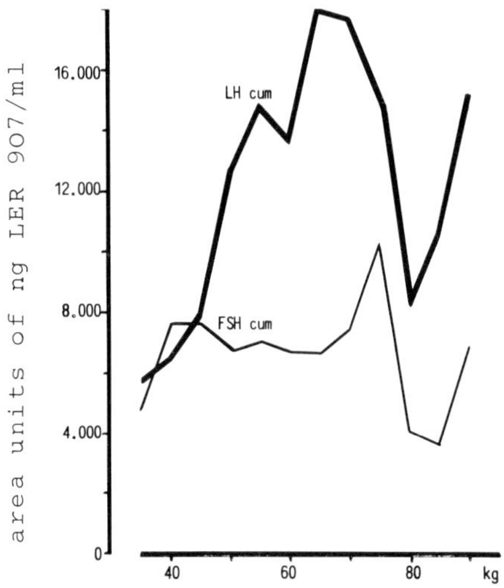

Fig.2. Mean luliberin (LH-RH) stimulated lutropin (LH) and follitropin (FSH) values (expressed in terms cumulative response - CUM) in correlation to body weight in n=490 women with functional amenorrhea.

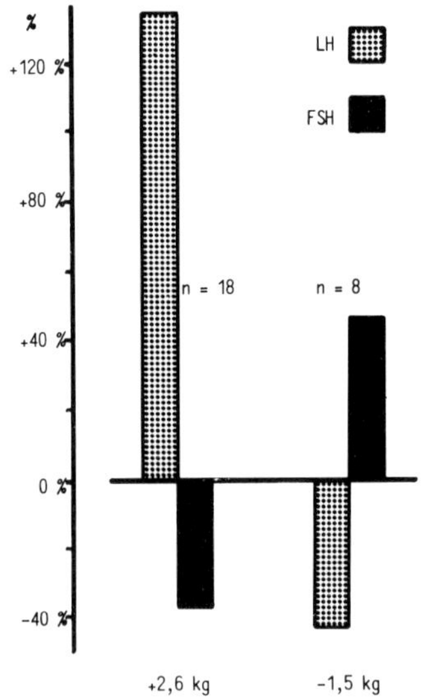

Fig.3. Percentage changes of pituitary gonadotropin responsiveness to luliberin (LH-RH) in n = 26 women with psychogenic amenorrhea (anorexia nervosa: n = 11; anorectic reaction: n = 15) after a nine months' interval of psychotherapy. Patients who responded to psychotherapy: n = 18 (left columns); patients who did not respond to psychotherapy: n = 8 (right columns).

raised with increasing body weight up to 80 kg - although the correlation coefficient is very poor (r = 0.076) while follitropin (FSH) levels did not reflect any characteristic change (r = - 0.057). The luliberin (LH-RH) - induced lutropin (LH) and follitropin (FSH) release values showed simular patterns in correlation to the body weight as the corresponding basal levels (Fig. 2). The correlation coefficients were for LH CUM r = 0.123 and for FSH CUM r = - 0.104.

In addition, in these patients no correlation to the body weight could be found for endogenous prolactin (r = 0.148) or estradiol (r = 0.053) levels.

Fig. 3 presents the percentage change of pituitary responsiveness to 25 µg luliberin (LH-RH) in n = 26 women with anorexia nervosa or anorectic reaction after a nine months' interval during which these patients were on psychotherapy. N = 18 women responded to psychotherapy with a weight gain of 2.6 kg while n = 8 did not respond. They even lost another 1.5 kg. In the group who responded to psychotherapy (n = 18) lutropin (LH) responsiveness increased by 135 % while follitropin (FSH) responsiveness decreased by 37 %. In contrast, the group who did not respond to psychotherapy (n = 8) reflected reverse response patterns: Lutropin (LH) responsiveness decreased by 43 % and follitropin (FSH) responsiveness increased by 47 %.

Discussion

Pituitary gonadotropin responsiveness to luliberin (LH-RH) varies widely but characteristically at different stages of the female sexual maturation process from childhood to adolescence: Before and after the onset of puberty reversed response patterns to luliberin (LH-RH) for follitropin (FSH) and lutropin (LH) are found. An initially exaggerated follitropin (FSH) responsiveness is followed by a continual decrease whereas lutropin (LH) responsiveness increases steadily during the hypothalamic-pituitary-ovarian maturation process (for review see: 9, 10, 11, 12). This maturation process is obviously reversible and was especially studied in women with anorexia nervosa or self-induced weight loss because of dietetic reasons. In the luliberin (LH-RH) test these patients often show prepubertal or pubertal response patterns (13, 14, 15, 16,17).

Fig. 1 and 2 demonstrate that patients with functional amenorrhea and low body weight also have low endogenous and impaired luliberin (LH-RH)-stimulated lutropin (LH) levels whereas follitropin (FSH), prolactin and estradiol remained unaffected: Thus, typical juvenile gonadotropin response patterns to luliberin (LH-RH) could not be found in those patients who had a low body weight which was comparable to that of patients with anorexia nervosa or anorectic reaction (35 - 45 kg).

In contrast, changes of gonadotropin responsiveness to luliberin (LH-RH) found in women with anorexia nervosa or anorectic reaction under successful psychotherapy correspond to those changes seen at various stages of the female sexual maturation process. From those patients who did not respond to psychotherapy it could be seen that deterioration of the illness caused a regression of luliberin (LH-RH)-induced gonadotropin responsiveness to response patterns seen in the early pubertal stages (Fig. 3). That means that in patients with psychogenic amenorrhea the maturity of the hypothalamic-pituitary-ovarian system might gradually disappear and then recur under successful therapy as during puberty (18).

The data presented indicate that the pituitary gonadotropin responsiveness to luliberin (LH-RH) in patients with psychogenic amenorrhea differs from the responsiveness seen in women with functional amenorrhea. This might be important for therapeutical approaches. Obviously, besides a critical body weight psychosexual factors are necessary for the female hypothalamic-pituitary-ovarian maturation process.

References

1. Frisch,R.E., Revelle,R.: Science 169, 397 - 399 (1971).
2. Frisch,R.E., Revelle,R.: Arch. Dis. Child 46, 695 - 701 (1971).
3. Keller,E., Friedrich,E., Schindler,A.E.: Endokrinologie des Ovars. In: Hormone im Kindesalter, ed. by Gupta,D., F.K. Schattauer Verlag, Stuttgart - New York 1980
4. Keller,E., Dahlen,H.G., Friedrich,E., Bohnet,H.G., Richter, R., Joel,E.W., Schubring,G., Klemt,W., Staemmler,H.J., Wyss,H.I., Schindler,A.E., Schneider.H.P.G.: J. Clin. Endocrinol. Metab. 40, 959 - 969 (1975).
5. Joel,E.W., Schönberg,D.K., Ilg,W., Keller,E.: Problems of Optimization of Double Antibody Radioimmunoassay of HLH, HFSH and HGH. In: Radioimmunoassay and Related Procedures in Medicine, ed. by IAEA, IAEA STI/PUB/350, Vienna 1974

6. Friedrich,E., Jaeger-Whitegiver,E.R., Bieder,M., Halverscheidt,H., Penke,B., Pallai,P., Keller,E., Schindler,A.E. Acta endocr. (Kbh.) Suppl. 184, 89 (1974).

7. Yen,S.S.C., Vandenberg,G., Rebar,R., Ehara,Y.: J. Clin. Endocrinol. Metab. 35, 931 - 934 (1972).

8. Franchimont,P., Gaspard,U., van Cauwenberge,J.R., Depas, E.: Some Aspects of Hypothalamus-Hypophyseal-Ovarian Feedback Mechanisms in Women. In: The Endocrine Function of the Human Ovary, ed. by James, V.H.T., Serio,M., Giusti, G., Academic Press, London - New York - San Francisco 1976

9. Schönberg,D.K.: LH-RH in Pediatrics. In: Hypothalamic Hormones - Structure, Synthesis, and Biological Activity, ed. : Gupta,D., Voelter,W., Verlag Chemie, Weinheim - New York 1975

10. Schneider,H.P.G., Keller,E., Hanker,J.P., Bohnet,H.G., Friedrich,E., Wyss,H.I., Schindler,A.E.: Clinical Use of LH-RH in Hypothalamic Amenorrhea. In: Progress of Reproductive Biology, vol. 2, ed. by Hubinont,P.O., L'Hermite, M., Robyn,C., Karger, Basel 1977

11. Job,J.C.: The Neuroendocrine System and Puberty. In: Clinical Neuroendocrinology, ed. by Martini,L., Besser,G.M., Academic Press, New York - San Francisco - London 1977

12. Styne,D.M., Grumbach,M.M.: Puberty in the Male and Female: Its Physiology and Disorders: In: Reproductive Endocrinology Physiology and Clinical Management, ed. by Yen, S.S. C., Jaffé, R.B., W.B. Saunders Company Philadilphia - London - Toronto 1978

13. Fries,H.: Secondary Amenorrhea, Self-induced Weight Reduction and Anorexia Nervosa. Acta Psychiatrica Scandinavia Suppl. 248, 1974.

14. Warren,M.P., Jewelewicz,R., Dyrenfurth, I., Ans, R., Khalaf,S., Vande Wiele, R.L.: J. Clin. Endocrinol. Metab. 40, 601 - 611 (1975).

15. Nillius,S.V., Wide,I.: The Pituitary Responsiveness to Acute and Chronic Administration of Gonadotropin-Releasing Hormone in Acute and Recovery Stages of Anorexia Nervosa. In: Anorexia Nervosa, ed. by Vigersky,R.A., Raven Press, New York 1977

16. Keller,E., Schier,I., Frick,V., Göser,R., Zubke,W., Friedrich,E., Joel,E.W., Schindler,A.E.: LH-RH Induced Gonadotropin Release in Women with Anorexia Nervosa and Anorectic Reaction Before and Under Psychotherapy. In: Human Fertilization, ed. by Ludwig,H. Tauber,P.F., Georg Thieme

Publishers, Stuttgart 1978

17. Yen,S.S.C.: Chronic Anovulation due to Inappropriate Feed-back System. In: Reproductive Endocrinology - Physiology, Pathophysiology and Clinical Management, ed. by Yen, S.S. C., Jaffé, R.B., W.B.Saunders Company Philadelphia - London - Toronto 1978

18. Bohnet,H.H., Dahlén,H.G., Keller,E., Friedrich,E., Schindler,A.E., Wyss,H.I., Schneider,H.P.G.: Clin. Endocrinol. 5, 25 - 36 (1976).

NATURAL PEPTIDE LACTONES AS ION CARRIERS IN MEMBRANES

Ilse Oberbäumer, Peter Feigl, Horst Ruf, Ernst Grell
Max-Planck-Institut für Biophysik
D-6000 Frankfurt 71, Germany

I. Introduction

The streptogramins are a well-known family of antibiotics that are produced by several streptomycetes (1, 2). According to their different structures these antibiotics have been classified into the two groups A and B. All members of group A are macrocyclic lactones while those of group B are peptide lactones containing six to seven amino acid residues in the ring. The structures of the three main components of group B are shown in Fig. 1.

It has been generally accepted that the streptogramins of both groups interact with the bacterial 50 S ribosomal subunit and block protein biosynthesis (1, 2, 3). The A components interfere with the ribosomal peptidyl transferase, whereas the effect of the B components appears to be rather indirect.

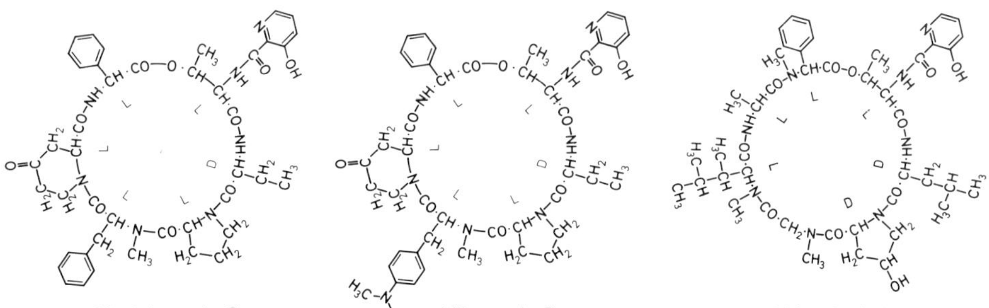

Virginiamycin S_1 Mikamycin B_1 Viridogrisein I

Fig. 1. Structures of streptogramin antibiotics of group B.

Furthermore, it is important to note that mixtures of antibiotics of both groups show a remarkable synergistic effect in vivo. Most of the results concerning the molecular structure and the antibacterial action of the streptogramins have been recently reviewed (3).

In addition to these main effects of the streptogramins, it has been reported that crude mixtures of these antibiotics inhibit respiration in rat liver and yeast mitochondria (4, 5). These results together with structural considerations of molecular models raised the question of whether the antibiotics are capable of passively modifying the cation permeabilities of membranes. The present article, dealing with the membrane activity and physico-chemical properties of the streptogramin antibiotics, will be restricted to their B components virginiamycin S_1, mikamycin B_1 and viridogrisein I (cf. Fig. 1). The A components have been discussed elsewhere (6).

II. Membrane Activity

In order to test this hypothesis, electrical measurements on black phospholipid bilayer membranes (≈ 60 Å thick) separating two aqueous compartments have been performed in the presence of the antibiotics of group B or relevant model compounds[a]. In fact, it could be shown that the antibiotic virginiamycin S_1

---

[a] The antibiotics have been isolated from the crude mixtures by specially developed chromatographic procedures (6). The preparation of virginiamycin S_1-acid, viridogrisein I-acid and of the derivatives of 3-hydroxypicolinic acid are also given in (6).
Crude mikamycin was obtained from Banyu Pharmaceutical Co., Ltd., Tokyo (Japan). Crude virginiamycin was kindly made available to us by R.I.T., Genval (Belgium) and the starting material for the purification of viridogrisein I was a gift of Parke, Davies & Co, Detroit (U.S.A) and Bristol Laboratories, Syracuse (U.S.A).

acts as a general cation carrier with the following specificities at pH 7.5 and 25°C: $H^+ >> Rb^+ > K^+ > Cs^+ > Na^+ \approx Ca^{2+}$ (7). However, virginiamycin S_1 apparently exhibits a special preference for protons.

In the case of the alkali ions, a cavity formed by the peptide lactone ring is suggested as the binding site. In principle, the peptide lactone can adopt such a conformation which is suitable for the coordination of alkali ions and which is similar to those of cyclic peptides and depsipeptides (e.g. valinomycin) exhibiting well-known alkali ion specificities.

All the results concerning the proton transport of virginiamycin S_1 can be related to the physico-chemical properties of the 3-hydroxypicolinyl residue of the antibiotic (6, 7). It is therefore very likely that this residue carries the proton across the membrane. This is consistent with the fact that the 3-hydroxypicolinyl moiety exhibits the properties of a dibasic acid (cf. Fig. 2a). The above interpretation is supported by preliminary results according to which not only the antibiotics mikamycin B_1 and viridogrisein I but also the simple model compound N-dodecyl-3-hydroxypicolinamide appear to act as proton carriers in phospholipid bilayer membranes.

For the transport of divalent cations (e.g. Ca^{2+}), it is assumed that the deprotonated 3-hydroxypicolinyl residue of the B components could act as binding site. However, contributions from coordinating groups located in the peptide-lactone ring might be expected.

In order to gain a more detailed knowledge of the processes that are relevant to the cation transport mediated by the group B streptogramins, it is important to investigate the protolysis and the complex formation with cations of these antibiotics in homogeneous solution.

III. Protolysis

The protolytic reactions of the 3-hydroxypicolinyl residue common to all the antibiotics of group B and the model compounds are illustrated in Fig. 2a, b. Only the macroscopic transition is shown for the dissociation of the amide proton in Fig. 2b. This system can be described by introducing the three apparent, macroscopic dissociation constants K'_I, K'_{II}

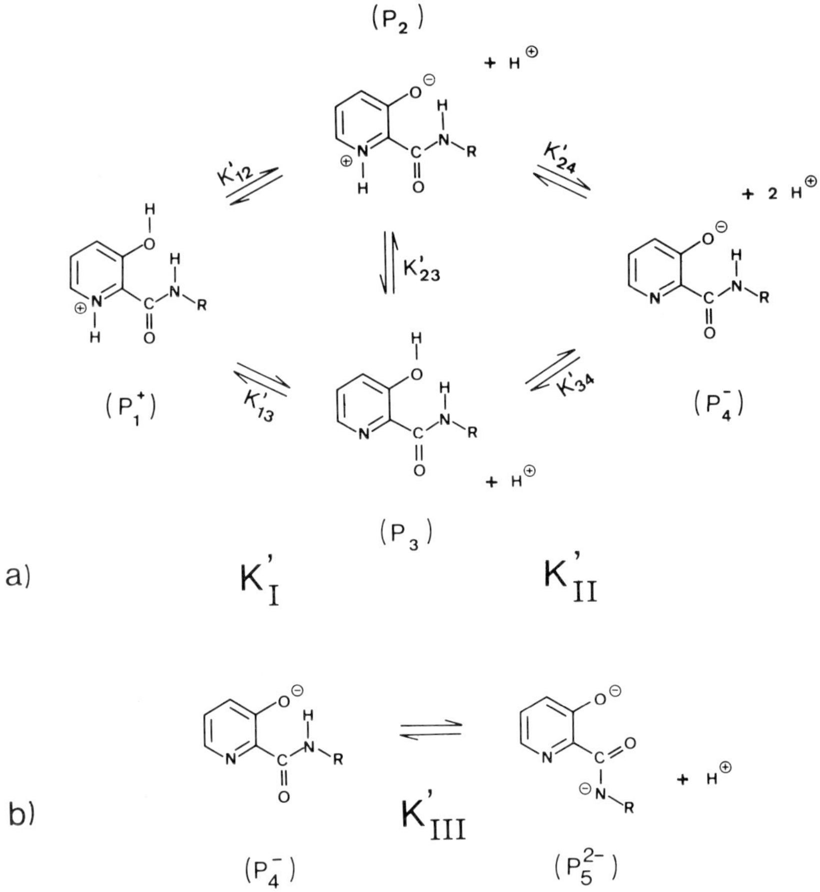

Fig. 2. Scheme of the protolytic reactions of group B antibiotics (R: peptide lactone residue) and of model compounds (R: H, CH_3, $CH_3(CH_2)_{11}$) indicating the macroscopic dissociation constants K'_I and K'_{II} (a) and K'_{III} (b).

and K'_{III}. Since only the 3-hydroxypicolinyl moiety absorbs in the wavelength region from 300 to 400 nm, it should be possible to determine the above dissociation constants by spectrophotometric titrations provided that each of the five protolytic states indicated in Fig. 2 exhibits a different absorption spectrum. A detailed study of the absorption spectra of the various states indicated that this condition is fulfilled (6). The wavelengths of the absorption maxima for the protolytic states follow the order $\lambda(P_2) > \lambda(P_4^-) > \lambda(P_5^{2-}) > \lambda(P_1^+) > \lambda(P_3)$. The spectrophotometric titrations have been carried out in aqueous solution and in 33.8 % water in methanol (v/v) at 25°C. The latter medium was chosen because the behaviour of virginiamycin S_1 in this medium closely resembles that of the membrane-bound antibiotic (7).

As the dissociation of the first proton (pK'_I) does not occur in the physiological pH range and the fully protonated state P_1^+ seems to contribute very little to the transport processes across the membrane, the pK'_I values have only been determined for some selected compounds. The same argument holds for the dissociation step of the third proton (pK'_{III}). Nevertheless, the protolytic state P_5^{2-} plays an important role with respect to heavy metal ion coordination (cf. section IV).

In Fig. 3b and c the spectrophotometric titration of mikamycin B_1 for the determination of the pK'_{II} value and the corresponding evaluation is shown. In the case of mikamycin B_1 an additional titration process is observed between pK'_I and pK'_{II} as indicated by the pronounced spectral change at 254 nm (cf. Fig. 3a, c). This additional titratable group which is not observed in the case of virginiamycin S_1 is due to the p-dimethylaminophenylalanine residue (cf. Fig. 1). The corresponding pK' value is denoted pK'_{IV}.

The results of similar titrations for the B components and of relevant model compounds are summarized in Tab. 1. It shows

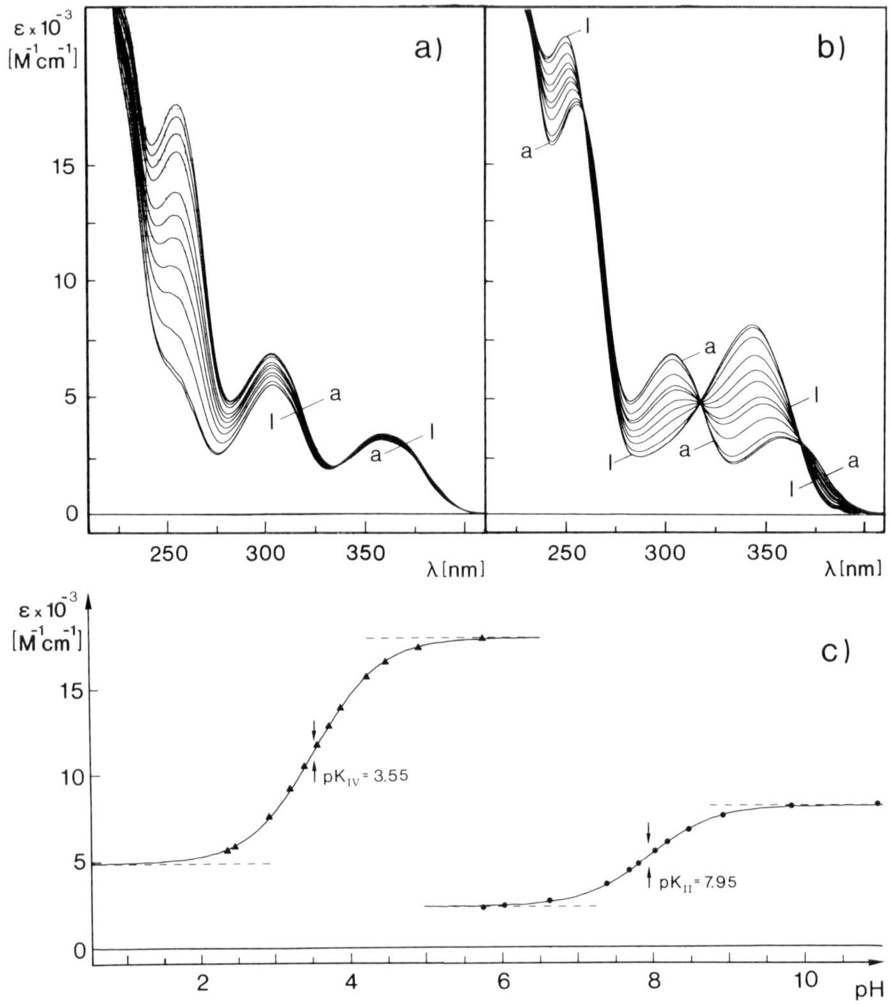

Fig. 3. Spectrophotometric pH-titration at 25°C of mikamycin B_1 ($2.31x10^{-5}$M) in 33.8% water in methanol (v/v) containing 0.1 M tetraethylammonium perchlorate,

a) with perchloric acid ranging from pH 5.76 (spectrum a) to pH 2.36 (spectrum l) and
b) with tetraethylammonium hydroxide ranging from pH 5.76 (spectrum a) to pH 11.0 (spectrum l).
c) Plot of extinction coefficient at 254 nm (▲) and 338 nm (●) vs. pH. The theoretical titration curves have been calculated for the indicated pK' values.

Tab. 1. Macroscopic pK'-values of B components of streptogramin antibiotics and of model compounds in water (medium A) and 33.8% water in methanol (v/v) (medium B) containing 0.1 M tetraethylammonium perchlorate at 25°C (experimental error ± 0.1).

Compound	pK'_{I}		pK'_{IV}		pK'_{II}		pK'_{III}
medium	A	B	A	B	A	B	A
virginiamycin S_1	1.25	0.90[a]			7.10	7.80	11.7
mikamycin B_1			4.40	3.55	7.25	7.95	
viridogrisein I	1.60				7.10	7.65	
virginiamycin S_1-acid	2.25				7.60		
viridogrisein I-acid	1.80				7.50		
3-hydroxypicolinamide	2.20				7.95	8.85	
N-methyl-3-hydroxy-picolinamide	3.85				8.00	8.90	
N-dodecyl-3-hydroxy-picolinamide						8.90	

[a] 40% water in methanol (v/v)

that the pK'_{II} values of all three main B components are nearly identical. The open chain compounds virginiamycin S_1-acid and viridogrisein I-acid which exhibit a hydrolyzed lactone bond and have no antibacterial activity in vivo are characterized by noticeably higher pK'_{II} values. Even higher pK'_{II} values than for the open chain compounds are found for 3-hydroxypicolinamide and its two substituted amide derivatives (cf. Tab.1). These differences clearly indicate that the nature of the threonine residue coupled to the 3-hydroxypicolinyl moiety as well as ring closure and ring conformation are responsible for the observed decrease in the pK'_{II} values compared with that of the simple model compounds. It is only because of this decrease in pK' value that the B components of the streptogramin antibiotics are capable of acting as efficient proton carriers in the physiological pH range. In the case of mikamycin B_1 it is furthermore assumed because of its low pK'_{IV} value, that the protolytic states of the antibiotic with the protonated p-dimethylamino group will barely have an effect on the proton transport properties.

The differences between the pK' values determined in pure aqueous solution and 33.8 % water in methanol (cf. Tab. 1) can mainly be attributed to the different K'_{23} values of the corresponding pH-independent equilibrium (cf. Fig. 2a) in the two media. The determination of the microscopic pK' values will not be discussed here (cf. 6).

IV. Complex Formation with Cations

In this section complex formation of the streptogramins of group B and of the model compounds 3-hydroxypicolinamide and N-methyl-3-hydroxypicolinamide with alkaline earth and heavy metal ions will be considered. Complex formation with these cations has been investigated by spectrophotometric titrations employing UV absorption and circular dichroism in the wave-

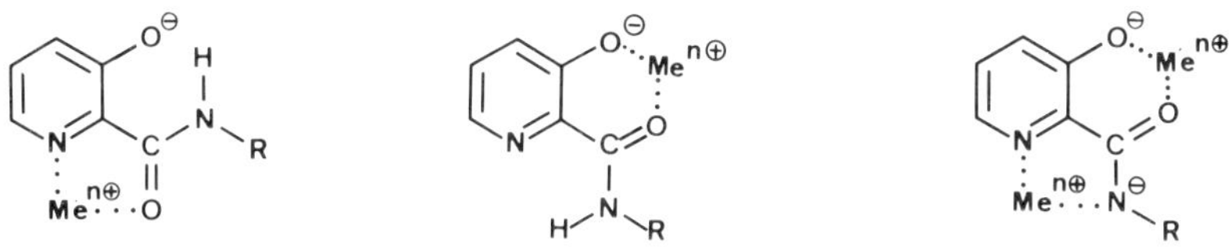

I II III

Fig. 4. Structures of cation complexes of the streptogramins of group B and model compounds respectively.

length region from 200 to 400 nm, as the spectral properties of the coordinated ligands differed from those of the uncomplexed ones. For the investigation of complex formation with alkali ions these methods appear to be less suitable. In the case of some heavy metal ions, the absorption of the complexes in the visible spectral range due to the metal ion could also be studied. The general aim of this approach is to elucidate the structures of the complexes in terms of protolytic states of the coordinated ligands and to determine the stoichiometry and the dissociation constants of the various complexes that are formed in solution. By comparison of the spectrophotometric titrations of the antibiotics with the corresponding ones of the model compounds it is also hoped to obtain information concerning the participation of groups from the peptide lactone ring of the antibiotics in the coordination of cations.

In Fig. 4 two possible 1:1 complexes of the 3-hydroxypicolinyl residue are shown, where coordination occurs either to the pyridine nitrogen (structure I) or to the deprotonated phenolic oxygen (structure II) with participation of the amide carbonyl group. According to these models similar complexes with 3:1 and 2:1 stoichiometry are also expected. With respect to complex formation of the antibiotics and the model compounds with the alkaline earth ions, only the 1:1 complexes are considered here. If the deprotonated protolytic state P_4^- is involved in coordination of the alkaline earth ions as depicted in Fig. 4 (structures I and II), a dependence of the stability

Tab. 2. pK' values (negative logarithm of dissociation constant of complex) for 1:1 complexes of virginiamycin S_1 and 3-hydroxypicolinamide with alkaline earth ions in different media at 25°C (experimental error ± 0.15).

Compound	Medium[a)]	Cation	pH	pK'
virginiamycin S_1	ethanol	Mg^{2+}	b)	3.80
	methanol	Ca^{2+}	b)	2.50
	40% water in methanol	Ca^{2+}	5.75	0.2
		Ca^{2+}	7.30	1.85
3-hydroxy-picolinamide	"	Ca^{2+}	7.70	1.50

a) the media contained 0.1 M tetraethylammonium perchlorate and 0.005 M N-ethylmorpholine or MES resp.

b) the media contained 0.0025 M perchloric acid

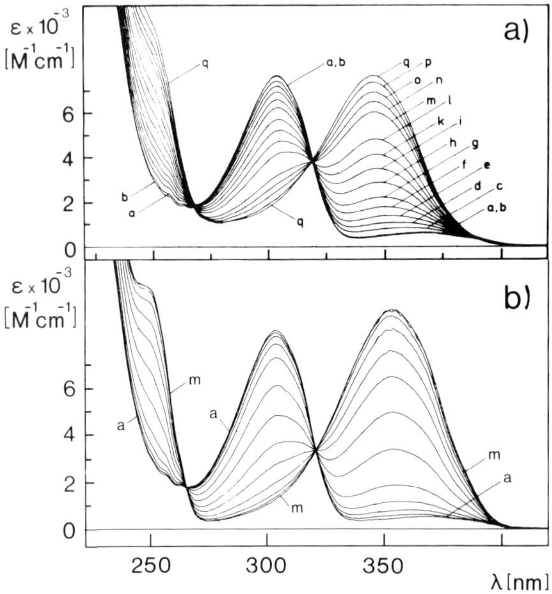

Fig. 5.
a) Spectrophotometric pH titration at 25°C of virginiamycin S_1 (2.31×10^{-5} M) in methanol containing 0.05 M tetramethylammonium chloride with tetramethylammonium hydroxide (in propanol) ranging from 6.67×10^{-5} M HCl (spectrum a) to 8.81×10^{-4} M OH^- (spectrum q).
b) Spectrophotometric titration at 25°C of virginiamycin S_1 (4.93×10^{-5} M) in ethanol containing 0.1 M tetraethylammonium perchlorate, 0.005 M N-ethylmorpholine and 0.0025 M perchloric acid (spectrum a) with $Mg(ClO_4)_2$ up to 9.07×10^{-3} M (spectrum m).

of the complex on pH must be expected. This is clearly concluded from the pK' values for the virginiamycin S_1-Ca^{2+} complex at different pH values as given in Tab. 2. The corresponding complex of the model compound 3-hydroxypicolinamide is even slightly weaker than that of the antibiotic but this is expected to be due to the higher pK'_{II} value of the model compound (cf. Tab. 1).

The structure of the alkaline earth complexes can be deduced in an illustrative way from Fig. 5. Fig. 5a shows the spectra corresponding to a titration of virginiamycin S_1 in methanol with tetraethylammonium hydroxide acting as base. A virtually identical spectroscopic behaviour is found for the titration of virginiamycin S_1 in ethanol with $Mg(ClO_4)_2$ as demonstrated in Fig. 5b. From this similarity it is concluded that the protolytic state P_4^- of the antibiotics is coordinated to Mg^{2+} and Ca^{2+} as indicated in Fig. 4 (structure II). The same conclusion

Tab. 3. pK' values for 1:1 and 1:2 complexes of virginiamycin S_1 and model compounds with heavy metal ions in different media containing 0.1 M tetraethylammonium perchlorate, 0.005 M N-ethylmorpholine- or MES-perchloric acid at 25°C (experimental error ± 0.15).

Compound	Medium	Cation	pH	pK'	Stoichiom.
3-hydroxy-picolinamide	water	Mn^{2+}	5.2	2.0	1:1
			5.8	2.8	
		Ni^{2+}	5.4	4.4	1:1
		Cu^{2+}	4.8	5.1	1:1
				2.85	1:2
			5.2	5.75	1:1
				3.45	1:2
N-methyl-3-hydroxypicolinamide	water	Cu^{2+}	5.2	5.55	1:1
virginiamycin S_1	40% water in methanol (v/v)	Mn^{2+}	5.6	2.15	1:1
		Ni^{2+}	5.45	5.05	

is valid for the model compounds indicating that no groups of the peptide lactone ring are likely to coordinate to the alkaline earth ions.

A considerably different behaviour is found for the complex formation between the ligands discussed above and heavy metal ions such as Mn^{2+}, Ni^{2+}, Cu^{2+}. In the case of some of the

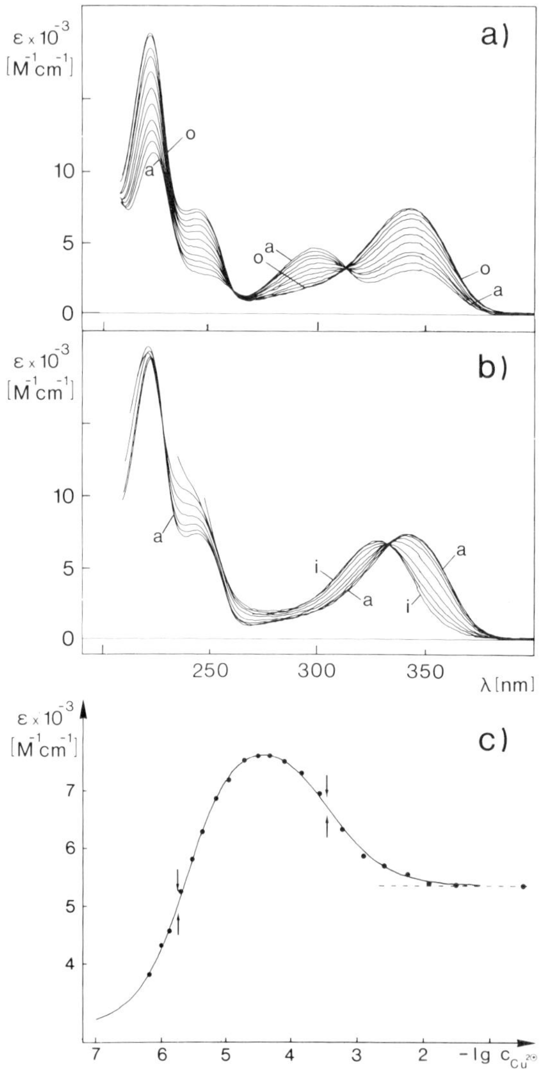

Fig. 6.
Spectrophotometric titration at 25°C of 3-hydroxypicolinamide (1.29 x10^{-6}M) in water containing 0.1 M tetraethylammonium perchlorate, 0.005 M MES-$HClO_4$ pH 5.2,

a) in the absence of $Cu(ClO_4)_2$ (spectrum a) and with $Cu(ClO_4)_2$ up to 3.15x10^{-5}M (spectrum o),

b) with $Cu(ClO_4)_2$ ranging from 3.15x10^{-5}M (spectrum a) to 5.72x10^{-3}M (spectrum i).

c) Plot of extinction coefficient at 340 nm vs. negative logarithm of Cu^{2+}-concentration. The theoretical titration curve has been fitted to the experimental data points.

heavy metal ions pK' values even higher than 5 are found for the 1:1 complexes (cf. Tab. 3). It is concluded that these cations exhibit much higher affinities to the model compounds and the B components than the alkaline earth ions. As a typical example complex formation between 3-hydroxypicolinamide and Cu^{2+} can be illustrated on the basis of a detailed spectrophotometric titration as depicted in Fig. 6. The transfer of the uncomplexed ligand to the 1:1 Cu^{2+} complex (spectrum o) is shown in Fig. 6a. The higher order complexes that are also formed during this titration step are neglected here. A further increase of the Cu^{2+} concentration leads to a second titration step (cf. Fig. 6b), indicating that the ligand is capable of coordinating a second Cu^{2+} ion, thus leading to the formation of a 1:2 complex. The pK' value for the coordination of this second cation is about two orders of magnitude lower compared with that of the 1:1 complex (cf. Tab. 3 and Fig. 6c). The pK' values for both complexes have been evaluated by a least square computer fit to the data points shown in Fig. 6c. The absorption spectrum of the 1:1 3-hydroxypicolinamide-Cu^{2+} complex is similar to that of the ligand in the protolytic state P_2 (cf. Fig. 2), whereas the spectrum of the corresponding 1:2 complex closely resembles that of the protolytic state P_5^{2-}. From this comparison it has to be concluded that the 1:1 complex corresponds to structure I in Fig. 4, where the metal ion is essentially coordinated to the pyridine nitrogen of the ligand. With regard to the 1:2 complex, structure III (Fig. 4) is suggested where the more strongly bound Cu^{2+} is coordinated to a binding site formed by the pyridine nitrogen and the deprotonated amide group of the ligand, while the second Cu^{2+} ion is coordinated to a binding site similar to that for the alkaline earth ions. For the model compound N-methyl-3-hydroxypicolinamide, the formation of a 1:2 Cu^{2+} complex is not observed under equivalent experimental conditions.

An even more complicated behaviour is found for the complex formation of Cu^{2+} with virginiamycin S_1, mikamycin B_1 and

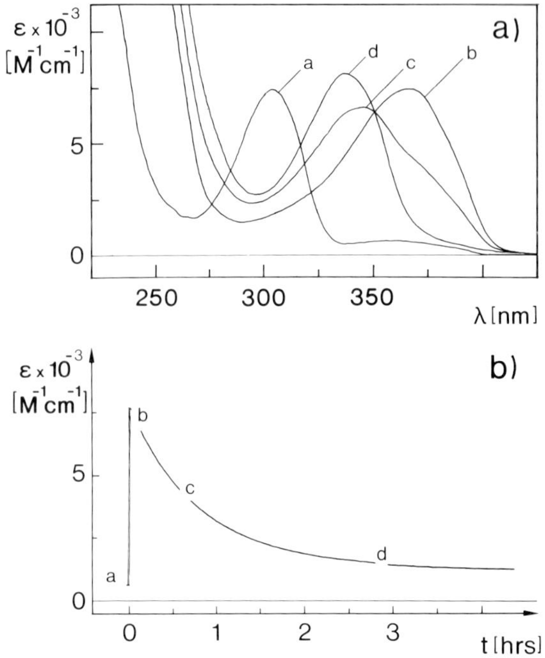

Fig. 7.
a) Absorption spectra at 25°C of virginiamycin S_1 (4.93×10^{-5}M) in methanol (spectrum a) and after addition of 1.48×10^{-4}M $Cu(ClO_4)_2$: 5 min (spectrum b), 38 min (spectrum c), 172 min (spectrum d).
b) Plot of extinction coefficient at 370 nm vs. time with the conditions given in a). The time for the spectra a to d is indicated.

viridogrisein I. Fig. 7a and 7b illustrate the change in ligand absorption for virginiamycin S_1 upon addition of three equivalents of Cu^{2+} as a function of time. Spectrum b which is obtained shortly after the addition of Cu^{2+} is typical for the formation of complex A in Fig. 8 (cf. structure I, Fig. 4). In contrast to the model compound 3-hydroxypicolinamide with the antibiotics the transition of the 1:1 complex to the 1:2 complex (structure B in Fig. 8) occurs in the time range of hours (cf. Fig. 7b) and is thus considerably slower. This reflects an alteration of the properties of the ligand due to the peptide lactone ring. The pK' values for these 1:1 and 1:2 Cu^{2+} complexes of the antibiotics can only be estimated. For virginiamycin S_1 in aqueous solution at pH 5.4 the lower limits for both pK' values must be at least 6 which implies that the 1:2 complex B (Fig. 8) is about three orders of magnitude more stable than the corresponding complex of the model compound. Again in contrast to the model compound 3-hydroxypicolinamide

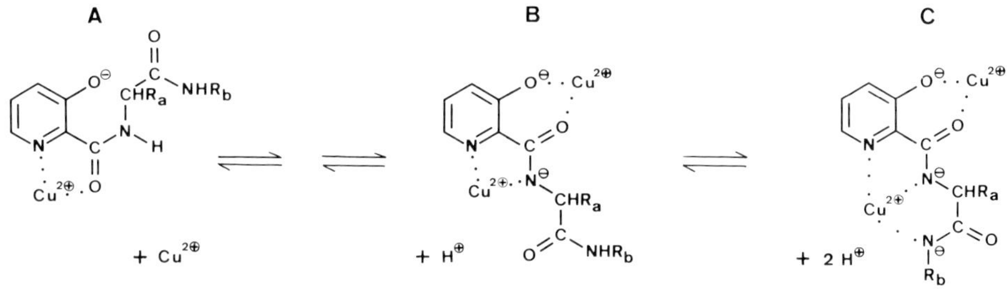

Fig. 8. Complex formation of streptogramins of group B with Cu^{2+} (R_a, R_b: part of the peptide lactone ring residue).

the formation of a second 1:2 virginiamycin S_1-Cu^{2+} complex is found. However, under the experimental conditions of the measurements illustrated in Fig. 7, the transition from the first to the second 1:2 complex would take several months. But if the experiment in Fig. 7 is modified by adding at least two equivalents of a base prior to the addition of Cu^{2+}, the formation of the second complex is nearly complete within 30 minutes. This indicates that the formation of this complex is also very sensitive to pH. The corresponding absorption spectrum of the complexed ligand differs only slightly from spectrum d in Fig. 7a. On the other hand, considerable changes occur in the circular dichroism of the ligand and the absorption spectra in the visible spectral range (cf. 5). The intense absorption band at 575 nm which is due to the coordinated Cu^{2+} is consistent with the coordination of an additional deprotonated amide group located in the peptide lactone ring according to structure C in Fig. 8. The scheme in Fig. 8 summarizes the complex formation between the antibiotics of group B and Cu^{2+} ions.

Although not all of the properties of the B components and model compounds with respect to their coordination of cations are yet understood, these studies clearly help to elucidate the possible binding sites of these antibiotics that are relevant for the transport of cations across membranes. According to the presented results, it seems likely that the biological action of the streptogramin antibiotics of group B might not be restricted to the inhibition of protein biosynthesis.

Acknowledgememts

The authors are very grateful to Prof. D. Margerum and Dr. H. Diebler for helpful and stimulating discussions.

References

1. Tanaka, N.: in "Antibiotics III", Corcoran, J.W., Hahn, F.E., eds., Springer, Berlin-Heidelberg-New York (1975), pp. 487-497.
2. Vazquez, D.: in "Antibiotics III", Corcoran, J.W., Hahn, F.E., eds., Springer, Berlin-Heidelberg-New York (1975), pp. 521-534.
3. Cocito, C.: Microbiol. Rev. 43, 145-198 (1979).
4. Dixon, H., Kellerman, G.M., Mitchell, C.H., Towers, N.H, Linnane, A.W.: Biochem. Biophys. Res. Commun. 43, 780-786 (1971).
5. Groot Obbink, D.J., Spithill, T.W., Maxwell, R.J., Linnane, A.W.: Molec. gen. Genet. 151, 127-136 (1977).
6. Oberbäumer, I.: Diplomarbeit, Göttingen (1975), Dissertation, Göttingen (1979).
7. Grell, E., Oberbäumer, I., Ruf, H., Zingsheim, H.P.: in "Biochemistry of Membrane Transport", Sememza, G., Carafoli, E., eds., Springer, Berlin-Heidelberg-New York (1977), pp. 147-178.

DIFFERENT EFFECTS OF THYMUS EXTRACTS ON HUMAN FIBROBLASTS AND CANCER CELLS IN CULTURE

Volker Paffenholz
Research Laboratories of the Vitorgan Pharmaceutical Company, D-7302 Ostfildern 1, FRG

Summary

We studied the influence of soluble extracts of fetal and juvenile calf thymus on the proliferative capacity of human cells in culture. Low doses of thymus extracts increase nucleoside incorporation into DNA and RNA of fibroblasts and activate cell growth comparable to the effect of potent stimulators like fetal bovine serum or somatotropin . In contrast to these results thymus extracts cause a varying inhibition of DNA synthesis in cancer cells which is intensified by the addition of hydrocortisone. The opposite reaction in the fibroblast and cancer cell system may reflect a selective influence of the thymus extracts on the various cell types.

Introduction

The involvment of thymus in maturation of precursor T lymphocytes and development of cell-mediated immunity raised renewed interest when experiments discovered that immunocompetence of neonatally thymectomized animals was restored by grafting thymus tissue within a cell-impermeable chamber that allowed the passage of macromolecules (1). More direct studies showed that immunologically deficient mice recovered the capacity to

Structure and Activity of Natural Peptides

induce a graft-versus-host reaction after injection of crude and partially purified cell-free thymus extracts (2). Furthermore, metabolic deficiencies as a result of thymectomy were reversed by the administration of thymus extracts of different animal sources (3). On the basis of these findings several thymus-dependent factors were detected in the serum of animals and humans active in the regulation of various physiological processes. The essential role of these hormone-like substances encouraged the medical use of thymus preparations which now gain increasing importance in many fields of cure and therapy. Applications of thymus substances or purified hormones like thymosin affect growth and metabolism and were confirmed as successful agents for treatment of various diseases like immune infirmity, cancer and senility (4,5). In our experimental approach we intended to evaluate the molecular effects of thymus extracts on human fibroblasts and cancer cells in culture.

Results

The experimental conditions applied were close to the procedures reported by Gospodarowicz and Moran (6). Monolayer cultures of human fibroblasts and cancer cells were maintained in a resting phase during a 72 hours incubation in serum-deprived minimal medium containing 0.25% FBS. The bioassay utilizes the fact that the cells are still fully responsive to growth factors or mitogens and may initiate DNA synthesis following stimulation by appropriate substances; on the other hand, the response may be as well diminished by mitotic inhibitors or toxic components. Cell-free thymus extracts were prepared of either native tissue or dilutions of lyophilized organ powder after centrifugation and sterile filtration. Changes of metabolic events in the cell cultures were calculated according to the rate of ^{3}H-thymidine incorporation after addition of the tissue extracts and compared with known proliferative stimulators like fetal bovine serum (FBS) and somatotropine (STH).

On the left side of Fig. 1 we see a striking increase of thymidine incorporation in MRC-5 cells (derived from human embryonic lung) after addition of native fetal thymus extracts (F.Th.). Although the final protein concentrations given in g/ml are very low the extract has a significantly higher stimulating activity than the comparatively submitted 1% of fetal bovine serum (final concentration 200 x 10^{-6} g/ml) or the 10^{-2} international units of somatotropin (a crude preparation with 50 x 10^{-6} g/ml). The initiation of DNA synthesis in these cells is not caused by unspecific reactions but exclusively depends on the addition of essential proteins or growth factors since solutions of bovine serum albumin (BSA) even in higher concentrations have no effect.

The right side of Fig. 1 demonstrates the influence on DNA synthesis of fibroblast cultures established from human skin biopsy (H.S.) which behave similar to the embryonic MRC-5 cells. We still find good results after incubation with lyophilized and redissolved organ powder of fetal thymus in the concentration of 10^{-6} g/ml but the influence of higher dilutions is less significant. Partial hydrolysis or denaturation of the organ substances reduces the cellular response which emphasizes the requirement of vital proteins for stimulation of DNA synthesis and metabolism in cell cultures of diploid state.

In contrast to these results are the effects of thymus extracts on cancer cells. In the case of Wish, a cell line derived from human amnion tissue, ^{3}H-thymidine incorporation after incubation with lyophilized and redissolved extracts of fetal (F.Th.) and of juvenile thymus (K.Th.) is significantly reduced (Fig. 2). Inhibition of Wish cell DNA synthesis is not expressed as a concentration-dependent reaction which may result of the nature of the organ solutions: Higher amounts lead to an increase of

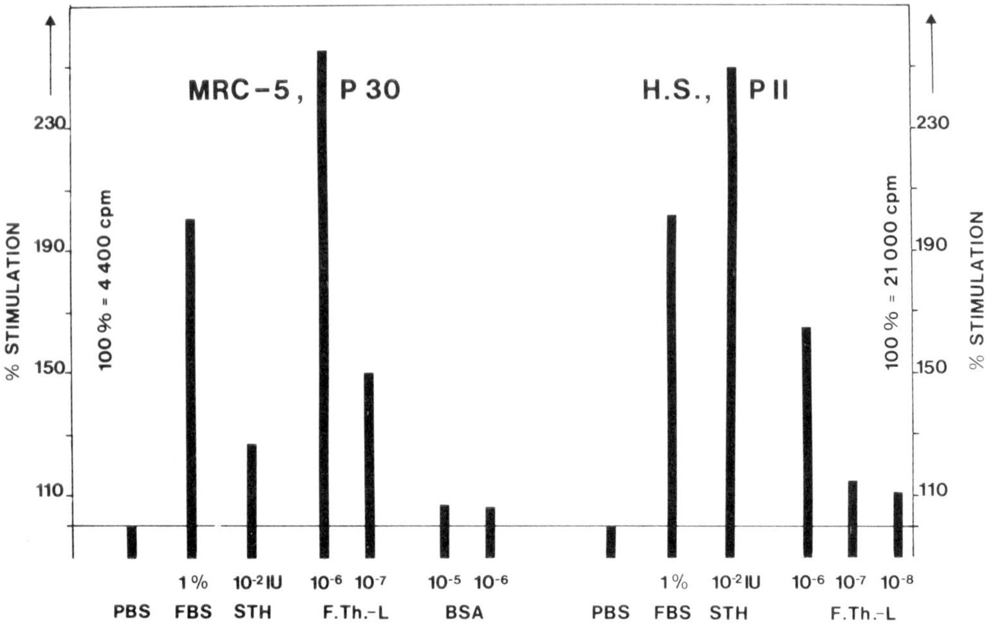

Fig. 1.
Stimulation of DNA synthesis in MRC-5 cells, passage 30, after incubation with phosphate-buffered saline (PBS = control), fetal bovine serum (FBS), somatotropin (STH), native fetal calf thymus extract (F.Th.) and bovine serum albumin (BSA). Stimulation of DNA synthesis in skin fibroblast cultures (H.S.), passage 11, after incubation with lyophilized and redissolved extract of fetal calf thymus (F.Th.). Final protein concentration given in g/ml.

nucleoside incorporation and to a stimulation of the cancer cells possibly due to the supply of nutritional substances. Additional dilutions (10^{-9} g/ml) of the tissue-specific metabolic inhibitors have little or no influence. Compared to the situation in diploid cell cultures the activation of the cancer

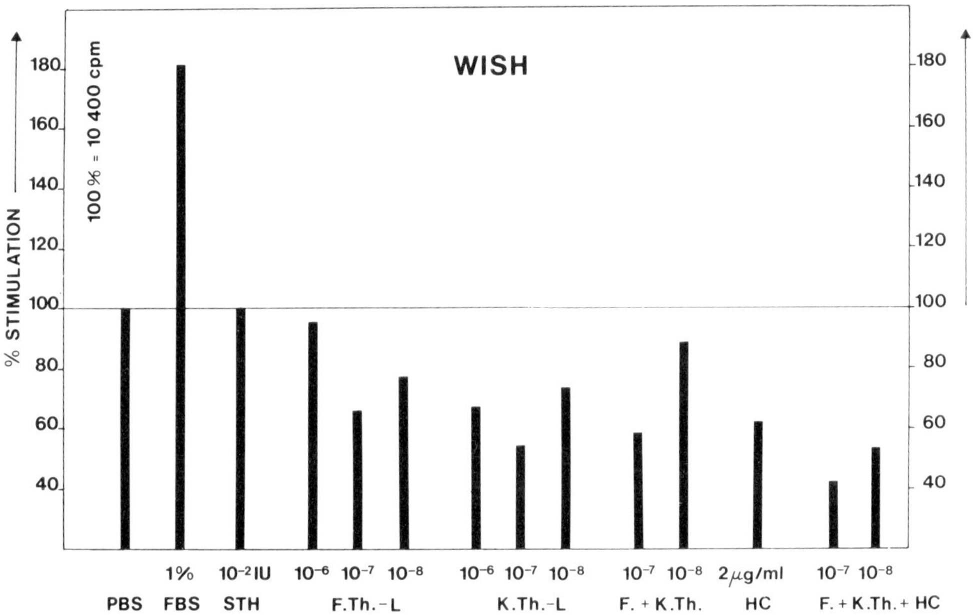

Fig. 2.

DNA synthesis in Wish cells after incubation with PBS, FBS, STH, lyophilized and redissolved extract of fetal calf thymus (F.Th.), juvenile calf thymus (K.Th.), mixtures of both extracts (F.+K.Th.) and hydrocortisone (HC). Final protein concentration given in g/ml.

cells by fetal bovine serum and somatotropin is less pronounced. The combination of the extract mixtures with hydrocortisone (HC) intensifies the inhibitory effect and decreases the ^{3}H-thymidine incorporation to values as low as 40 - 50% of the untreated control.

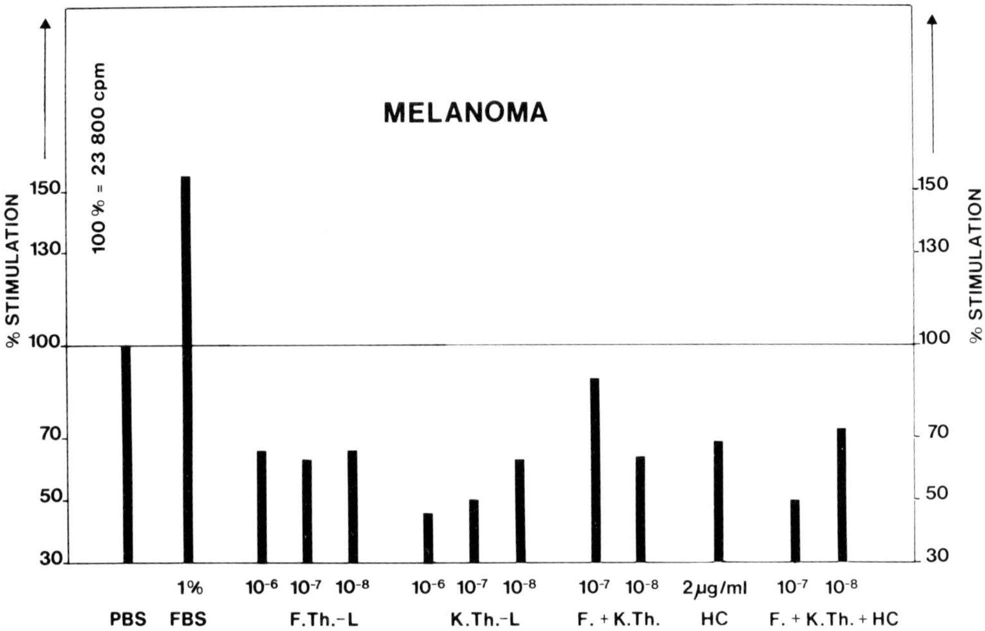

Fig. 3.

DNA synthesis in a human melanoma cell line after incubation with PBS, FBS, STH, lyophilized and redissolved extract of fetal calf thymus (F.Th.), juvenile calf thymus (K.Th.), mixtures of both extracts (F.+K.Th.) and hydrocortisone (HC). Final protein concentration given in g/ml.

Similar results were seen after incubation of a human melanoma cell line with lyophilized and redissolved thymus extracts (Fig. 3). An average of 40% reduction of DNA synthesis was obtained with fetal and juvenile thymus. The inhibition is effective at a broad concentration range but less amplified in the combination of the extract mixtures with hydrocortisone.

Discussion

Although it is well-known that thymectomy or age-associated deficiencies of thymus function increase the development of neoplasia and even stimulate tumor growth, only few reports exist concerning the influence of thymus factors on human cells in culture. Data published mostly deal with maturation studies of precursor T cells into immunologically competent T lymphocytes or the expression of T cell surface antigens.

Our results illustrate the action of calf thymus extracts on different cell types and demonstrate that the thymus gland contains and/or secretes several biologically active factors which may play an important role in the physiological regulation of cells and organs not involved in immunological processes.

Already at low extract concentrations we detected a stimulation of DNA synthesis and metabolism in diploid human cells of different origin. On the other hand, thymus substances in a concentration range of 10^{-6} - 10^{-8} g/ml exerted a significant decline on DNA synthesis in 2 human cancer cell lines which may provide evidence for the existence of tissue-specific factors with opposite activities. In contrast to the stimulation of diploid cell growth which resulted in a better multiplication and higher cumulative cell number we were not able to influence cancer cell proliferation permanently. Reports on in vivo studies with thymus substances revealed inhibitory effects to chemically induced tumors with incidence of volume reduction and of malignancy (3). Based on successful approaches of clinical applications and therapy with thymus substances we may conclude that thymus factors provoke immune reactivity, influence metabolism and neoplastic growth and restore functions in patients with thymus-dependent immunological deficiencies.

References

1. Osoba, D.: J. Exp. Med. 122, 633-650 (1965).
2. Goldstein, A.L., White, A.: In:Biochemical Actions of Hormones, I., Litwach, G. (ed.), Academic Press, New York, 465-502 (1970).
3. Milcu, S.M., Potop, I.: In:Thymic Hormones, Luckey, T.D. (ed.), Urban und Schwarzenberg, München, 97-134 (1973).
4. Goldstein, A.L., Thurman, G.B., Cohen, G.H., Hooper, J.A.: In:The Biological Activity of Thymic Hormones, Van Bekkum, D.W. (ed.), Kooyker Scientific Publication, Rotterdam, 173-197 (1975).
5. Weksler, M.E., Innes, J.B., Goldstein, G.: J. Exp. Med. 148, 996-1006 (1978).
6. Gospodarowicz, D., Moran, J.S.: Exp. Cell Res. 90, 279-284 (1975).

STUDIES WITH MELANOSTATIN (MIF) AND BROMOERGOCRYPTINE IN PATIENTS WITH AMENORRHEA AND HYPERPROLACTINEMIA

Adolf E. Schindler, Cornelia Hofstätter, Erich Keller, Reinhold Göser, Gernot Harrant

Universitäts-Frauenklinik, D-7400 Tübingen, FRG

Introduction

The isolation and structure analysis of melanostatin (MIF) was reported in 1971 (1) and revealed a tripeptide (L-prolyl-L-leucyl-glycine amide). This peptide was found to potentiate L-dopa effect in animals and man (2,3,4) and apomorphine action in rats by modifying the dopamine receptor (5) improving perhaps responsiveness to the natural neurotransmitter dopamine. Influence on growth hormone secretion of this peptide was demonstrated (6,9) and suppression of ACTH in patients with Morbus Addison or adrenalectomized bilaterally because of Cushing's disease was found (7,8). In these patients also a decrease of Prl was noted (8). Therefore, the aim of study was to evaluate the effect of MIF on various plasma hormones in amenorrheic patients with hyperprolactinemia. For comparison these patients were also treated with the dopamine agonist bromoergocryptine.

Material and methods

The study group consisted of six patients, age 28 - 39 (mean 33.6 years), with secondary amenorrhea of 3 - 28 years duration (mean 10.8 years) and hyperprolactinemia of 42 - 148 ng/ml (mean 100.3 ng/ml). In two patients sella enlargement was

Structure and Activity of Natural Peptides

found. Synthetic L-Pro-L-Leu-Gly-NH_2 , Serono , and bromoergocryptine - as comercially available (Pravidel[R], Sandoz) - were used in this investigation.
The experimental design for the application of MIF is demonstrated in Figure 1.

Fig. 1. Time and dose schedule of the intravenous application of melanostatin (MIF).

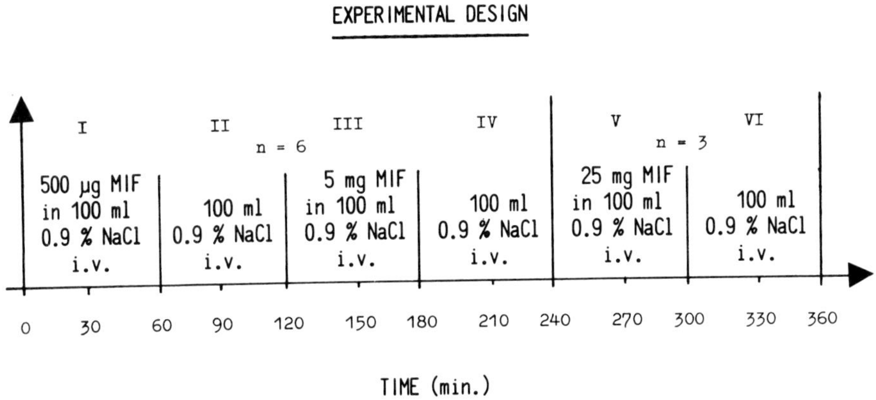

Bromoergocryptine (Pravidel[R]) was given orally p.o. at a dose of 2 x 2.50 mg daily continuously. This drug was started on the day after the study with MIF.

Types of hormones measured and methods used were as follows: Prolactin (Prl)-RIA Serono, hLH- RIA Serono and NIH standard LER 907 ng/ml, hFSH— RIA Serono and NIH standard LER 907 ng/ ml, hGH - RIA Serono, hTSH - RIA Serono, insulin - RIA Serono, T_3 - RIA Serono and estradiol -17ß (E_2) - RIA with a specific antiserum raised against E_2-6-oxo:BSA.

Results

The data for the six patients up to 5 mg MIF with 2 control periods with a total of 240 minutes are listed in Table 1.

The concentration of hGH was very variable in the volunteers and demonstrated inconsistant changes throughout the period of study. No effect of MIF could be detected.

Three patients were also infused with 25 mg MIF during a one hour period followed by a one hour control period. Again, no characteristic changes of the measured hormones could be detected. Only FSH exhibited a consistent decrease throughout the entire course of the study, but this change was not significant. All other hormones did not reveal major and consistent responses. During the study pulse and blood pressure did not change significantly. From the three patients given 25 mg of MIF various complains such as headaches and hot flushes were recorded. The leucocyte counts changed from an average of 5 167 leucocytes/ml to 7 875. The liver enzymes, electrolytes, bilirubin, cholesterol, triglycerides, creatinine and blood urea nitrogen remained unaltered.

Five of the six patients were treated with bromoergocryptine. Within 4 weeks using 5 mg daily the Prl concentration decreased from 97.8 $\pm$ 43.0 ng/ml to 34.3 $\pm$ 13.1 ng/ml (mean $\pm$ SD). Resumption of ovulatory cycles could be established in all patients but the dose of bromoergocryptine had to be further increased in four. Three of the five treated women conceived.

Discussion

Studies with MIF have resulted in a number of conflicting results. This is reflected in the publications reporting on the treatment of parkinsonism (4). Differences in results are also apperent for growth hormone (6,9), ACTH (6,7,8,11) and

Table 1. Plasma hormones (mean ± SD) during infusion of 0.5 mg and 5.0 mg MIF and corresponding control periods.

Time of studies in months	Type of infusion	Prl ng/ml	FSH ng LER 907/ml	LH ng LER 907/ml	E_2 pg/ml	Insulin uIU/ml	FSH uIU/ml	T_3 ng/ml
0 30 60	0.5mg MIF 0.9% NaCl	100 ± 45 97 ± 39 90 ± 36	251 ± 45 246 ± 47 238 ± 38	44 ± 16 35 ± 9 34 ± 10	34 ± 14 34 ± 13 34 ± 16	24 ± 6 20 ± 5 19 ± 4	3.1 ± 1.0 2.8 ± 0.9 2.6 ± 0.7	0.7 ± 0.1 0.8 ± 0.1 0.7 ± 0.1
90 120	0.9% NaCl	95 ± 36 93 ± 49	241 ± 41 232 ± 48	31 ± 9 31 ± 9	36 ± 13 36 ± 14	17 ± 5 19 ± 6	2.4 ± 6.8 2.3 ± 0.9	0.7 ± 0.1 0.7 ± 0.1
150 180	0.5mg MIF 0.9% NaCl	94 ± 44 97 ± 43	224 ± 49 226 ± 43	30 ± 10 32 ± 10	36 ± 13 35 ± 14	19 ± 4 18 ± 4	2.5 ± 0.8 2.3 ± 0.8	0.7 ± 0.08 0.7 ± 0.1
210 240	0.9% NaCl	96 ± 43 90 ± 37	217 ± 46 227 ± 50	31 ± 10 28 ± 14	34 ± 14 37 ± 16	16 ± 5 17 ± 3	2.2 ± 0.7 2.3 ± 0.8	0.7 ± 0.1 0.7 ± 0.1

Prl (6,8). So far, available data have been obtained from normoprolactinemic patients. Therefore, the present investigation was based upon hyperprolactinemic patients. Similar to the data of FAGLIA et al. (6), significant changes of Prl under MIF infusion,when compared with periods of 0.9% NaCl infusion, could not be detected. However, in patients with Morbus Cushing a decrease of Prl with a similar dose of MIF and with an increase after withdrawal of MIF was found (8). Furthermore, in this study significant changes were also absent for hFSH, hLH, hTSH, hGH, insulin, E_2 and T_3. Lack of changes of hFSH, hLH, hTSH and insulin were previously published using comparable doses (6,8). No consistent effect on hGH was found during the entire period of this investigation in contrast to the effect in agromegalics seen by others (6).

Since MIF has a short biological half-life and is degraded enzymatically rapidly (12), results might be dependent upon the experimental design. Furthermore, the period of time in which a certain dose of MIF is given probably influences the dose which effectively reaches the target tissues. Taking intravasal and parenchymal inactivation as well as the blood brain barrier into consideration, the diversity of results so far available could perhaps be caused. Otherwise, one has to conclude that up to the present time and with the doses so far applied no major impact on the endocrine status in the human by MIF has been established.

Using the dopamine agonist bromoergocryptine these patients treated with the drug responded as indicated by the change in Prl levels, resumption of menstruation and in part conception reflecting the influence of the dopaminergic pathways and supporting previously published experience (10).

Overall, this study did not demonstrate an influence of MIF on the lactotroph of patients with amenorrhea and hyperprolactinemia inspite of the reported dopaminergic action of the compound, while bromoergocryptine proved to be an effective Prl inhibitor in these patients.

Summary

Melanostatin (MIF) did not effect circulating levels of Prl, FSH, LH, TSH, GH, insulin, E_2 and T_3 by i.v. infusion of 0.5, 5 and 25 mg MIF/hr in patients with secondary amenorrhea and hyperprolactinemia, while bromoergocryptine was effective in these women normalizing the raised Prl levels and restoring menstrual cycle function.

Acknowledgements

Melanostatin (MIF) was kindly supplied by Serono, Freiburg (FRG). The authors are grateful for materials provided by the National Pituitary Agency NIH Bethesda for some of the RIA for protein hormones.

References

1. Nair,R.M.G., Kastin,A.J., Schally,A.V.: Biochem. Biophys. Res. Commun. 43, 1376 - 1381 (1971).

2. Plotnikoff,N.P., Kastin,A.J., Anderson,M.S., Schally,A.V.: Life Scienes 10, 1279 - 1283 (1971).

3. Barbeau,A.: Lancet II, 683 - 684 (1975).

4. Caroli,A., Palladrin,G., Margotta,V.: Melanocyte-Stimulating Hormone Release-Inhibiting Factor (MIF): Appraisal of Nervous and Endocrine Effects, in: Polleri,A., McLeod, R.M. (eds): Neuroendocrinology: Biological and Clinical Aspects, Serono Symposion 20, 55 - 58 (1979).

5. Kostrzewa,R.W., Kastin,A.J., Solbrain,S.K.: Pharmacology, Biochemistry and Behaviour 9, 375 - 378 (1978).

6. Faglia,G., Parachi,A., Sterari,C., Beck-Peccoz,P., Ambrosi,B., Travaglini,P., Spadan,A., Oliver,C.: J. Clin. Endocrin. Metab. 42, 991 - 994 (1976).

7. Voigt,K.H., Fehm,H.L., Lang,R.E., Beinert,K.E., Pfeiffer, E.F.: Horm. Metab. Res. 9, 991 - 994 (1976).

8. Althoff,P. , Rosak,C., Heitmann,A., Schulz,F., Usadel, K.-H., Neubauer,M.: Acta Endocrin. 94, Suppl. 234, 51 - 52 (1980).

9. Panerei,A.A., Cozchi,D., Gil-ad,I., Locatelly,V., Rossi, G.I., Müller,E.E.: Clinical Endocrinology 5, 717 - 722 (1976).

10. Schindler,A.E.: Bromoergocryptine for Ovulation Induction, in: Hafez,E.S.E. (ed.): Human Ovulation, North Holland, Amsterdam 405 - 412 (1979).

11. Ambrosi,B., Gaggini,M., Secchi,F., Faglia,G.: Acta Endocrin. 91, Suppl. 225, 67 (1979).

12. Redding,T.W., Kastin,A.J., Gonzales-Barcena,D., Coy,D.H., Hirotsu,Y., Ruelas,J., Schally,A.V.: Neuroendocrinology 16, 119 - 126 (1974).

AUTHOR INDEX

SUBJECT INDEX

Reagenzien
MERCK
Stabil · wasserlöslich · wirtschaftlich
H
C=O
C=O
H
Fluescin®
398
Fluoreszenzreagenz auf Aminosäuren nach Roth
Zur qualitativen Erfassung und quantitativen Bestimmung von Aminosäuren auch im Sub-Nanomol-Bereich durch Fluorimetrie. Für manuelle Methoden und als Detektionsreagenz in automatischen Aminosäureanalysatoren.
E. Merck, Darmstadt
Erhältlich in 2 Formen:
Art.-Nr. 11455 Fluescin® Fluoreszenzreagenz auf Aminosäuren nach Roth: eine gebrauchsfertige Reagenzzubereitung, die vor Verwendung in automatischen Aminosäureanalysatoren nur noch mit 0,1 M Boratpuffer pH 9–10 verdünnt wird.
Art.-Nr. 11452 Phthaldialdehyd für die Fluoreszenzanalyse: eine speziell aufgereinigte Grundsubstanz zur Herstellung von Fluoreszenzreagenzien nach eigenen Rezepturen.
Literatur: M. Roth, A. Hampaï, Journal of Chromatography 83 (1973), 353–356
Bitte verlangen Sie weitere Informationen